T0136078

About Island Press

Since 1984, the nonprofit Island Press has been stimulating, shaping, and communicating the ideas that are essential for solving environmental problems worldwide. With more than 800 titles in print and some 40 new releases each year, we are the nation's leading publisher on environmental issues. We identify innovative thinkers and emerging trends in the environmental field. We work with world-renowned experts and authors to develop cross-disciplinary solutions to environmental challenges.

Island Press designs and implements coordinated book publication campaigns in order to communicate our critical messages in print, in person, and online using the latest technologies, programs, and the media. Our goal: to reach targeted audiences—scientists, policymakers, environmental advocates, the media, and concerned citizens—who can and will take action to protect the plants and animals that enrich our world, the ecosystems we need to survive, the water we drink, and the air we breathe.

Island Press gratefully acknowledges the support of its work by the Agua Fund, Inc., Annenberg Foundation, The Christensen Fund, The Nathan Cummings Foundation, The Geraldine R. Dodge Foundation, Doris Duke Charitable Foundation, The Educational Foundation of America, Betsy and Jesse Fink Foundation, The William and Flora Hewlett Foundation, The Kendeda Fund, The Andrew W. Mellon Foundation, The Curtis and Edith Munson Foundation, Oak Foundation, The Overbrook Foundation, the David and Lucile Packard Foundation, The Summit Fund of Washington, Trust for Architectural Easements, Wallace Global Fund, The Winslow Foundation, and other generous donors.

The opinions expressed in this book are those of the author(s) and do not necessarily reflect the views of our donors.

Foundations of
Ecological Resilience

Foundations of Ecological Resilience

EDITED BY

LANCE H. GUNDERSON,

CRAIG R. ALLEN,

AND C. S. HOLLING

Washington | Covelo | London

Copyright © 2010 Island Press

All rights reserved under International and Pan-American Copyright Conventions. No part of this book may be reproduced in any form or by any means without permission in writing from the publisher: Island Press, 1718 Connecticut Avenue NW, Suite 300, Washington, DC 20009, USA.

Island Press is a trademark of The Center for Resource Economics.

A list of permissions and original sources appears on pages 451–52.

LIBRARY OF CONGRESS CATALOGING-IN-PUBLICATION DATA

Foundations of ecological resilience / edited by Lance H. Gunderson, Craig R. Allen, and C.S. Holling.

     p. cm.

   Includes bibliographical references and index.

   ISBN-13: 978-1-59726-510-2 (cloth : alk. paper)

   ISBN-10: 1-59726-510-1 (cloth : alk. paper)

   ISBN-13: 978-1-59726-511-9 (pbk. : alk. paper)

   ISBN-10: 1-59726-511-X (pbk. : alk. paper)

   1. Ecosystem management. 2. Ecosystem health. I. Gunderson, Lance H. II. Allen, Craig R. III. Holling, C. S.

   QH75.F68 2009

   577—dc22

   2008048428

Printed on recycled, acid-free paper

Manufactured in the United States of America

10 9 8 7 6 5 4 3 2 1

Keywords: ecological resilience, panarchy, perturbation, regime shift, resilience, resilience theory, threshold

Contents

Note from the Publisher

THIS VOLUME IS A COLLECTION of previously published seminal papers in the developing field of resilience theory over the past thirty years. The volume editors have grouped them according to theme and provided commentary explaining their relevance and contribution to this important and fast-growing field. We have set the articles in a consistent format and type but have otherwise retained the text, style, documentation, and idiosyncrasies of the originals. Full publishing formation can be found on the opening page of each article.

Why Resilience?
Why Now?

LANCE H. GUNDERSON AND CRAIG R. ALLEN

LANCE H. GUNDERSON, *Department of Environmental Studies, Emory University, Atlanta*

CRAIG R. ALLEN, *Nebraska Cooperative Fish & Wildlife Research Unit, University of Nebraska, Lincoln*

PERHAPS ECOLOGY, rather than economics, should be called "the dismal science" because the popular application of the information generated by ecologists is generally bad news. Since the 1960s, prominent ecologists have been referred to by the popular press as the "New Jeremiahs," because some have interpreted their ecological research of human impacts on ecosystems as prophecies of doom and gloom. Indeed, many of their successors continue the trend of documenting the increasing size and magnitude of the human footprint on the planet. A recent Millennium Ecosystem Assessment (2005) determined that global and regional ecosystems have been altered by human activity more in the past fifty years than at any time in history, a trend that is likely to continue.

As we begin the twenty-first century, the rate (and spatial scale) of ecological change is accelerating. Across the planet, natural disasters are severely affecting individual and collective societies. Over time, humans adapt and learn how to cope with these events. Yet the abrupt and often unpredictable dynamics associated with these events have led to greater uncertainty in spite of technological and scientific advances. An increasing human population and anthropogenic land use and land cover change have left humankind more vulnerable to these events (Kasperson et al. 1995) and the potential loss of ecosystem goods and services (Vitousek et al. 1997, Millenium Ecosystem Assessment 2005). The root of the word *disaster* literally means "bad star," suggesting an extraplanetary origin of these events but perhaps better describing the large uncertainties associated with them. Such large-scale disasters include outbreaks of disease, such as influenza (Barry 2004), AIDS, and hantavirus; recurring tsunamis (Adger et al. 2005); tropical cyclones, such as Hurricane Katrina in 2005 (Bohannan 2005); and global climate change (Steffen et al. 2004), to name but a few.

Resilience theory (Walker and Salt 2006) was developed by ecologists over the past three decades to explain surprising and nonlinear dynamics of complex adaptive systems (Gunderson and Holling 2002, Walker et al. 2004). Moreover, resilience theory is the basis for adaptive management, which embraces uncertainty of complex resource systems (Holling 1978, Walters 1986, 1997). The following section presents a review of the concepts of resilience as developed by ecological theorists and applied ecologists.

Ecological Resilience

Development of ecological resilience theory began in the 1960s with attempts to mathematically model dynamic ecosystems. Early models (Lewontin 1969, May 1977) focused on the stability of systems—that is, the processes that control or mediate the persistence of ecological structures. Part of the construct of modeling systems was to create boundaries that identified the system. In doing so, processes were categorized as internal or external to the system, with a focus on interactions between the external and internal. Disturbances were viewed as external drivers or perturbations. Human interventions, such as the harvest of natural resources (e.g., fishing, lumbering), were included in the category of disturbances to an ecological system. A major assumption for theory and practice was that ecological systems were stable. The stability assumptions rested on two additional assumptions: (1) that a system would generally persist in form and function (unless, of course, humans perturbed it) and (2) that a system would recover to its former equilibrium state after disturbances. Implicit to this was the notion of global or stable equilibrium, such as population or ecosystem carrying capacity. Although rates of recovery might vary, perturbed systems would eventually recover to a predisturbance state. When applied to renewable resources, these theories led to policies of maximum sustained yield, or harvest rates that assumed whatever was being harvested would be replaced. Hence, the idea of optimal control of harvests was introduced and persists.

Holling (1973) introduced the word *resilience* to describe three aspects of changes that occur in an ecosystem over time. The first was to describe the "persistence of relationships within a system" and the "ability of systems to absorb changes of state variables, driving variables and parameters, and still persist." The second concept recognized the occurrence of alternative and multiple states as opposed to the assumption of a single equilibrium and global stability; hence, resilience was "the size of a stability domain or the amount of disturbance a system could take before it shifted into alternative configuration." The third insight was the surprising and discontinuous nature of change, such as the collapse of fish stocks or the sudden outbreak of spruce budworms in forests. These insights altered the way in which theorists perceived ecological systems and how practitioners have attempted to manage these systems.

The community of ecologists has viewed and defined resilience in multiple ways, each based on different assumptions. One interpretation of the idea of resilience is rooted in the etymology of the word, which has been traced to the Latin word *resilire*, meaning to "leap back." Some ecologists (Pimm 1991) define resilience as how fast a variable that has been displaced from equilibrium returns to that equilibrium. This is essentially a return time, or time of recovery, which can be mathematically defined but is based on an assumption of behavior around a single equilibrium. Ludwig et al. (1997) assert that this perspective is applicable to a linear system or to the behavior of a nonlinear system in the vicinity of a local equilibrium that can approximated by a linear function.

Holling (1996) distinguished two types of resilience: engineering and ecological. Engineering resilience is defined as the rate or speed of recovery of a system following a shock. Ecological resilience, on the other hand, assumes multiple states (or "regimes") and is defined as the magnitude of a disturbance that triggers a shift between alternative states (Holling 1973, 1996). In this sense, a regime shift occurs when the controlling variables in a system (including feedbacks) result in a qualitatively different set of structures and dynamics of these systems (Walker et al. 2004). Whether or not alternative regimes or states exist in ecological systems has been the subject of debate over the past three decades.

One of the first demonstrations of alternative states and regime shifts came from an international modeling project conducted by the International Institute for Applied Systems Analysis (IIASA) on the outbreaks of spruce budworms in the boreal forests of Canada (Ludwig et al. 1978, Holling 1978, Clark et al. 1979). This project modeled pest outbreaks by using a small number of variables (budworm population, predation effectiveness, and the volume of forest canopy) and the interactions or relationships among these variables. The model depicted two states. One state (no outbreak) is characterized by low numbers of budworm and young, fast-growing trees, and the other state (outbreak) is characterized by high numbers of budworm and old, senescent trees. The shift between regimes occurs when the growing trees result in an increase in canopy volume such that the set of birds that eat budworms can no longer control the pest populations (Holling 1988).

A similar model has been used to explain the outbreak of such human diseases as influenza, hantavirus, or malaria (Janssen and Martens 1997). One state is an outbreak state, when the infection is rapidly spreading among susceptible humans. The other is a dormant or inactive state, when either no or few cases of infection are present. As documented in many cases of regime shifts, the transition among states or regimes is mediated by the interaction between slower and faster components in ecosystems (Holling and Gunderson 2002). In this model, the faster variables include the population or numbers of disease organisms, the slower variables include susceptibility of the hosts and disease vectors, and the slowest variables are the mutation rates of the disease and the size of the human population (May 1977). Some of these factors, such as disease vectors and host densities, may be managed by humans, whereas others, such as mutation rates, may not.

The demonstration that ecosystems exhibited alternative states or regimes gained momentum in the 1980s and 1990s, but not all studies supported the idea. Sousa and Connell (1985) examined a multiple-decade time series data set of marine populations and found no evidence for alternate regimes or states. Other ecologists, working in disturbance-driven ecosystems, found that resilience theory was the only theory that helped explain the complex changes that they were studying. Walker (1981), Westoby et al. (1989), and Dublin et al. (1990) studied semiarid rangelands and found dramatic shifts between grass-dominated and shrub-dominated systems. Those shifts were mediated by interactions among herbivores, fires, and drought cycles. Scheffer (1998) and Scheffer et al. (2001) described two alternative states (clear water with rooted aquatic vegetation and turbid water with phytoplankton) in shallow lake systems. Coral reef systems may shift from a coral-dominated state to a macroalgae-dominated state (Hughes 1994, McClanahan et al. 2002, Hughes et al. 2003, Bellwood et al. 2004). Multiple pathways have been documented for this transition, including overfishing and resulting population declines of key grazing species, increases in nutrients, and shifts in recruitment patterns (Jackson et al. 2001). Estes and Duggins (1995) and Steneck et al. (2004) have shown how near-shore temperate marine systems shift between dominance by kelp or dominance by sea urchins as a function of the density of sea otters and other grazers.

Regime shifts have been observed in hundreds of different ecosystems, including marine, freshwater, and terrestrial (Gunderson and Pritchard 2002, Scheffer et al. 2001, Folke et al. 2004). In all of these systems, the transitions among regimes, and the resilience of the system, can be traced to a small number of variables, including biological and physical controls and recurring larger-scale perturbations (or disturbances). A key insight is that ecological resilience is mediated and lost due to the interaction of variables that operate at distinctive scales of space and time (Holling 1986, Holling and Gunderson 2002).

Resilience and Scale

Complex ecological systems operate across wide ranges of scales of space and time. An example is global influenza outbreaks, as occurred with the "Spanish" flu in the early part of the twentieth century (Barry 2004). These outbreaks involve structures and dynamics that range from interactions among microscopic organisms to the scale of the planet. Populations of bacteria in host organisms operate on time scales of minutes to weeks, suggested by the normal time course of the disease within a host. Numbers of infected or susceptible humans can literally cover the planet. The many pathways (especially air travel) by which humans move can accelerate rates of spread, so that the disease can flare up in one place and within days be spread around the planet. As argued by Barry (2004), ignorance of the vectors of spread lead to greater concentrations of humans and larger outbreaks.

Another example of a natural disaster that indicates the cross-scale nature of dynamics is from Florida Bay, a shallow, subtropical marine ecosystem located at the southern terminus of the Florida peninsula. For most of the twentieth century, the bay had clear water and a bottom covered by sea grass. Around 1990, the sea grass began dying over most of the bay. The die-off was viewed as an ecological crisis and created great political, social, and economic turmoil. Because much of the bay is in Everglades National Park, the social objectives of conservation in the park were brought into question: Would the grass return? What (if any) management actions led to the die-off? Sport fishing and tourism relied on the clear-water state of the bay. Many wealthy people (including the U.S. president at that time) used the bay for recreation, so the crisis became instantly politicized.

The sea-grass die-off resulted in the system "flipping" from a clear-water, grass-dominated system to one with muddy water and recurring algae blooms. Living sea grass plants store nutrients, and their root systems stabilize sediments. The loss of sea-grass released nutrients into the water column and allowed sediments to become suspended in the water column by wind-generated currents. But much wrangling and discussion went into trying to understand what caused the shift in ecological regimes.

A number of hypotheses were proposed to explain the system shift (Gunderson 2001). These included hypersalinity resulting from a decrease in freshwater flow and altered water circulation, an increase in nutrient inputs from surrounding urban and agricultural areas, a lack of hurricanes, a loss of herbivores (turtles and manatees), disease, and temperature stress. The most plausible explanation involved a spatially homogenous stand of high sea grass biomass (probably related to a lack of disturbances, such as storms and grazers). The stress caused by high temperatures caused local die-offs as photosynthesis could not produce enough oxygen to keep up with respiratory demand. Because the beds contained high biomass, the die-off spread as dead material in the water column further depressed photosynthesis. Without sea grass, the sediments and nutrients became suspended in the water column, leading to algae blooms and muddy water. Both of these factors inhibited subsequent sea grass establishment. Hence, loss of ecological resilience (the amount of disturbance that the system can absorb without changing state) was related to the slowly changing variable of sea grass biomass. The regime shift (or state change) was related not to a single stressor but to a small set of factors, including sea grass biomass, disturbance regimes, and sediment stability.

The cross-scale nature of natural disasters requires that assessment and monitoring should cover multiple scales, up to the scale of the planet. Local, regional, national, and international agencies and institutions are not sufficient by themselves. Integration and communication are key to management and require novel approaches to dealing with the cross-scale issues. Longstaff and Yang (2008) describe the role of trust in leadership following natural disasters. Scale issues apply to policy, management, and leadership. In the best cases, leadership spans social-political scales; one person can lead for a time, but several are better locally, regionally, and politically (Westley et al. 2002; Folke et al. 2002, 2005).

Adaptive Management

The dynamics of complex adaptive systems following large-scale disasters, such as disease outbreaks, flood and drought cycles, hurricanes, and cyclones, present problems of predictability for management (Holling 1978, Walters 1986, Gunderson 1999, 2003). Yet planning and management require some estimate of future conditions. Certainly, many things are known, especially the broad and the general. In August 2005, it was well known, at least three days prior to landfall, that Hurricane Katrina would strike the Gulf Coast of the United States (with a given probability), but the specific impacts and the exact location of landfall could not be predetermined. In the case of influenza, similar levels of uncertainty abound. Certainly, as described elsewhere, the broad and the general forms of an influenza outbreak can be estimated and imagined, but the particulars, such as where and when a large-scale outbreak will occur, cannot. Our predictive abilities of such systems are limited for many reasons, including the evolutionary, adaptive, and cross-scale nature of complex systems as well as the lack of data to monitor and to test ideas about system dynamics across scales.

Adaptive management is an approach to natural resource management that was developed from theories of resilience (Holling 1978). Adaptive management acknowledges the deep uncertainties of resource management and attempts to winnow those uncertainties over time by using management actions as experiments to test policy (Walters 1986). Management must confront various sources of complexity in systems, including the ecological, economic, social, political, and organizational components of these systems (Holling and Gunderson 2002, Westley 2002), as well as the interactions among system components. The difficulties in managing the interacting aspects of social-ecological systems have led some to call them wicked problems (Rittel and Webber 1973, Ludwig et al. 2001). Developers of adaptive management approaches (Holling 1978, Walters 1986) acknowledged the complex multidimensionality of natural resource issues but focused on analytic approaches primarily in the ecological and economic domains from a systems perspective. Lee (1993) was the first to separate these issues into scientific (primarily ecological) and social (political) components. Adaptive management attempts to bring together disciplinary approaches for analysis and assessment and then integrate those ideas with policy and governance in the social arenas in a framework some describe as adaptive governance (Folke et al. 2005).

Adaptive governance is an emergent framework for managing complex environmental issues. Dietz et al. (2003) used the phrase to describe the social and human context for applying adaptive management. Folke et al. (2005) describe this form of governance as necessary for the management of complex ecosystems, particularly when change is "abrupt, disorganizing, or turbulent." Brunner et al. (2006) provide a rich set of examples to illustrate the emergence of adaptive governance as a method for solving problems created by top-down control of decision making and attempts at implementing myopic scientific and technical solutions that are bereft of political considerations. They describe adaptive governance as operating in situations where the science is contextual, knowledge is incomplete, multiple ways of knowing and understanding are present, policy is implemented in modest steps, and unintended consequences and decision making are both top-down (although fragmented) and bottom-up. As such, adaptive governance is meant to integrate science, policy, and decision making in systems that assume and manage for change rather than against change (Gunderson et al. 1995).

Foundations of Resilience

The purpose of this volume is to synthesize the key scientific papers that led to our current understanding of resilience. In these papers lies the corpus of thought and understanding of resilience theory and thinking (Walker and Salt 2006). Our focus is specifically on ecological resilience. The concept of ecological resilience is often applied to social-ecological systems, but the foundations are in ecology. This book is organized both chronologically and categorically, focusing on theory, examples, and models.

The articles in this volume were chosen using a combination of factors. The first criteria for selection was articles that have shaped or defined the concepts and theories of resilience. We chose to include key papers that broke new conceptual ground and contributed a novel idea, or expansion of ideas, to our collective understanding of resilience, including the first formal descriptions of resilience theory from the early 1970s. The second criteria was to include examples that demonstrated ecological resilience in a range of ecosystems—that is, we included a small set of exemplar ecosystems where ecological resilience has been observed and reported. The third criteria for selection was a set of articles that provided the practical and methodological advances in understanding resilience. In essence, these factors became the groupings by which the volume is organized: concepts, examples, and models. The articles selected for each of these groups are described in the following paragraphs.

The first section of the collection includes the papers that are most often cited as the origins of the descriptions of ecological resilience and development of the fundamental concepts. These include the article that started the notion: C. S. Holling's 1973 article in the *Annual Review of Ecology and Systematics*. Holling expanded on these ideas in 1986, introducing the concept of an adaptive cycle—a general heuristic of ecological dynamics over time, which is a prototheory of nonlinear dynamics in complex systems of people and ecosystems. Although Holling introduced the idea (and importance) of scale in 1986, the idea was developed first by Folke and colleagues in 1996, along with the entwined arguments of the relationship between diversity and resilience. Holling wrote a short chapter in 1996 to contrast and highlight at least two different ways that ecologists use and define resilience. To round out the section on theory and concepts, we include the annual review article by Folke and colleagues (2004), which is an outstanding summary of the knowledge of ecological resilience that had accumulated over three decades.

The second section of this volume includes classic papers from ecology that reported on regime shifts in very different ecosystems. All were the first publication for the specific ecosystem type that used the idea of alternative regimes to describe the resource dynamics that they observed. The first is by Terry Hughes (1994), who describes phase shifts in coral reefs of the Caribbean. The second is by James Estes and Donald Duggins (1995), who describe trophic-regulated regime shifts in the northern Pacific involving sea otters, sea urchins, and kelp forests. The third, by Craig Allen, Beth Forys, and C. S. Holling (1999), examines regime shifts in animal communities in transforming landscapes.

The final section of the book presents methodological and practical implications of resilience, especially how human intervention erodes or enhances resilience. We include the article on resource science by Holling and Chambers (1973) because it introduces the notion that ecological resilience is most often revealed in managed resource systems, or systems that have a large degree of human intervention. That article also introduced the use of models to help understand the complex dynamics exhibited by such resource systems. We also include two articles that each describe different aspects of forest pest management. Don Ludwig and colleagues (1978) provide elegant quantitative and qualitative analyses of outbreak dynamics. Clark, Jones, and Holling (1979) describe the practical implications of resilience for policy development and implementation.

We hope that this collection is a contribution to our collective understanding of resilience. We have likely overlooked some key articles and perhaps included some that are repetitive. In the end, those reading this volume will be the judge of the collection's adequacy. In either case, we invite the reader to read or reread what we consider to be the foundational articles to that collective understanding.

Literature Cited

Adger, W. N., T. P. Hughes, C. Folke, S. R. Carpenter, and J. Rockström. 2005. Social-ecological resilience to coastal disasters. *Science* 309:1036–39.

Barry, J. 2004. *The great influenza: The epic story of the deadliest plague in history*. New York: Viking Adult.

Bellwood, D. R., T. P. Hughes, C. Folke, and M. Nystrom. 2004. Confronting the coral reef crisis. *Nature* 429:827–33.

Bohannon, J. 2005. Disasters: Searching for lessons from a bad year. *Science* 310:1883.

Brunner, R. D., T. D. Steelman, L. Coe-Juell, C. M. Cromley, C. M. Edwards, and D. W. Tucker. 2006. *Adaptive governance: Integrating science policy and decision making*. New York: Columbia University Press.

Clark, W. C., D. D. Jones, and C. S. Holling. 1979. Lessons for ecological policy design: A case study of ecosystem management. *Ecological Modeling* 7:1–52.

Dietz T., E. Ostrom, and P. C. Stern. 2003. The struggle to govern the commons. *Science* 302:1902–12.

Dublin, H. T., A. R. E. Sinclair, and J. McGlade. 1990. Elephants and fire as causes of multiple stable states in the Serengeti-Mara woodlands. *Journal of Animal Ecology* 59:1147–64.

Estes, J. A., and D. O. Duggins. 1995. Sea otters and kelp forests in Alaska: Generality and variation in a community ecological paradigm. *Ecological Monographs* 65: 75–100.

Folke, C., S. Carpenter, T. Elmqvist, L. Gunderson, C. S. Holling, and B. Walker. 2002. Resilience and sustainable development: Building adaptive capacity in a world of transformations. *Ambio* 31:437–40.

Folke, C., S. Carpenter, B. Walker, M. Scheffer, T. Elmqvist, L. Gunderson, and C. S. Holling. 2004. Regime shifts, resilience and biodiversity in ecosystem management. *Annual Review of Ecology and Systematics* 35:557–81.

Folke, C., T. Hahn, P. Olsson, and J. Norberg. 2005. Adaptive governance of social-ecological systems. *Annual Review of Environment and Resources* 30:441–73.

Gunderson, L. H. 1999. Resilience, flexibility and adaptive management. *Conservation Ecology* 3 (1): 7. http://www.consecol.org/vol3/iss1/art7/.

———. 2001. Managing surprising ecosystems in southern Florida. *Ecological Economics* 37:371–78.

———. 2003. Adaptive dancing. In *Navigating social-ecological systems: Building resilience for complexity and change*, ed. Fikret Berkes, Johan Colding, and Carl Folke, 33–53. Cambridge: Cambridge University Press.

Gunderson, L. H., and C. S. Holling. 2002. *Panarchy: Understanding transformations in human and natural systems*. Washington, DC: Island Press.

Gunderson, L. H., C. S. Holling, and S. S. Light. 1995. *Barriers and bridges to the renewal of ecosystems and institutions*. New York: Columbia University Press.

Gunderson, L. H., and L. Pritchard. 2002. *Resilience and the behavior of large scale systems*. Washington, DC: Island Press.

Holling, C. S. 1973. Resilience and stability of ecological systems. *Annual Review of Ecology and Systematics* 4:1–23.

———. 1978. *Adaptive environmental assessment and management.* Caldwell, NJ: Blackburn.

———. 1986. The resilience of terrestrial ecosystems: Local surprise and global change. In *Sustainable development of the biosphere,* ed. W. C. Clark and R. E. Munn, 292–317. Cambridge: Cambridge University Press.

———. 1988. Temperate forest insect outbreaks, tropical deforestation and migratory birds. *Memoirs Entomological Society of Canada* 146:22–32.

———. 1996. Engineering resilience vs. ecological resilience. In *Engineering within ecological constraints,* ed. P. C. Schulze, 31–43. Washington, DC: National Academy Press.

Holling, C. S., and L. H. Gunderson. 2002. Resilience and adaptive cycles. In *Panarchy: Understanding transformations in human and ecological systems,* ed. L. H. Gunderson and C. S. Holling, 25–62. Washington, DC: Island Press.

Hughes, T. P. 1994. Catastrophes, phase shifts, and large-scale degradation of a Caribbean coral reef. *Science* 265:1547–51.

Hughes, T. P., A. H. Baird, D. R. Bellwood, M. Card, S. R. Connolly, C. Folke, R. Grosberg, et al. 2003. Climate change, human impacts, and the resilience of coral reefs. *Science* 301:929–33.

Jackson, J. B. C., M. X. Kirby, W. H. Berger, K. A. Bjorndal, L. W. Botsford, L. W. Bourque, R. H. Bradbury, et. al. 2001. Historical overfishing and the recent collapse of coastal ecosystems. *Science* 293:629–38.

Janssen, M. A., and W. J. M. Martens. 1997. Modeling malaria as a complex adaptive system. *Artificial Life* 3:213–36.

Kasperson, J. X., R. E. Kasperson, and B. L. Turner II. 1995. *Regions at risk: Comparisons of threatened environments.* Tokyo: United Nations University Press.

Lee, K. 1993. *Compass and gyroscope: Integrating science and politics for the environment.* Washington, DC: Island Press.

Lewontin, R. C. 1969. The meaning of stability. *Brookhaven Symposium in Biology* 22:13–24.

Longstaff, P. H., and S. Yang. 2008. Communication management and trust: Their role in building resilience to "surprises" such as natural disasters, pandemic flu, and terrorism. *Ecology and Society* 13 (1): 3. http://www.ecologyandsociety.org/vol13/iss1/art3/.

Ludwig, D., D. D. Jones, and C. S. Holling. 1978. Qualitative analysis of insect outbreak systems: The spruce budworm and forest. *Journal of Animal Ecology* 47:315–32.

Ludwig, D., B. Walker, and C. S. Holling. 1997. Sustainability, stability, and resilience. *Conservation Ecology* 1 (1): 7. http://www.consecol.org/vol1/iss1/art7/.

Ludwig, D., M. Mangel, and B. Haddad. 2001. Ecology, conservation, and public policy. *Annual Review of Ecology and Systematics* 32:481–517.

May, R. M. 1977. Thresholds and breakpoints in ecosystems with a multiplicity of stable states. *Nature* 269:471–77.

McClanahan, T., N. Polunin, and T. Done. 2002. Resilience of coral reefs. In *Resilience and behavior of large-scale systems,* ed. L. H. Gunderson and L. Pritchard Jr., 111–64. Washington, DC: Island Press.

Millennium Ecosystem Assessment. 2005. *Ecosystems and human well-being: Current state and trends findings of the condition and trends working group.* Washington, DC: Island Press.

Pimm, S. L. 1991. *The balance of nature?* Chicago: University of Chicago Press.

Rittel, H., and M. Webber. 1973. Dilemmas in a general theory of planning. *Policy Sciences* 4:155–69.

Scheffer, M. 1998. *The ecology of shallow lakes.* London: Chapman and Hall.

Scheffer, M., S. R. Carpenter, J. Foley, C. Folke, and B. Walker. 2001. Catastrophic shifts in ecosystems. *Nature* 413:591–696.

Sousa, W. P., and J. H. Connell. 1985. Further comments on the evidence for multiple stable points in natural communities. *American Naturalist* 125:612–15.

Steffen, W., A. Sanderson, J. Jäger, P. D Tyson, B. Moore III, P. A. Matson, K. Richardson, F. Oldfield, H. J. Schellnhuber, B. L. Turner II, and R. J. Wasson. 2004. *Global change and the earth system: A planet under pressure.* Heidelberg: Springer-Verlag.

Steneck, R. S., J. Vavrinec, and A. Leland. 2004. Accelerating trophic level dysfunction in kelp forest ecosystems of the western North Atlantic. *Ecosystems* 7:323–31.

Vitousek, P. M., H. A. Mooney, J. Lubchenco, and J. M. Melillo. 1997. Human domination of earth's ecosystems. *Science* 277: 494–99.

Walker, B., C. S. Holling, S. R. Carpenter, and A. Kinzig. 2004. Resilience, adaptability and transformability in social–ecological systems. *Ecology and Society* 9 (2): 5. http://www.ecologyandsociety.org/vol9/iss2/art5/.

Walker, B., and D. Salt. 2006. *Resilience thinking: Sustaining ecosystems and people in a changing world.* Washington, DC: Island Press.

Walker, B. H. 1981. Is succession a viable concept in African savanna ecosystems? In *Forest succession: Concepts and application,* ed. D. C. West, H. H. Shugart, and D. B. Botkin, 431–47. New York: Springer-Verlag.

Walters, C. J. 1986. *Adaptive management of renewable resources.* New York: McGraw-Hill.

———. 1997. Challenges in adaptive management of riparian and coastal ecosystems. *Conservation Ecology* 1 (1). http://www.consecol.org/vol1/iss2/art1.

Westley, F. 2002. The devil in the dynamics: Adaptive management on the front lines. In *Panarchy: Understanding transformations in human and natural systems,* ed. L. H. Gunderson and C. S. Holling, 333–60. Washington, DC: Island Press.

Westley, F., S. R. Carpenter, W. A. Brock, C. S. Holling, and L. H. Gunderson. 2002. Why systems of people and nature are not just social and ecological systems. In *Panarchy: Understanding transformations in human and natural systems,* ed. L. H. Gunderson and C. S. Holling, 103–19. Washington, DC: Island Press.

Westoby, M., B. H. Walker, and I. Noyr-Meir. 1989. Opportunistic management for rangelands not at equilibrium. *Journal of Rangeland Management* 42:266–74.

Concepts and Theory

Commentary on
Part One Articles

CRAIG R. ALLEN, LANCE H. GUNDERSON, AND C. S. HOLLING

WHILE SOME IN COMMON PARLANCE use the word *theory* to mean something that is speculative or opinion, we use it in a scientific sense to indicate "an explanation based on observation and reasoning," which is consistent with its meaning for over four hundred years (Harper 2001). As such, resilience and panarchy are an integral part of ecological theory because of their application in understanding and explaining commonalities in patterns of change in complex systems. Those systems can be ecosystems or human or social systems, or combinations thereof (Gunderson and Holling 2002, Walker et al. 2004). This section presents six papers that collectively describe much of the theory of ecological resilience. They include the first article to use the word *resilience* to describe a new way of conceptualizing ecosystem dynamics.

There are two views of human interactions with, and management of, the world. In one, efforts are focused on maintaining a degree of constancy by reducing natural variability. In the other, focus is on maintaining the "consistency of relationships" among various parts of the ecological system in question. The former comes from traditions of physics, engineering, and other similar quantitative sciences, while the latter focuses on qualitative properties of systems. The first focuses on equilibrium states; the latter, on the persistence of function and structure. Both approaches are useful, under the right circumstances.

Systems have a capacity to absorb disturbances, but this capacity has limits and bounds, and when these limits are exceeded the system may rapidly transform. Holling was the first to recognize the significance of thresholds in ecological systems, and the importance of avoiding them. Holling (1973) outlines how the response of systems can exhibit threshold behavior. Changes in either driving or state variables may cause collapses. Often, the system provides no warning, and collapse follows an unexpected, but often inevitable, event. More recently, increasing variability in some variables has been suggested as an indicator of impending collapse of a system as it approaches the limits of its resilience (Carpenter and Brock 2006, Wardwell and Allen 2009).

Knowledge of the form of critical population processes (particularly predation) suggests that ecological systems have more than one stable state (*multiple stable states, multiple equilibria,* and *meta-stable states* are all words used to describe such behavior). Variables move between some of those states, and that variability both results in and is caused by diversity in space, time, and species. Abrupt jumps in variables are the rule, not the exception.

Resilience is described here as the property that allows the fundamental functions of an ecosystem to persist in the face of extremes of disturbance. It can be measured by the size of the viable stability domains. Stability, in contrast to resilience, is used in a narrow sense of elasticity. It is the property that resists departure from equilibrium and that maximizes the speed of return to the equilibrium following small disturbances. Resilience focuses on the role of positive feedbacks, of behavior far from steady states and with internally generated variability. Stability, in the narrow sense above, deals with negative feedback, of behavior near steady states, and with constancy. Different views and definitions are still being used to distinguish between resilience and stability. The view expressed above is generally used by those ecologists who develop theory empirically, who often use simulation models, and who conduct their science integrated with policy and ecological management. They are typically trained within a biological tradition.

Those who define resilience differently use a measure of elasticity or return time, a definition that is the opposite of that above. It is a definition that implicitly assumes there is only one equilibrium state. Scientists holding this view tend to be more deductive in their formation of theory or are influenced by an engineering and applied mathematical tradition. They tend to apply theory to practice rather than to develop theory empirically as part of practice.

One empirically based critique of multi-stable states has been influential (Sousa and Connell 1985). But this critique is inadequate because it relies on only existing published time series data and ignores any kind of analysis of causation. It is a phenomenological investigation, not causal. As a result, behavior is seen as being determined or explainable by only one variable, the time horizon for the variable is too short, and multiscale interactions between variables of different speeds are ignored. This reflects a common limitation of many population studies.

Holling (1973) documents stability domains using empirical evidence from numerous studies. Stability is defined as the return of a system to an equilibrium state following disturbance, and resilience is defined as a measure of a system's persistence and its ability to absorb change and disturbance but still maintain the same relationships among population or state variables. A system can be highly unstable but very resilient. In fact, a key insight of this paper is that instability may create highly resilient systems (e.g., grassland persistence is reliant on frequently occurring fires). Managing for stability, as humans so often do, has the unexpected outcome of reducing a system's resilience (Holling and Meffe 1996, Allen and Holling 2008). An equilibrium-focused view is attractive to humans, who often focus on optimizing single elements of systems, but it fails to capture the behavior of complex systems.

By the early 1990s, many authors in the ecological literature (O'Neill et al. 1986, Pimm 1984, Tilman and Downing 1994) had applied the word *resilience* as the speed or time of return of an ecological system to an equilibrium following a disturbance. This was part of a multifaceted definition outlined by Holling in 1973, but it is a narrow definition and ignores the presence of alternative states. In response, Holling (1996) explicitly contrasts and compares two primary definitions of resilience, which he describes as *engineering resilience* and *ecological resilience*. Ecological systems differ from engineering ones in that change is not continuous but, rather, discontinuous; ecological change is characterized by surprising events (such as hurricanes, fires, or pest outbreaks) that open windows of opportunities for establishing new combinations of species and ecological processes (Allen and Holling 2008). Ecological attributes are also distributed discontinuously in space, across scales. Ecosystems don't have single equilibria; rather, they have multiple equilibrium and are often far from equilibrium and are on dynamic trajectories. Like the location of electrons about an atom, an ecological system has changed by the time it can be measured, and optimal approaches to ecosystem management are prone to failure. Additionally, management actions and policies focused on constant yields and the reduction of variability reduce the resilience of a system. Because these systems are "moving targets," management needs to be flexible, adaptive, and experimental and must recognize the multiple critical scales characterizing a given system.

Engineering resilience focuses on equilibrium states and is measured simply as the return time following disturbance. This definition, often used by population biologists, is analogous to the intrinsic rate of increase of a species (r). Engineering resilience focuses on stability, in the sense of elasticity. It is the ability of a system to resist departure from an equilibrium following disturbances and to return to the same equilibrium when sufficiently perturbed.

Ecological resilience focuses on conditions far from equilibrium, when abrupt shifts between multiple stable states are possible. Here, the measurement of resilience is the magnitude or amount of disturbance that can be absorbed without undergoing the shift to an alternative stable state characterized by changes in controlling variables and processes and their dominant scales. More recently, stable states have been characterized by their process regimes, and the term *regime shift* has been used (Scheffer et al. 2001). Ecological resilience can be measured by the size of the stability domains.

The two differing definitions of resilience lead to grossly different strategies for managing systems and responding to surprise. Managing for stability is suggested by engineering resilience; however, this often has demonstrable negative consequences in the long run. Productivity or yield is often increased over short time periods due to management efficiency and optimization but suffers in the long run as ecological surprises exceed the diminished resilience of the system. This happens because managing for reduced variability in one or a small number of variables alters competitive interactions and the buildup of capital (such as fuel for fires) and leads to the loss of important processes and functions. The reduction in variability means that key structuring variables and processes are lost or greatly diminished (e.g., pest outbreaks or fires).

The changes that occur when the resilience of a system is exceeded can lead to an undesirable, but highly resilient, system state. Reversing the system can be very difficult because undesirable systems can be extremely resilient and the regime shifts may exhibit hysteresis. Concomitant with the reduction in resilience when management attempts to reduce variability is an increasingly rigid management bureaucracy and ever more dependent—and vulnerable—human societies. An increasing reliance by humans on systems where management has reduced variability often means that ecological shifts have ever greater and negative implications for human economies and societies.

Holling (1996) also begins to formulate a model of the relationship among ecological diversity, resilience, and scale (formally conceptualized in Peterson et al. 1998). He notes that resilient systems have multiple controls that are most efficient on different scales, and that the distribution of diversity within and across scales is what matters.

Holling (1986) is part of a groundbreaking volume that was one of the first works to build and synthesize understanding around themes of sustainability and global environmental change (Clark and Munn 1986). This was years before global climate change was a widespread research topic or undertaken by large international research bodies. As part of this work, Holling (1986) applied the concept of resilience to help understand how a wide range of ecosystems would respond to broad-scale environmental (climatic) change. Ecological systems exhibit a diverse array of responses to global changes—a characteristic that is inherent in their resilience. Nonadaptive systems with little flexibility in response to perturbation and disturbance would be in a constant state of flux and disarray. This paper provides an early recognition of discontinuous and nonlinear response in ecological systems and represents an early attempt to link ecological and social systems across scales. Holling (1986) recognized that positive feedbacks are responsible for maintaining the systems on which humanity relies—for example, feedbacks between the atmosphere and vegetation.

In many ways, this paper was an early warning of the very real possibility that the resilience of global systems could be exceeded, resulting in very sudden and effectively irreversible regime shifts. Because many of the anticipated changes are global, rather than local, in nature, adaptation to changes caused by the human footprint will need to occur not only within individuals but within institutions and social systems as well. Twenty years later, approaches linking social-economic-ecological systems are commonplace and viewed as the frontier in global change and resilience research (Walker and Salt 2006).

A theme of the Holling (1986) paper is recognizing the inevitability of "surprises"— unexpected outcomes with causes and responses very different from those anticipated, or results or behaviors that are induced by human actions but which are very different than expected. Such thoughts were given a wide public airing when Donald Rumsfeld, secretary of defense under U.S. president George W. Bush, stated the following at a Department of Defense news briefing on February 12, 2002: "Because as we know, there are known knowns; there are things we know we know. We also know there are known unknowns; that is to say we know there are some things we do not know. But there are also unknown unknowns—the ones we don't know we don't know."

This quote is often offered as an example of incompetence and incoherence, but it is actually an explicit acknowledgment of the inevitability of surprise, even when every envisioned contingency has been planned for. That nearly all of the global public did not understand it suggests that humanity still cherishes the idea of predictability and linear response. However, an increasing emphasis on adaptive management in the United States and elsewhere—for example, the shift of the U.S. Department of Interior to an adaptive management paradigm (Williams et al. 2007) with an emphasis on identifying and reducing uncertainty—offers some hope for the future of resource management. The viewpoint of linear causation and response, and constancy as a management goal, is still common and suggests to policy makers and the public alike that mistakes can be made but that the affected system, be it local or global, will eventually return to equilibrium. The two viewpoints—that of discontinuous change and surprise versus that of continuous change and predictability—lead to very different approaches to policy and institutions. For looming global changes, such as those potentially wrought by climate change, the latter suggests that there is plenty of time for social and ecological adaptation to change, while the former suggests sudden and perhaps catastrophic change exceeding the limits of human adaptability.

Importantly, Holling (1986) describes how a resilient system might be self-maintaining. First, self-organizing processes within and across scales provide positive feedbacks. Second, instabilities and variability experienced by the system, if not of a magnitude sufficient to exceed the systems' resilience, help to strengthen structures and encourage adaptation and may generate novelty (Allen and Holling 2008).

Scaling, and the characteristic time and space domains in systems, is described as discontinuous and critical to understanding the organization, and thus the evolution and resilience, of systems. This description and related understanding builds directly on hierarchy theory (Allen and Starr 1982) and itself has evolved to the theory of complex adaptive systems (Arthur et al. 1997, Levin 1998) and complexity theory. This insight, though not unique, was critical in developing resilience theory. It suggests that systems are characterized by having discrete structures and functions and processes at each time and space domain present (Holling 1992); that scale-specific interactions and positive feedbacks maintain scale-specific structures and functions; that changes between scales are discontinuous; that the number of scales present is finite and limited; and that when the resilience of a system is exceeded, reorganization rapidly occurs and the scale-specific structure and function present in the new system may be vastly different from the old.

Holling introduced the adaptive cycle in the 1986 article, which was a foundation of panarchy theory (Gunderson, Holling, and Light 1995). The adaptive cycle consists of four phases of system state: exploitation, conservation, creative destruction, and renewal. These phases are also described for human institutions. For ecosystems and institutions, increasing stored capital and connectedness are described as accidents waiting to happen. Panarchy theory took an additional decade to develop fully. Ideas of discontinuous change and scaling developed sooner (Holling 1992), but it took another fifteen or more years for ecologists and others to recognize the ubiquity of discontinuities in ecosystems (Ludwig 2008).

Holling (1986) provides a number of examples where human management in the form of reducing or controlling variability in the system made those systems less resilient in the long run and vulnerable to catastrophic change. Usually, management has been in the form of command and control (Holling and Meffe 1996) and eschewed the tenets of adaptive management (Holling 1978, Walters 1986).

Stability and resilience may have an inverse relationship. Systems with high variability may have high resilience but low stability. Holling (1986) describes a pathology of managed ecosystems that occurs when people successfully control and reduce the variability of key ecological processes, which in turn leads to a decrease in ecological resilience. Rigid institutions reinforce the fragility introduced to ecological systems by command-and-control management with the result that both the ecological and institutional systems become ever more vulnerable to the inevitable surprise.

The proposition that ecosystems exhibit multiple stable states was posited by Holling in 1973, but that idea, and resilience theory itself, gained little immediate traction. In 1985, Sousa and Connell challenged the idea by reviewing a number of studies that they considered reasonably long term and concluding that multiple stable states were, at best, the rare exception. The long-term studies they reviewed were, in fact, too small spatially and too short temporally to identify even decadal shifts and cycles, but the data they analyzed were some of the best available. In the more than two decades since their review, evidence has accumulated that many, and in fact most, ecological systems exhibit multiple stable states. Currently, this is an expanding avenue of research, and the rapid transitions between stable states are referred to as regime shifts (Scheffer et al. 2001). These shifts are characterized by a rapid reorganization of form and structure controlled by new or altered processes. New stable states (or regimes) produce different ecological goods and services, which may be beneficial or harmful for humans and which may have low or high resilience (Gunderson 2000, Folke et al. 2004).

Folke et al. (2004) document the evidence of regime shifts. They also document the increasing prevalence of such shifts and suggest that this is due to human activities that erode resilience. Because of rapid land use and land cover alterations at scales from the local to the global, and because of current and anticipated anthropogenic climate change, understanding the causes of regime shifts and, inversely, the components of ecological systems that create and maintain their resilience is critical.

Often the disruption or alteration of a key structuring process (or variable) is sufficient to trigger a regime shift. Examples include phosphorus cycling in lakes (Carpenter 2003) and fire periodicity in rangelands (Dublin et al. 1990). The transition of state variables following a change in key processes can be sudden—as is the case with lake ecosystems—or may exhibit pronounced time lags—as with rangelands. Regardless of the response time of state variables, after the key structuring processes have been altered the eventual change in structure is inevitable.

Folke et al. (2004) forward a refined definition of resilience that more explicitly recognizes within- and across-scale linkages and adaptability. They suggest that "resilience reflects the degree to which a complex adaptive system is capable of self-organization ... and the degree to which the system can build and increase the capacity for learning and adaptation." This definition explicitly considers that ecosystems self-organize, that self-organization (i.e., positive feedbacks) occurs at a distinct but limited scale of space and time, that biodiversity has a critical role, and that these scales are nested. This nested structure within ecosystems has been described as a panarchy. Panarchies are largely constrained by top-down controls, and they are structured hierarchically, but reorganization may also be triggered at smaller scales (at lower levels in the hierarchy). The increasing frequency of regime shifts should be of great concern if this view of ecological organization is true. Increasing transformations of ecosystems may be predicted to eventually cause global transformations with consequences that are highly uncertain.

Decreasing resilience in a system by, for example, reducing biological diversity, and thus ecological function, or by removing too much biomass makes systems ever more sensitive to smaller and smaller external forces that can trigger regime shifts. Folke et al. (2004) argue that it is human actions that have increased the prevalence of regime shifts, and that such shifts often transform ecosystems to a less desirable state while also reducing the provision of ecological goods and services to humans. Human actions have been both top down and bottom up (e.g., nutrient cycling) or via alteration of disturbance regimes. Following regime shifts, the new ecological organization is more likely to include more weedy, colonizing, or invasive species and the new systems may be more prone to surprise, at least at human time scales.

Although humans are now the predominant drivers of ecological organization, and thus regime shifts, across the globe, humans also have the capacity to manage ecosystems in a manner that can build or enhance resilience. This can be done by managing to avoid catastrophic thresholds rather than managing for reduced variability in some state variables (Biggs and Rogers 2003). Resilience-building or transforming approaches to ecosystem management will need to be flexible and open to learning. Adaptive management approaches are designed to increase learning and to reduce and identify key sources of uncertainty.

The relationship between species richness (biological diversity) and measures of "stability" has been a subject of debate and theory for well over a hundred years (Darwin 1859). Modeling (Drake et al. 1996), mesocosms (Naeem et al. 1994), and small-scale experiments (Tilman 1996) have focused the debate toward the role of functional diversity in maintaining stability. Those studies have established that increasing functional diversity increases stability in the face of natural variation. Sources of variation include temperature, rainfall, and the influx of new species (invasions). Despite this research, much conservation effort is still expended to protect the largest number of individual taxa regardless of functional importance or population viability, in static reserves where species are preserved and humans excluded, regardless of the rapidly transforming landscapes within which they are embedded.

Folke et al. (2004) advocate a conservation approach that creates incentives for humanity to "work in harmony" with the ecosystems they inhabit. Because humanity relies on ecosystems for an enormous quantity of services, many of which are simply taken for granted (e.g., photosynthesis), conservation should focus on maintaining those services and thus the functional groups that create and maintain them. Ecosystems usually maintain their essential functions even when species are lost. This suggests that functions are much more robust than species richness. Holling (1992) presented the Textural Discontinuity Hypothesis, which posits that ecosystems are structured by a relatively few key processes, each operating at distinct temporal and spatial scales. Thus, maintaining those key processes should be a conservation priority. And those key processes are entwined in scale-specific positive biotic and abiotic feedbacks — that is, the structure and functions self-organize within ranges of scale — and are responsible for maintaining the overall resilience of an ecological systems (Peterson et al. 1998, Elmqvist et al. 2003, Allen et al. 2005).

Currently, especially in the developed world, humans are disconnected from ecological systems at many levels. The ecosystems humanity inhabits are not recognized as absolutely critical to humanity's survival: a dominance-of-nature viewpoint still reigns in the twenty-first century. Human economies are disconnected from ecosystems, and ecosystems are simply economic externalities, rarely addressed in mainstream economics. Government policies often subsidize the exploitation of natural resources. This disconnect is a primary driver of species loss and management strategies that attempt to reduce variability in a few economically important variables, such as deer populations or board feet of aspen. Both result in reduced resilience. Because the loss of resilience often is not manifest with an immediate catastrophe, this lack of connection between humans and nature is further reinforced. Because the loss of resilience in ecological systems increases the probability of a regime shift, the services that humanity depends on will change. Regime shifts often lead to less desirable system states and fewer or less reliable ecological services of benefit to humans, or at least vastly different arrays of services. Therefore, it is in humanity's self-interest to maintain the resilience and adaptive capacity of ecosystems by managing for complex systems of humans and nature, rather than managing humans and managing nature.

Models of the relationship between biological diversity and ecological stability all explicitly or implicitly focus on species' functions. Most have considered that increasing species richness increases functional representation and thus stability. Most authors have focused on stability—that is, the return rate of ecosystems following disturbance. More stable systems return to their equilibrium state more quickly.

The model of Darwin (1859; formalized by MacArthur 1955) considered implicitly that all species provided an equivalent amount and intensity of function. Thus, adding more species increases the functions present and the redundancy of function. That all species provide equivalent degrees of function, and that each functional group provides equally important functions, is unlikely. Lawton (1994) presented a model that considered that the intensity and amount (breadth) of ecological function provided by each species varies. Therefore, adding more species to an ecological system may trivially or substantially increase stability. Moreover, because species interact, sometimes strongly, the idiosyncratic nature of those interactions influences ecological stability more than adding new or redundant functions does. Thus, adding species may actually decrease stability although the overall relationship is one of increasing stability with increasing species richness.

Ehrlich and Ehrlich (1981) presented a model that explicitly incorporated redundancy. Because the number of ecological functions is finite, the relationship between species richness and stability eventually asymptotes. Ehrlich and Ehrlich's model considered that the functions provided by species were equivalent in terms of breadth and intensity of function. Walker's model (1992, 1995), on the other hand, like Lawton's (1994), considered variable breadth and intensity of function among species. He termed species with strong functions *driver species* and those performing weak functions *passenger species*. In his model, adding passengers to the system has little effect on stability but adding drivers has a strong effect.

All but the earliest models consider that functional redundancy exists among species. Redundancy of function adds to the stability of systems because losing one or a few species from a system will have little or no effect on stability as long as at least one species providing that function is present.

The model presented by Peterson et al. (1998) enriches our understanding of the role of biodiversity on at least two fronts. First, the authors focus on the property of resilience rather than stability. Today, the use of the term *stability* in describing ecological systems is often inappropriate. Ecological systems are best described as occupying a trajectory, characterized by nonlinear and non-optimal dynamics. Indeed, attempts at optimal control are doomed to failure because by the time the required measurements are made the system is in a different state. If ecological systems can exist in alternative dynamic states, then resilience is the most appropriate measure of system behavior. In this view, stability is a system property to be avoided because stable systems are unlikely to be adaptive and are more likely to experience catastrophic failure. Second, the authors incorporate scale into their model of the relationship between biodiversity and resilience.

In the Peterson et al. model (1998), the relationship between species richness and resilience depends on the distribution of functions present both within and across scales. Within a single range of scale, resilience is increased when a diversity of functions is present and when the breadth of those functions present is high (Elmqvist et al. 2003). This allows a system to respond to a great variety of perturbations that may occur. Across scales, a redundancy of function increases resilience. However, the authors term this characteristic *cross-scale reinforcement* because a function occurring at different scales is not truly redundant. For example, the function of seed dispersal occurs at very small scales via ants, at meso scales via birds, and at large scales via mammals. Though the function is the same, it is not redundant because it is occurring at grossly different scales. This redundancy of function is important because it provides a robust control of perturbations that can scale up, such as pest outbreak (Holling 1988). The loss of seemingly redundant functions that occur at different scales may not have an immediate impact on a system (i.e., its stability appears unchanged), but its resilience is reduced—a fact that could well remain undetected until just the wrong disturbance at the wrong scale occurs.

Determining the scales present in systems is important to resilience and panarchy theory. Holling (1992) applied a crude method based on a split moving window analysis of animal body masses. Methods for the detection of discontinuities that bound scale-specific regimes (scales of positively reinforced interactions) are still problematic but have made some strides (Stow et al. 2007). The model of Peterson et al. (1998) has been tested (Fischer et al. 2007, Fischer et al. 2008, Wardwell et al. 2008), and the quantification of the distribution of function within and across scales has been forwarded as a method of assessing the relative resilience of ecosystems (Allen et al. 2005).

Literature Cited

Allen, C. R., L. Gunderson, and A. R. Johnson. 2005. The use of discontinuities and functional groups to assess relative resilience in complex systems. *Ecosystems* 8:958–66.

Allen, C. R., and C. S. Holling, eds. 2008. *Discontinuities in ecosystems and other complex systems*. New York: University of Columbia Press.

Allen, T. F. H., and T. B. Starr. 1982. *Hierarchy: Perspectives for ecological complexity*. Chicago: University of Chicago Press.

Arthur, W. B., S. N. Durlauf, and D. Lane. 1997. Introduction. In *The economy as an evolving complex system II*, ed. W. B. Arthur, S. N. Durlauf, and D. Lane, 1–14. Reading, MA: Addison-Wesley.

Biggs, H. C., and K. H. Rogers. 2003. An adaptive system to link science, monitoring, and management in practice. In *The Kruger experience: Ecology and management of savanna heterogeneity*, ed. J. T. du Toit, K. H. Rogers, and H. G. Biggs, 59–80. Washington, DC: Island Press.

Carpenter, S. R. 2003. *Regime shifts in lake ecosystems: Pattern and variation*. Vol. 15, Excellence in Ecology Series. Oldendorf/Luhe, Ger.: Ecology Institute.

Carpenter, S. R., and W. A. Brock. 2006. Rising variance: A leading indicator of ecological transition. *Ecology Letters* 9:311–18.

Clark, W. C., and R. E. Munn, eds. 1986. The resilience of terrestrial ecosystems: Local surprise and global change. In *Sustainable development of the biosphere*. Cambridge: Cambridge University Press.

Darwin, C. 1859. *On the origin of species by means of natural selection or the preservation of favoured races in the struggle for life*. Reprint, 1964, Cambridge, MA: Harvard University Press.

Drake, J. A., G. R. Huxel, and C. L. Hewitt. 1996. Microcosms as models for generating and testing community theory. *Ecology* 77:670–77.

Dublin, H. T, A. R. E. Sinclair, and J. McGlade. 1990. Elephants and fire as causes of multiple stable states in the Serengeti–Mara woodlands. *Journal of Animal Ecology* 59:1147–64.

Ehrlich, P. R., and A. H. Ehrlich. 1981. *Extinction: The causes and consequences of the disappearance of species*. New York: Random House.

Elmqvist, T., C. Folke, M. Nystrom, G. Peterson, J. Bengtsson, B. Walker, and J. Norberg. 2003. Response diversity, ecosystem change, and resilience. *Frontiers in Ecology and the Environment* 1:488–94.

Fischer, J., D. B. Lindenmayer, S. P. Blomberg, R. Montague-Drake, A. Felton, and J. A. Stein. 2007. Functional richness and relative resilience of bird communities in regions with different land use intensities. *Ecosystems* 10:964–74.

Fischer, J., D. B. Lindenmayer, and R. Montague-Drake. 2008. The role of landscape texture in conservation biogeography: A case study on birds in south-eastern Australia. *Diversity and Distributions* 14:38–46.

Folke, C., S. Carpenter, M. Scheffer, T. Elmqvist, L. Gunderson, and C. S. Holling. 2004. Regime shifts, resilience and biodiversity in ecosystem management. *Annual Review of Ecology and Systematics* 35:557–81.

Gunderson, L. H. 2000. Resilience in theory and practice. *Annual Review of Ecology and Systematics* 31:425–39.

Gunderson, L. H., and C. S. Holling. 2002. *Panarchy: Understanding transformations in human and natural systems*. Washington, DC: Island Press.

Gunderson, L. H., C. S. Holling, and S. Light, eds. 1995. *Barriers and bridges to renewal of ecosystems and institutions*. New York: Columbia University Press.

Harper, D. 2001. *Online etymology dictionary.* http://www.etymonline.com.

Holling, C. S. 1973. Resilience and stability of ecological systems. *Annual Review of Ecological Systems* 4:1–23.

———, ed. 1978. *Adaptive environmental assessment and management.* Chichester: Wiley.

———. 1986. The resilience of terrestrial ecosystems: Local surprise and global change. In *Sustainable development of the biosphere*, ed. W. C. Clark and R. E. Munn, 292–317. Cambridge: Cambridge University Press.

———. 1988. Temperate forest insect outbreaks, tropical deforestation and migratory birds. *Memoirs of the Entomological Society of Canada* 146:21–32.

———. 1992. Cross-scale morphology, geometry, and dynamics of ecosystems. *Ecological Monographs* 62:447–502.

———. 1996. Engineering resilience versus ecological resilience. In *Engineering within ecological constraints*, ed. P. C. Schulze, 31–44. Washington, DC: National Academies Press.

Holling, C. S., and G. K. Meffe. 1996. Command and control and the pathology of natural resource management. *Conservation Biology* 10:328–37.

Lawton, J. H. 1994. What do species do in ecosystems? *Oikos* 71:367–74.

Levin, S. A. 1998. Ecosystems and the biosphere as complex adaptive systems. *Ecosystems* 1:431–36.

Ludwig, D. 2008. Synthesis. In *Discontinuities in ecosystems and other complex systems*, ed. C. R. Allen and C. S. Holling, 234–39. New York: University of Columbia Press.

MacArthur, R. H. 1955. Fluctuations of animal populations and a measure of community stability. *Ecology* 36:533–36.

Naeem, S., L. J. Thompson, S. P. Lawler, J. H. Lawton, and R. M. Woodfin. 1994. Declining biodiversity can alter the performance of ecosystems. *Nature* 368:734–37.

O'Neill, R. V., D. L. DeAngelis, J. B. Waide, and T. F. H. Allen. 1986. *A hierarchical concept of ecosystems.* Princeton, NJ: Princeton University Press.

Peterson, G., C. R. Allen, and C. S. Holling. 1998. Ecological resilience, biodiversity and scale. *Ecosystems* 1:6–18.

Pimm, S. L. 1984. The complexity and stability of ecosystems. *Nature* 307:321–26.

Scheffer, M., S. Carpenter, J. A. Foley, C. Folke, and B. Walker. 2001. Catastrophic shifts in ecosystems. *Nature* 413:591–96.

Sousa, W. P., and J. H. Connell. 1985. Further comments on the evidence for multiple stable points in natural communities. *American Naturalist* 125:612–15.

Stow, C. A., C. R. Allen, and A. S. Garmestani. 2007. Evaluating discontinuities in complex systems: Toward quantitative measures of resilience. *Ecology and Society* 12 (1): 26. http://www.ecologyandsociety.org/vol12/iss1/art26/.

Tilman, D. 1996. Biodiversity: Population versus ecosystem stability. *Ecology* 77:350–63.

Tilman, D., and J. A. Downing. 1994. Biodiversity and stability in grasslands. *Nature* 367:363–65.

Walker, B. 1992. Biological diversity and ecological redundancy. *Conservation Biology* 6:18–23.

———. 1995. Conserving biological diversity through ecosystem resilience. *Conservation Biology* 9:747–52.

Walker, B., C. S. Holling, S. R. Carpenter, and A. Kinzig. 2004. Resilience, adaptability and transformability in social–ecological systems. *Ecology and Society* 9 (2): 5. http://www.ecologyandsociety.org/vol9/iss2/art5/.

Walker, B., and D. Salt. 2006. *Resilience thinking: Sustaining ecosystems and people in a changing world*. Washington, DC: Island Press.

Walters, C. 1986. *Adaptive management of renewable resources*. New York: Macmillan.

Wardwell, D., and C. R. Allen. 2009. Variability in population abundance is associated with thresholds between scaling regimes. *Ecology and Society* (in press).

Wardwell, D., C. R. Allen, G. D. Peterson, and A. D. Tyre. 2008. A test of the cross-scale resilience model: Functional richness in Mediterranean-climate ecosystems. *Ecological Complexity* 5:165–82.

Williams, B. K., R. C. Szaro, and C. D. Shapiro. 2007. *Adaptive management: The U.S. Department of Interior technical guide*. Washington, DC: Adaptive Management Working Group, U.S. Department of the Interior.

Resilience and Stability of Ecological Systems

C. S. HOLLING

C. S. HOLLING, *Institute of Resource Ecology, University of British Columbia, Vancouver, Canada.*

Annual Review of Ecology and Systematics 1973.4:1–23.
Originally Published by *Annual Reviews.*

Introduction

INDIVIDUALS DIE, POPULATIONS DISAPPEAR, and species become extinct. That is one view of the world. But another view of the world concentrates not so much on presence or absence as upon the numbers of organisms and the degree of constancy of their numbers. These are two very different ways of viewing the behavior of systems and the usefulness of the view depends very much on the properties of the system concerned. If we are examining a particular device designed by the engineer to perform specific tasks under a rather narrow range of predictable external conditions, we are likely to be more concerned with consistent nonvariable performance in which slight departures from the performance goal are immediately counteracted. A quantitative view of the behavior of the system is, therefore, essential. With attention focused upon achieving constancy, the critical events seem to be the amplitude and frequency of oscillations. But if we are dealing with a system profoundly affected by changes external to it, and continually confronted by the unexpected, the constancy of its behavior becomes less important than the persistence of the relationships. Attention shifts, therefore, to the qualitative and to questions of existence or not.

Our traditions of analysis in theoretical and empirical ecology have been largely inherited from developments in classical physics and its applied variants. Inevitably, there has been a tendency to emphasize the quantitative rather than the qualitative, for it is important in this tradition to know not just that a quantity is larger than another quantity, but precisely how much larger. It is similarly important, if a quantity fluctuates, to know its amplitude and period of fluctuation. But this orientation may simply reflect an analytic approach developed in one area because it was useful and then transferred to another where it may not be.

Our traditional view of natural systems, therefore, might well be less a meaningful reality than a perceptual convenience. There can in some years be more owls and fewer mice and in others, the reverse. Fish populations wax and wane as a natural condition, and insect populations can range over extremes that only logarithmic transformations can easily illustrate. Moreover, over distinct areas, during long or short periods of time, species can completely disappear and then reappear. Different and useful insight might be obtained, therefore, by viewing the behavior of ecological systems in terms of the probability of extinction of their elements, and by shifting emphasis from the equilibrium states to the conditions for persistence.

An equilibrium-centered view is essentially static and provides little insight into the transient behavior of systems that are not near the equilibrium. Natural, undisturbed systems are likely to be continually in a transient state; they will be equally so under the influence of man. As man's numbers and economic demands increase, his use of resources shifts equilibrium states and moves populations away from equilibria. The present concerns for pollution and endangered species are specific signals that the well-being of the world is not adequately described by concentrating on equilibria and conditions near them. Moreover, strategies based upon these two different views of the world might well be antagonistic. It is at least conceivable that the effective and responsible effort to provide a maximum sustained yield from a fish population or a nonfluctuating supply of water from a watershed (both equilibrium-centered views) might paradoxically increase the chance for extinctions.

The purpose of this review is to explore both ecological theory and the behavior of natural systems to see if different perspectives of their behavior can yield different insights useful for both theory and practice.

Some Theory

Let us first consider the behavior of two interacting populations: a predator and its prey, a herbivore and its resource, or two competitors. If the interrelations are at all regulated we might expect a disturbance of one or both populations in a constant environment to be followed by fluctuations that gradually decrease in amplitude. They might be represented as in Figure 1, where the fluctuations of each population over time are shown as the sides of a box. In this example the two populations in some sense are regulating each other, but the lags in the response generate a series of oscillations whose amplitude gradually reduces to a constant and sustained value for each population. But if we are also concerned with persistence we would like to know not just how the populations behave from one particular pair of starting values, but from all possible pairs since there might well be combinations of starting populations for which ultimately the fate of one or other of the populations is extinction. It becomes very difficult on time plots to show the full variety of responses possible, and it proves convenient to plot a trajectory in a phase plane. This is shown by the end of the box in Figure 1 where the two axes represent the density of the two populations.

The trajectory shown on that plane represents the sequential change of the two populations at constant time intervals. Each point represents the unique density of each population at a particular point in time and the arrows indicate the direction of change over time. If oscillations are damped, as in the case shown, then the trajectory is represented as a closed spiral that eventually reaches a stable equilibrium.

We can imagine a number of different forms for trajectories in the phase plane (Figure 2). Figure 2a shows an open spiral which would represent situations where fluctuations gradually increase in amplitude. The small arrows are added to suggest that this condition holds no matter what combination of populations initiates the trajectory. In Figure 2b

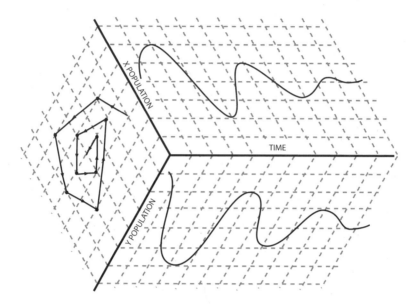

FIGURE 1: *Derivation of a phase plane showing the changes in numbers of two populations over time.*

the trajectories are closed and given any starting point eventually return to that point. It is particularly significant that each starting point generates a unique cycle and there is no tendency for points to converge to a single cycle or point. This can be termed "neutral stability" and it is the kind of stability achieved by an imaginary frictionless pendulum.

Figure 2c represents a stable system similar to that of Figure 1, in which all possible trajectories in the phase plane spiral into an equilibrium. These three examples are relatively simple and, however relevant for classical stability analysis, may well be theoretical curiosities in ecology. Figures 2d–2f add some complexities. In a sense Figure 2d represents a combination of a and c, with a region in the center of the phase plane within which all possible trajectories spiral inwards to equilibrium. Those outside this region spiral outwards and lead eventually to extinction of one or the other populations. This is an example of local stability in contrast to the global stability of Figure 2c. I designate the region within which stability occurs as the domain of attraction, and the line that contains this domain as the boundary of the attraction domain.

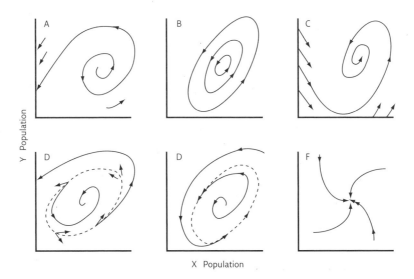

FIGURE 2: *Examples of possible behaviors of systems in a phase plane; (a) unstable equilibrium, (b) neutrally stable cycles, (c) stable equilibrium, (d) domain of attraction, (e) stable limit cycle, (f) stable node.*

The trajectories in Figure 2e behave in just the opposite way. There is an internal region within which the trajectories spiral out to a stable limit cycle and beyond which they spiral inwards to it. Finally, a stable node is shown in Figure 2f in which there are no oscillations and the trajectories approach the node monotonically. These six figures could be combined in an almost infinite variety of ways to produce several domains of attraction within which there could be a stable equilibrium, a stable limit cycle, a stable node, or even neutrally stable orbits. Although I have presumed a constant world throughout, in the presence of random fluctuations of parameters or of driving variables (Walters 39), any one trajectory could wander with only its general form approaching the shape of the trajectory shown. These added complications are explored later when we consider real systems. For the moment however, let us review theoretical treatments in the light of the possibilities suggested in Figure 2.

The present status of ecological stability theory is very well summarized in a number of analyses of classical models, particularly May's (23–25) insightful analyses of the Lotka-Volterra model and its expansions, the graphical stability analyses of Rosenzweig (33, 34), and the methodological review of Lewontin (20).

May (24) reviews the large class of coupled differential equations expressing the rate of change of two populations as continuous functions of both. The behavior of these models results from the interplay between (a) stabilizing negative feedback density-dependent responses to resources and predation, and (b) the destabilizing effects produced by the way individual predators attack and predator numbers respond to prey density [termed the functional and numerical responses, as in Holling (11)]. Various forms have been given to these terms; the familiar Lotka-Volterra model includes the simplest and least realistic, in which death of prey is caused only by predation, predation is a linear function of the product of prey and predator populations, and growth of the predator population is linearly proportional to the same product. This model generates neutral stability as in Figure 2b, but the assumptions are very unrealistic since very few components are included, there are no explicit lags or spatial elements, and thresholds, limits, and nonlinearities are missing.

These features have all been shown to be essential properties of the predation process (Holling 12, 13) and the effect of adding some of them has been analyzed by May (24). He points out that traditional ways of analyzing the stability properties of models using analytical or graphical means (Rosenzweig & MacArthur 33, Rosenzweig 34, 35) concentrate about the immediate neighborhood of the equilibrium. By doing this, linear techniques of analysis can be applied that are analytically tractable. Such analyses show that with certain defined sets of parameters stable equilibrium points or nodes exist (such as Figure 2c), while for other sets they not, and in such cases the system is, by default, presumed to be unstable, as in Figure 2a. May (24), however, invokes a little-used theorem of Kolmogorov (Minorksy 26) to show that all these models have either a stable equilibrium point or a stable limit cycle (as in Figure 2e). Hence he concludes that the conditions presumed by linear analysis are unstable, and in fact must lead to stable limit cycles. In every instance, however, the models are globally rather than locally stable, limiting their behavior to that shown in either Figures 2c or 2e.

There is another tradition of models that recognizes the basically discontinuous features of ecological systems and incorporates explicit lags. Nicholson and Bailey initiated this tradition when they developed a model using the output of attacks and survivals within one generation as the input for the next (29). The introduction this explicit lag generates oscillations that increase in amplitude until one or other of the species becomes extinct (Figure 2a). Their assumptions are as unrealistically simple as Lotka's and Volterra's; the instability results because the number of attacking predators at any moment is so much a consequence of events in the previous generation that there are "too many" when prey are declining and "too few" when prey are increasing. If a lag is introduced into the Lotka-Volterra formulation (Wangersky & Cunningham 40) the same instability results.

The sense one gains, then, of the behavior of the traditional models is that they are either globally unstable or globally stable, that neutral stability is very unlikely, and that when the models are stable a limit cycle is a likely consequence.

Many, but not all, of the simplifying assumptions have been relaxed in simulation models, and there is one example (Holling & Ewing 14) that joins the two traditions initiated by Lotka-Volterra and Nicholson and Bailey and, further, includes more realism in the operation of the stabilizing and destabilizing forces. These modifications are described in more detail later; the important features accounting for the difference in behavior result from the introduction of explicit lags, a functional response of predators that rises monotonically to a plateau, a nonrandom (or contagious) attack by predators, and a minimum prey density below which reproduction does not occur. With these changes a very different pattern emerges that conforms most closely to Figure 2d. That is, there exists a domain of attraction within which there is a stable equilibrium; beyond that domain the prey population becomes extinct. Unlike the Nicholson and Bailey model, the stability becomes possible, although in a limited region, because of contagious attack. [Contagious attack implies that for one reason or another some prey have a greater probability of being attacked than others, a condition that is common in nature (Griffiths & Holling 9).] The influence of contagious attack becomes significant whenever predators become abundant in relation to the prey, for then the susceptible prey receive the burden of attention, allowing more prey to escape than would be expected by random contact. This "inefficiency" of the predator allows the system to counteract the destabilizing effects of the lag.

If this were the only difference the system would be globally stable, much as Figure 2c. The inability of the prey to reproduce at low densities, however, allows some of the trajectories to cut this reproduction threshold, and the prey become extinct. This introduces a lower prey density boundary to the attraction domain and, at the same time, a higher prey density boundary above which the amplitudes of the oscillations inevitably carry the population below the reproduction threshold. The other modifications in the model, some of which have been touched on above, alter this picture in degree only. The essential point is that a more realistic representation of the behavior of interacting populations indicates the existence of at least one domain of attraction. It is quite possible, within this domain, to imagine stable equilibrium points, stable nodes, or stable limit cycles. Whatever the detailed configuration, the existence of discrete domains of attraction immediately suggests important consequences for the persistence of the system and the probability of its extinction.

Such models, however complex, are still so simple that they should not be viewed in a definitive and quantitative way. They are more powerfully used as a starting point to organize and guide understanding. It becomes valuable, therefore, to ask what the models leave out and whether such omissions make isolated domains of attraction more or less likely.

Theoretical models generally have not done well in simultaneously incorporating realistic behavior of the processes involved, randomness, spatial heterogeneity, and an adequate number of dimensions or state variables. This situation is changing very rapidly as theory and empirical studies develop a closer technical partnership. In what follows I refer to real world examples to determine how the four elements that tend to be left out might further affect the behavior of ecological systems.

Some Real World Examples

Self-Contained Ecosystems

In the broadest sense, the closest approximation we could make of a real world example that did not grossly depart from the assumptions of the theoretical models would be a self-contained system that was fairly homogenous and in which climatic fluctuations were reasonably small. If such systems could be discovered they would reveal how the more realistic interaction of real world processes could modify the patterns of systems behavior described above. Very close approximations to any of these conditions are not likely to be found, but if any exist, they are apt to be fresh water aquatic ones. Fresh water lakes are reasonably contained systems, at least within their watersheds; the fish show considerable mobility throughout, and the properties of the water buffer the more extreme effects of climate. Moreover, there have been enough documented manmade disturbances to liken them to perturbed systems in which either the parameter values or the levels of the constituent populations are changed. In a crude way, then, the lake studies can be likened to a partial exploration of a phase space of the sorts shown in Figure 2. Two major classes of disturbances have occurred: first, the impact of nutrient enrichment from man's domestic and industrial wastes, and second, changes in fish populations by harvesting.

The paleolimnologists have been remarkably successful in tracing the impact of man's activities on lake systems over surprisingly long periods. For example, Hutchinson (17) has reconstructed the series of events occurring in a small crater lake in Italy from the last glacial period in the Alps (2000 to 1800 BC) to the present. Between the beginning of the record and Roman times the lake had established a trophic equilibrium with a low level of productivity which persisted in spite of dramatic changes in surroundings from *Artemesia* steppe, through grassland, to fir and mixed oak forest. Then suddenly the whole aquatic system altered. This alteration towards eutrophication seems to have been initiated by the construction of the Via Cassia about 171 BC, which caused a subtle change in the hydrographic regime. The whole sequence of environmental changes can be viewed as changes in parameters or driving variables, and the long persistence in the face of these major changes suggests that natural systems have a high capacity to absorb change without dramatically altering. But this resilient character has its limits, and when the limits are passed, as by the construction of the Roman highway, the system rapidly changes to another condition.

More recently the activities of man have accelerated and limnologists have recorded some of the responses to these changes. The most dramatic change consists of blooms of algae in surface waters, an extraordinary growth triggered, in most instances, by nutrient additions from agricultural and domestic sources.

While such instances of nutrient addition provide some of the few examples available of perturbation effects in nature, there are no controls and the perturbations are exceedingly difficult to document. Nevertheless, the qualitative pattern seems consistent, particularly in those lakes (Edmundson 4, Hasler 10) to which sewage has been added for a time and then diverted elsewhere. This pulse of disturbance characteristically triggers periodic algal blooms, low oxygen conditions, the sudden disappearance of some plankton species, and appearance of others. As only one example, the, nutrient changes in Lake Michigan (Beeton 2) have been accompanied by the replacement of the cladoceran *Bosmina coregoni* by *B. Longirostris, Diaptomus oregonensis* has become an important copepod species, and a brackish water copepod *Eurytemora affinis* is a new addition to the zooplankton.

In Lake Erie, which has been particularly affected because of its shallowness and intensity of use, the mayfly *Hexagenia*, which originally dominated the benthic community, has been almost totally replaced by oligochetes. There have been blooms of the diatom *Melosira binderana*, which had never been reported from the United States until 1961 but now comprises as much as 99% of the total phytoplankton around certain islands. In those cases where sewage has been subsequently diverted there is a gradual return to less extreme conditions, the slowness of the return related to the accumulation of nutrients in sediments.

The overall pattern emerging from these examples is the sudden appearance or disappearance of populations, a wide amplitude of fluctuations, and the establishment of new domains of attraction.

The history of the Great Lakes provides not only some particularly good information on responses to man-made enrichment, but also on responses of fish populations to fishing pressure. The eutrophication experience touched on above can be viewed as an example of systems changes in driving variables and parameters, whereas the fishing example is more an experiment in changing state variables. The fisheries of the Great Lakes have always selectively concentrated on abundant species that are in high demand. Prior to 1930, before eutrophication complicated the story, the lake sturgeon in all the Great Lakes, the lake herring in Lake Erie, and the lake whitefish in Lake Huron were intensively fished (Smith 37). In each case the pattern was similar: a period of intense exploitation during which there was a prolonged high level harvest, followed by a sudden and precipitous drop in populations. Most significantly, even though fishing pressure was then relaxed, none of these populations showed any sign of returning to their previous levels of abundance. This is not unexpected for sturgeon because of their slow growth and late maturity, but it is unexpected for herring and white-fish. The maintenance of these low populations in recent times might be attributed to the increasingly unfavorable chemical or biological envi-ronment, but in the case of the herring, at least, the declines took place in the early 1920s before the major deterioration in environment occurred. It is as if the population had been shifted by fishing pressure from a domain with a high equilibrium to one with a lower one. This is clearly not a condition of neutral stability as suggested in Figure 2b since once the populations were lowered to a certain point the decline continued even though fishing pressure was relaxed. It can be better interpreted as a variant of Figure 2d where populations have been moved from one domain of attraction to another.

Since 1940 there has been a series of similar catastrophic changes in the Great Lakes that has led to major changes in the fish stocks. Beeton (2) provides graphs summarizing the catch statistics in the lakes for many species since 1900. Lake trout, whitefish, herring, walleye, sauger, and blue pike have experienced precipitous declines of populations to very low values in all of the lakes. The changes generally conform to the same pattern. After sustained but fluctuating levels of harvest the catch dropped dramatically in a span of a very few years, covering a range of from one to four orders of magnitude. In a number of examples partic-ularly high catches were obtained just before the drop. Although catch statistics inevitably exaggerate the step-like character of the pattern, populations must have generally behaved in the way described.

The explanations for these changes have been explored in part, and involve various combinations of intense fishing pressure, changes in the physical and chemical environment, and the appearance of a foreign predator (the sea lamprey) and foreign competitors (the alewife and carp). For our purpose the specific cause is less interest than the infer-ences that can be drawn concerning the resilience of these systems and their stability behavior. The events in Lake Michigan provide a typical example of the pattern in other lakes (Smith 37). The catch of lake trout was high, but fluctuated at around six million pounds annually from 1898 to 1940. For four years catches increased noticeably and then suddenly collapsed to near extinction by the 1950s due to a complete failure of natural reproduction. Lake herring and whitefish followed a similar pattern (Beeton 2: Figure 7). Smith (37) argues that the trigger for the lake trout collapse was the appearance of the sea lamprey that had spread through the Great Lakes after the construction of the Welland Canal. Although lamprey populations were extremely small at the time of the collapse, Smith argues that even a small mortality, added to a commercial harvest that was probably at the maximum for sustained yield, was sufficient to cause the collapse. Moreover, Ricker (31) has shown that fishing pressure shifts the age structure of fish populations towards younger ages. He demonstrates that a point can come where only slight increases in mortality can trigger a collapse of the kind noted for lake trout. In addition, the lake trout was coupled in a network of competitive and predatory interconnections with other species, and pressures on these might have contributed as well.

Whatever the specific causes, it is clear that the precondition for the collapse was set by the harvesting of fish, even though during a long period there were no obvious signs of problems. The fishing activity, however, progressively reduced the resilience of the system so that when the inevitable unexpected event occurred, the populations collapsed. If it had not been the lamprey, it would have been something else: a change in climate as part of the normal pattern of fluctuation, a change in the chemical or physical environment, or a change in competitors or predators. These examples again suggest distinct domains of attraction in which the populations forced close to the boundary of the domain can then flip over it.

The above examples are not isolated ones. In 1939 an experimental fishery was started in Lake Windermere to improve stocks of salmonids by reducing the abundance of perch (a competitor) and pike (a predator). Perch populations were particularly affected by trapping and the populations fell drastically in the first three years. Most significantly, although no perch have been removed from the North Basin since 1947, populations have still not shown any tendency to return to their previous level (Le Cren et al 19).

The same patterns have even been suggested for terrestrial systems. Many of the arid cattle grazing lands of the western United States have gradually become invaded and dominated by shrubs and trees like mesquite and cholla. In some instances grazing and the reduced incidence of fire through fire prevention programs allowed invasion and establishment of shrubs and trees at the expense of grass. Nevertheless, Glendening (8) has demonstrated, from data collected in a 17-year experiment in which intensity of grazing was manipulated, that once the trees have gained sufficient size and density to completely utilize or materially reduce the moisture supply, elimination of grazing will not result in the grassland reestablishing itself. In short, there is a level of the state variable "trees" that, once achieved, moves the system from one domain of attraction to another. Return to the original domain can only be made by an explicit reduction of the trees and shrubs.

These examples point to one or more distinct domains of attraction in which the important point is not so much how stable they are within the domain, but how likely it is for the system to move from one domain into another and so persist in a changed configuration.

This sampling of examples is inevitably biased. There are few cases well documented over a long period of time, and certainly some systems that have been greatly disturbed have fully recovered their original state once the disturbance was removed. But the recovery in most instances is in open systems in which reinvasion is the key ingredient. These cases are discussed below in connection with the effects of spatial heterogeneity. For the moment I conclude that distinct domains of attraction are not uncommon within closed systems. If such is the case, then further confirmation should be found from empirical evidence of the way processes which link organisms operate, for it is these processes that are the cause of the behavior observed.

Process Analysis

One way to represent the combined effects of processes like fecundity, predation, and competition is by using Ricker's (30) reproduction curves. These simply represent the population in one generation as a function of the population in the previous generation, and examples are shown in Figures 3a, c, and e. In the simplest form, and the one most used in practical fisheries management (Figure 3a), the reproduction curve is dome-shaped. When it crosses a line with slope 1 (the straight line in the figures) an equilibrium condition is possible, for at such cross-overs the

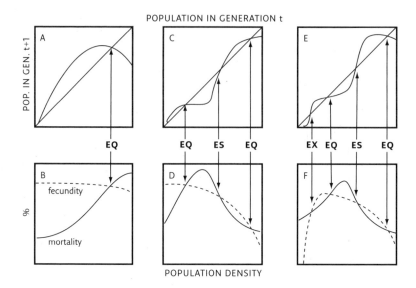

FIGURE 3: *Examples of various reproduction curves (a, c, and e) and their derivation from the contributions of fecundity and mortality (b, d, and f).*

population in one generation will produce the same number in the next. It is extremely difficult to detect the precise form of such curves in nature, however; variability is high, typically data are only available for parts of any one curve, and the treatment really only applies to situations where there are no lags. It is possible to deduce various forms of reproduction curves, however, by disaggregating the contributions of fecundity and mortality. The three lower graphs in Figure 3b, 3d, and 3f represent this disaggregation of their counterpart reproduction curves. The simplest types of reproduction curve (Figure 3a) can arise from a mortality that regularly increases with density and either a constant fecundity or a declining one. With fecundity expressed as the percentage mortality necessary to just balance reproduction, the cross-over point of the curves represents the equilibrium condition. But we know that the effects of density on fecundity and mortality can be very much more complicated.

Mortality from predation, for example, has been shown to take a number of classic forms (Holling 11, 13). The individual attack by predators as a function of prey density (the functional response to prey density) can increase with a linear rise to a plateau (type 1), a concave or negatively accelerated rise to a plateau (type 2), or an S-shaped rise to a plateau (type 3). The resulting contribution to mortality from these responses can therefore show ranges of prey density in which there is direct density dependence (negative feedback from the positively accelerated portions of the type 3 response), density independence (the straight line rise of type 1), and inverse dependence (the positive feedback from the negatively accelerated and plateau portions of the curves). There are, in addition, various numerical responses generated by changes in the number of predators as the density of their prey increases. Even for those predators whose populations respond by increasing, there often will be a limit to the increase set by other conditions in the environment. When populations are increasing they tend to augment the negative feedback features (although with a delay), but when populations are constant, despite increasing prey density, the percent mortality will inevitably decline since individual attack eventually saturates at complete satiation (the plateau of all three functional responses). In Figures 3d and 3f the mortality curves shown summarize a common type. The rising or direct density-dependent limb of the curve is induced by increasing predator populations and by the reduced intensity of attack at low densities, shown by the initial positively accelerated portion of the S-shaped type 3 response. Such a condition is common for predators with alternate prey, both vertebrates (Holling 14) and at least some invertebrates (Steele 38). The declining inverse density dependent limb is induced by satiation of the predator and a numerical response that has been reduced or stopped.

Fecundity curves that decline regularly over a very wide range of increasing population densities (as in Figure 3d) are common and have been referred to *Drosophila*-type curves (Fujita 6). This decline in fecundity is caused by increased competition for oviposition sites, interference with mating, and increased sterility. The interaction between a dome-shaped mortality curve and a monotonically decreasing fecundity curve can generate equilibrium conditions (Figure 3d). Two stable equilibria are possible, but between these two is a transient equilibrium designated as the escape threshold (ES in Figure 3). Effects of random changes on populations or parameters could readily shift densities from around the lower equilibrium to above this escape threshold, and in these circumstances populations would inevitably increase to the higher equilibrium.

The fecundity curves are likely to be more complex, however, since it seems inevitable that at some very low densities fecundity will decline because of difficulties in finding mates and the reduced effect of a variety of social facilitation behaviors. We might even logically conclude that for many species there is a minimum density below which fecundity is zero. A fecundity curve of this Allee-type (Fujita 6) has been empirically demonstrated for a number of insects (Watt 42) and is shown in Figure 3f. Its interaction with the dome-shaped mortality curve can add another transient equilibrium, the extinction threshold (EX in Figure 3f). With this addition there is a lower density such that if populations slip below it they will proceed inexorably to extinction. The extinction threshold is particularly likely since it has been shown mathematically that each of the three functional response curves will intersect with the ordinate of percent predation at a value above zero (Holling 13).

Empirical evidence, therefore, suggests that realistic forms to fecundity and mortality curves will generate sinuous reproduction curves like those in Figures 3c and 3e with the possibility of a number of equilibrium states, some transient and some stable. These are precisely the conditions that will generate domains of attraction, with each domain separated from others by the extinction and escape thresholds. This analysis of process hence adds support to the field observations discussed earlier.

The behavior of systems in phase space cannot be completely understood by the graphical representations presented above. These graphs are appropriate only when effects are immediate; in the face of the lags that generate cyclic behavior the reproduction curve should really produce two values for the population in generation $t + 1$ for each value of the population in generation t. The graphical treatment of Rosenzweig & MacArthur (33) to a degree can accommodate these lags and cyclic behavior. In their treatment they divide phase planes of the kind shown in Figure 2 into various regions of increasing and decreasing x and y populations. The regions are separated by two lines, one representing the collection of points at which the prey population does not change in density ($dx/dt = 0$, the prey isocline) and one in which the predator population does not so change ($dy/dt = 0$, the predator isocline). They deduce that the prey isocline will be dome-shaped for much the same reason as described for the fecundity curves of Figure 3f. The predator isocline, in the simplest condition, is presumed to be vertical, assuming that only one fixed level of prey is necessary to just maintain the predator population at a zero instantaneous rate of change.

Intersection of the two isoclines indicates a point where both populations are at equilibrium. Using traditional linear stability analysis one can infer whether these equilibrium states are stable (Figure 2c) or not (Figure 2a). Considerable importance is attached to whether the predator isocline intersects the rising or falling portion of the prey isocline. As mentioned earlier these techniques are only appropriate near equilibrium (May 24), and the presumed unstable conditions in fact generate stable limit cycles (Figure 2e). Moreover, it is unlikely that the predator isocline is a vertical one in the real world, since competition between predators at high predator densities would so interfere with the attack process that a larger number of prey would be required for stable predator populations. It is precisely this condition that was demonstrated by Griffiths & Holling (9) when they showed that a large number of species of parasites distribute their attacks contagiously. The result is a "squabbling predator behavior" (Rosenzweig 34, 35) that decreases the efficiency of predation at high predator/prey ratios. This converts an unstable system (Figure 2a) to a stable one (Figure 2c); it is likely that stability is the rule, rather than the exception, irrespective of where the two isoclines cross.

The empirical evidence described above shows that realistic fecundity and mortality (particularly predation) processes will generate forms that the theorists might tend to identify as special subsets of more general conditions. But it is just these special subsets that separate the real world from all possible ones, and these more realistic forms will modify the general conclusions of simpler theory. The ascending limb of the Allee-type fecundity curve will establish, through interaction with mortality, a minimum density below which prey will become extinct. This can at the same time establish an upper prey density above which prey will become extinct because the amplitude of prey fluctuations will eventually carry the population over the extinction threshold, as shown in the outer trajectory of Figure 2d. These conditions alone are sufficient to establish a domain of attraction, although the boundaries of this domain need not be closed. Within the domain the contagious attack by predators can produce a stable equilibrium or a stable node. Other behaviors of the mortality agents, however, could result in stable limit cycles.

More realistic forms of functional response change this pattern in degree only. For example, a negatively accelerated type of functional response would tend to make the domain of attraction somewhat smaller, and an S-shaped one larger. Limitations in the predator's numerical response and thresholds for reproduction of predators, similar to those for prey, could further change the form of the domain. Moreover, the behaviors that produce the sinuous reproduction curves of Figures 3c and 3e can add additional domains. The essential point, however, is that these systems are not globally stable but can have distinct domains of attraction. So long as the populations remain within one domain they have a consistent and regular form of behavior. If populations pass a boundary to the domain by chance or through intervention of man, then the behavior suddenly changes in much the way suggested from the field examples discussed earlier.

The Random World

To this point, I have argued as if the world were completely deterministic. In fact, the behavior of ecological systems is profoundly affected by random events. It is important, therefore, to add another level of realism at this point to determine how the above arguments may be modified. Again, it is applied ecology that tends to supply the best information from field studies since it is only in such situations that data have been collected in a sufficiently intensive and extensive manner. As one example, for 28 years there has been a major and intensive study of the spruce budworm and its interaction with the spruce-fir forests of eastern Canada (Morris 27). There have been six outbreaks of the spruce budworm since the early 1700s (Baskerville 1) and between these outbreaks the budworm has been an exceedingly rare species. When the outbreaks occur there is major destruction of balsam fir in all the mature forests, leaving only the less susceptible spruce, the nonsusceptible white birch, and a dense regeneration of fir and spruce. The more immature stands suffer less damage and more fir survives. Between outbreaks the young balsam grow, together with spruce and birch, to form dense stands in which the spruce and birch, in particular, suffer from crowding. This process evolves to produce stands of mature and overmature trees with fir a predominant feature.

This is a necessary, but not sufficient, condition for the appearance of an outbreak; outbreaks occur only when there is also a sequence of unusually dry years (Wellington 43). Until this sequence occurs, it is argued (Morris 27) that various natural enemies with limited numerical responses maintain the budworm populations around a low equilibrium. If a sequence of dry years occurs when there are mature stand of fir, the budworm populations rapidly increase and escape the control by predators and parasites. Their continued increase eventually causes enough tree mortality to force a collapse of the populations and the reinstatement of control around the lower equilibrium. The reproduction curves therefore would be similar to those in Figures 3c or 3e.

In brief, between outbreaks the fir tends to be favored in its competition with spruce and birch, whereas during an outbreak spruce and birch are favored because they are less susceptible to budworm attack. This interplay with the budworm thus maintains the spruce and birch which otherwise would be excluded through competition. The fir persists because of its regenerative powers and the interplay of forest growth rates and climatic conditions that determine the timing of budworm outbreaks.

This behavior could be viewed as a stable limit cycle with large amplitude, but it can be more accurately represented by a distinct domain of attraction determined by the interaction between budworm and its associated natural enemies, which is periodically exceeded through the chance consequence of climatic conditions. If we view the budworm only in relation to its associated predators and parasites we might argue that it is highly unstable in the sense that populations fluctuate widely. But these very fluctuations are essential features that maintain persistence of the budworm, together with its natural enemies and its host and associated trees. By so fluctuating, successive generations of forests are replaced, assuring a continued food supply for future generations of budworm and the persistence of the system.

Until now I have avoided formal identification of different kinds of behavior of ecological systems. The more realistic situations like budworm however, make it necessary to begin to give more formal definition to their behavior. It is useful to distinguish two kinds of behavior. One can be termed stability, which represents the ability of a system to return to an equilibrium state after a temporary disturbance; the more rapidly it returns and the less it fluctuates, the more stable it would be. But there is another property, termed resilience, that is a measure of the persistence of systems and of their ability to absorb change and disturbance and still maintain the same relationships between populations or state variables. In this sense, the budworm forest community is highly unstable and it is because of this instability that it has an enormous resilience. I return to this view frequently throughout the remainder of this paper.

The influence of random events on systems with domains of attraction is found in aquatic systems as well. For example, pink salmon populations can become stabilized for several years at very different levels, the new levels being reached by sudden steps rather than by gradual transition (Neave 28). The explanation is very much the same as that proposed for the budworm, involving an interrelation between negative and positive feedback mortality of the kinds described in Figures 3d and 3f, and random effects unrelated to density. The same pattern has been described by Larkin (18) in his simulation model of the Adams River sockeye salmon. This particular run of salmon has been characterized by a regular four-year periodicity since 1922, with one large or dominant year, one small or subdominant, and two years with very small populations. The same explanation as described above has been proposed with the added reality of a lag. Essentially, during the dominant year limited numerical responses produce an inverse density-dependent response as in the descending limb of the mortality curves of Figure 3d and 3f. The abundance of the prey in that year is nevertheless sufficient to establish populations of predators that have a major impact on the three succeeding low years. Buffering of predation by the smolts of the dominant year accounts for the larger size of the subdominant. These effects have been simulated (Larkin 18), and when random influences are imposed in order to simulate climatic variations the system has a distinct probability of flipping into another stable configuration that is actually reproduced in nature by sockeye salmon runs in other rivers. When subdominant escapement reaches a critical level there is about an equal chance that it may become the same size as the dominant one or shrivel to a very small size.

Random events, of course, are not exclusively climatic. The impact of fires on terrestrial ecosystems is particularly illuminating (Cooper 3) and the periodic appearance of fires has played a decisive role in the persistence of grasslands as well as certain forest communities. As an example, the random perturbation caused by fires in Wisconsin forests (Loucks 21) has resulted in a sequence of transient changes that move forest communities from one domain of attraction to another. The apparent instability of this forest community is best viewed not as an unstable condition alone, but as one that produces a highly resilient system capable of repeating itself and persisting over time until a disturbance restarts the sequence.

In summary, these examples of the influence of random events upon natural systems further confirm the existence of domains of attraction. Most importantly they suggest that instability, in the sense of large fluctuations, may introduce a resilience and a capacity to persist. It points out the very different view of the world that can be obtained if we concentrate on the boundaries to the domain of attraction rather than on equilibrium states. Although the equilibrium-centered view is analytically more tractable, it does not always provide a realistic understanding of the systems' behavior. Moreover, if this perspective is used as the exclusive guide to the management activities of man, exactly the reverse behavior and result can be produced than is expected.

The Spatial Mosaic

To this point, I have proceeded in a series of steps to gradually add more and more reality. I started with self-contained closed systems, proceeded to a more detailed explanation of how ecological processes operate, and then considered the influence of random events, which introduced heterogeneity over time.

The final step is now to recognize that the natural world is not very homogeneous over space, as well, but consists of a mosaic of spatial elements with distinct biological, physical, and chemical characteristics that are linked by mechanisms of biological and physical transport. The role of spatial heterogeneity has not been well explored in ecology because of the enormous logistic difficulties. Its importance, however, was revealed in a classic experiment that involved the interaction between a predatory mite, its phytophagous mite prey, and the prey's food source (Huffaker et al. 15). Briefly, in the relatively small enclosures used, when there was unimpeded movement throughout the experimental universe, the system was unstable and oscillations increased in amplitude. When barriers were introduced to impede dispersal between parts of the universe, however, the interaction persisted. Thus populations in one small locale that suffer chance extinctions could be reestablished by invasion from other populations having high numbers—a conclusion that is confirmed by Roff's mathematical analysis of spatial heterogeneity (32).

There is one study that has been largely neglected that is, in a sense, a much more realistic example of the effects of both temporal and spatial heterogeneity of a population in nature (Wellington 44, 45). There is a peninsula on Vancouver Island in which the topography and climate combine to make a mosaic of favorable locales for the tent caterpillar. From year to year the size of these locales enlarges or contracts depending on climate; Wellington was able to use the easily observed changes in cloud patterns in any year to define these areas. The tent caterpillar, to add a further element of realism, has identifiable behavioral types that are determined not by genetics but by the nutritional history of the parents. These types represent a range from sluggish to very active, and the proportion of types affects the shape of the easily visible web the tent caterpillars spin. By combining these defined differences of behavior with observations on changing numbers, shape of webs, and changing cloud patterns, an elegant story of systems behavior emerges. In a favorable year locales that previously could not support tent caterpillars now can, and populations are established through invasion by the vigorous dispersers from other locales. In these new areas they tend to produce another generation with a high proportion of vigorous behavioral types. Because of their high dispersal behavior and the small area of the locale in relation to its periphery, they then tend to leave in greater numbers than they arrive. The result is a gradual increase in the proportion of more sluggish types to the point where the local population collapses. But, although its fluctuations are considerable, even under the most unfavorable conditions there are always enclaves suitable for the insect. It is an example of a population with high fluctuations that can take advantage of transient periods of favorable conditions and that has, because of this variability, a high degree of resilience and capacity to persist.

A further embellishment has been added in the study of natural insect populations by Gilbert & Hughes (7). They combined an insightful field study of the interaction between aphids and their parasites with a simulation model, concentrating upon a specific locale and the events within it under different conditions of immigration from other locales.

Again the important focus was upon persistence rather than degree of fluctuation. They found that specific features of the parasite-host interaction allowed the parasite to make full use of its aphid resources just short of driving the host to extinction. It is particularly intriguing that the parasite and its host were introduced into Australia from Europe and in the short period that the parasite has been present in Australia there have been dramatic changes in its developmental rate and fecundity. The other major difference between conditions in Europe and Australia is that the immigration rate of the host in England is considerably higher than in Australia. If the immigration rate in Australia increased to the English level, then, according to the model the parasite should increase its fecundity from the Australian level to the English to make the most of its opportunity short of extinction. This study provides, therefore, a remarkable example of a parasite and its host evolving together to permit persistence, and further confirms the importance of systems resilience as distinct from systems stability.

Synthesis

Some Definitions
Traditionally, discussion and analyses of stability have essentially equated stability to systems behavior. In ecology, at least, this has caused confusion since, in mathematical analyses, stability has tended to assume definitions that relate to conditions very near equilibrium points. This is a simple convenience dictated by the enormous analytical difficulties of treating the behavior of nonlinear systems at some distance from equilibrium. On the other hand, more general treatments have touched on questions of persistence and the probability of extinction, defining these measures as aspects of stability as well. To avoid this confusion I propose that the behavior of ecological systems could well be defined by two distinct properties: resilience and stability.

Resilience determines the persistence of relationships within a system and is a measure of the ability of these systems to absorb changes of state variables, driving variables, and parameters, and still persist. In this definition resilience is the property of the system and persistence or probability of extinction is the result. Stability, on the other hand, is the ability of a system to return to an equilibrium state after a temporary disturbance. The more rapidly it returns, and with the least fluctuation, the more stable it is. In this definition stability is the property of the system and the degree of fluctuation around specific states the result.

Resilience versus Stability

With these definitions in mind a system can be very resilient and still fluctuate greatly, i.e. have low stability. I have touched above on examples like the spruce budworm forest community in which the very fact of low stability seems to introduce high resilience. Nor are such cases isolated ones, as Watt (41) has shown in his analysis of thirty years of data collected for every major forest insect throughout Canada by the Insect Survey program of the Canada Department of the Environment. This statistical analysis shows that in those areas subjected to extreme climatic conditions the populations fluctuate widely but have a high capability of absorbing periodic extremes of fluctuation. They are, therefore, unstable using the restricted definition above, but highly resilient. In more benign, less variable climatic regions the populations are much less able to absorb chance climatic extremes even though the populations tend to be more constant. These situations show a high degree of stability and a lower resilience. The balance between resilience and stability is clearly a product of the evolutionary history of these systems in the face of the range of random fluctuations they have experienced.

In Slobodkin's terms (36) evolution is like a game, but a distinctive one in which the only payoff is to stay in the game. Therefore, a major strategy selected is not one maximizing either efficiency or a particular reward, but one which allows persistence by maintaining flexibility above all else. A population responds to any environmental change by the initiation of a series of physiological, behavioral, ecological, and genetic changes that restore its ability to respond to subsequent unpredictable environmental changes. Variability over space and time results in variability in numbers, and with this variability the population can simultaneously retain genetic and behavioral types that can maintain their existence in low populations together with others that can capitalize on chance opportunities for dramatic increase. The more homogeneous the environment in space and time, the more likely is the system to have low fluctuations and low resilience. It is not surprising, therefore, that the commercial fishery systems of the Great Lakes have provided a vivid example of the sensitivity of ecological systems to disruption by man, for they represent climatically buffered, fairly homo-geneous and self-contained systems with relatively low variability and hence high stability and low resilience. Moreover, the goal of produc-ing a maximum sustained yield may result in a more stable system of reduced resilience.

Nor is it surprising that however readily fish stocks in lakes can be driven to extinction, it has been extremely difficult to do the same to insect pests of man's crops. Pest systems are highly variable in space and time; as open systems they are much affected by dispersal and therefore have a high resilience. Similarly, some Arctic ecosystems thought of as fragile may be highly resilient, although unstable. Certainly this is not true for some subsystems in the Arctic, such as Arctic frozen soil, self-contained Arctic lakes, and cohesive social populations like caribou, but these might be exceptions to a general rule.

The notion of an interplay between resilience and stability might also resolve the conflicting views of the role of diversity and stability of eco-logical communities. Elton (5) and MacArthur (22) have argued cogently from empirical and theoretical points of view that stability is roughly proportional to the number of links between species in a trophic web. In essence, if there are a variety of trophic links the same flow of energy and nutrients will be maintained through alternate links when a spe-cies becomes rare. However, May's (23) recent mathematical analyses of models of a large number of interacting populations show that this relation between increased diversity and stability is not a mathemati-cal truism. He shows that randomly assembled complex systems are in general less stable, and never more stable, than less complex ones. He points out that ecological systems are likely to have evolved to a very small subset of all possible sets and that MacArthur's conclusions, therefore, might still apply in the real world. The definition of stability used, however, is the equilibrium-centered one. What May has shown is that complex systems might fluctuate more than less complex ones. But if there is more than one domain of attraction, then the increased variability could simply move the system from one domain to another. Also, the more species there are, the more equilibria there may be and, although numbers may thereby fluctuate considerably, the overall per-sistence might be enhanced. It would be useful to explore the possibility that instability in numbers can result in more diversity of species and in spatial patchiness, and hence in increased resilience.

Measurement

If there is a worthwhile distinction between resilience and stability it is important that both be measurable. In a theoretical world such measurements could be developed from the behavior of model systems in phase space. Just as it was useful to disaggregate the reproduction curves into their constituent components o f mortality and fecundity, so it is useful to disaggregate the information in a phase plane. There are two components that are important: one that concerns the cyclic behavior and its frequency and amplitude, and one that concerns the configuration of forces caused by the positive and negative feedback relations.

To separate the two we need to imagine first the appearance of a phase space in which there are no such forces operating. This would produce a referent trajectory containing only the cyclic properties of the system. If the forces were operating, departure from this referent trajectory would be a measure of the intensity of the forces. The referent trajectories that would seem to be most useful would be the neutrally stable orbits of Figure 2b, for we can arbitrarily imagine these trajectories as moving on a flat plane. At least for more realistic models parameter values can be discovered that do generate neutrally stable orbits. In the complex predator-prey model of Holling (14), if a range of parameters is chosen to explore the effects different degrees of contagion of attack, the interaction is unstable when attack is random and stable when it is contagious. We have recently shown that there is a critical level of contagion between these extremes that generates neutrally stable orbits. These orbits, then, have a certain frequency and amplitude and the departure of more realistic trajectories from these referent ones should allow the computation of the vector of forces. If these were integrated a potential field would be represented with peaks and valleys. If the whole potential field were a shallow bowl the system would be globally stable and all trajectories would spiral to the bottom of the bowl, the equilibrium point. But if, at a minimum there were a lower extinction threshold for prey then, in effect, the bowl would have a slice taken out of one side, as suggested in Figure 4. Trajectories that initiated far up on the side of the bowl would have amplitude that would carry the trajectory over the slice cut out of it. Only those trajectories that just avoided the lowest point of the gap formed by the slice would spiral in to the bowl's bottom. If we termed the bowl the basin of attraction (Lewontin 20) then the domain of attraction

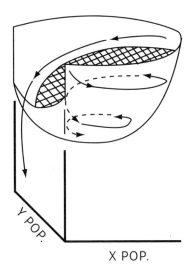

FIGURE 4: *Diagrammatic representation showing the feedback forces as a potential field upon which trajectories move. The shaded portion is the domain of attraction.*

would be determined by both the cyclic behavior and the configuration of forces. It would be confined to a smaller portion of the bottom of the bowl, and one edge would touch the bottom portion of the slice taken out of the basin.

This approach, then, suggests ways to measure relative amounts of resilience and stability. There are two resilience measures: Since resilience is concerned with probabilities of extinction, firstly, the overall area of the domain of attraction will in part determine whether chance shifts in state variables will move trajectories outside the domain. Secondly, the height of the lowest point of the basin of attraction (e.g. the bottom of the slice described above) above equilibrium will be a measure of how much the forces have to be changed before all trajectories move to extinction of one or more of the state variables.

The measures of stability would be designed in just the opposite way from those that measure resilience. They would be centered on the equilibrium rather than on the boundary of the domain, and could be represented by a frequency distribution of the slopes of the potential field and by the velocity of the neutral orbits around the equilibrium.

But such measures require an immense amount of knowledge of a system and it is unlikely that we will often have all that is necessary. Hughes & Gilbert (16), however, have suggested a promising approach to measuring probabilities of extinction and hence of resilience. They were able to show in a stochastic model that the distribution of surviving population sizes at any given time does not differ significantly from a negative binomial. This of course is just a description, but it does provide a way to estimate the very small probability of zero, i.e. of extinction, from the observed mean and variance. The configuration of the potential field and the cyclic behavior will determine the number and form of the domains of attraction, and these will in turn affect the parameter values of the negative binomial or of any other distribution function that seems appropriate. Changes in the zero class of the distribution, that is, in the probability of extinction, will be caused by these parameter values, which can then be viewed as the relative measures of resilience. It will be important to explore this technique first with a number of theoretical models so that the appropriate distributions and their behavior can be identified. It will then be quite feasible, in the field, to sample populations in defined areas, apply the appropriate distribution, and use the parameter values as measures of the degree of resilience.

Application

The resilience and stability viewpoints of the behavior of ecological systems can yield very different approaches to the management of resources. The stability view emphasizes the equilibrium, the maintenance of a predictable world, and the harvesting of nature's excess production with as little fluctuation as possible. The resilience view emphasizes domains of attraction and the need for persistence. But extinction is not purely a random event; it results from the interaction of random events with those deterministic forces that define the shape, size, and characteristics of the domain of attraction. The very approach, therefore, that assures a stable maximum sustained yield of a renewable resource might so change these deterministic conditions that the resilience is lost or reduced so that a chance and rare event that previously could be absorbed can trigger a sudden dramatic change and loss of structural integrity of the system.

A management approach based on resilience, on the other hand, would emphasize the need to keep options open, the need to view events in a regional rather than a local context, and the need to emphasize heterogeneity. Flowing from this would be not the presumption of sufficient knowledge, but the recognition of our ignorance; not the assumption that future events are expected, but that they will be unexpected. The resilience framework can accommodate this shift of perspective, for it does not require a precise capacity to predict the future, but only a qualitative capacity to devise systems that can absorb and accommodate future events in whatever unexpected form they may take.

Literature Cited

1.   Baskerville, G. L. 1971. *The Fir-Spruce-Birch Forest and the Budworm.* Forestry Service, Canada Dept. Environ., Fredericton, N. B. Unpublished.
2.   Beeton, A. D. 1969. Changes in the environment and biota of the Great Lakes. *Eutrophication: Causes, Consequences, Correctives.* Washington DC: Nat. Acad. Sci.
3.   Cooper, C. F. 1961. The ecology of fire. *Sci. Am.* 204:150–6, 158, 160.
4.   Edmondson, W. T. 1961. Changes in Lake Washington following increase in nutrient income. *Verh. Int. Ver. Limnol.* 14:167–75.
5.   Elton, C. S. 1958. *The Ecology of Invasions by Animals and Plants.* London: Methuen.
6.   Fujita, H. 1954. An interpretation of the changes in type of the population density effect upon the oviposition rate. *Ecology* 35:253–7.
7.   Gilbert, N., Hughes, R. D. 1971. A model of an aphid population—three adventures. *J. Anim. Ecol.* 40:525–34.
8.   Glendening, G. 1952. Some quantitative data on the increase of mesquite and cactus on a desert grassland range in southern Arizona. *Ecology* 33:319–28.
9.   Griffiths, K. J., Holling, C. S. 1969. A competition submodel for parasites and predators. *Can. Entomol.* 101:785–818.
10.   Hasler, A. D. 1947. Eutrophication of lakes by domestic sewage. *Ecology* 28:383–95.
11.   Holling, C. S. 1961. Principles of insect predation. *Ann. Rev. Entomol.* 6:163–82.
12.   Holling, C S. 1966. The functional response of invertebrate predators to prey density. *Mem. Entomol. Soc. Can.* 48:1–86.
13.   Holling, C. S. 1965. The functional response of predators to prey density and its role in mimicry and population regulations. *Mem. Entomol. Soc. Can.* 45:1–60.
14.   Holling, C. S., Ewing, S. 1971. Blind man's buff: exploring the response space generated by realistic ecological simulation models. *Proc. Int. Symp. Statist. Ecol.* New Haven, Conn.: Yale Univ. Press 2:207–29.
15.   Huffaker, C. D., Shea, K. P., Herman, S. S. 1963. Experimental studies on predation. Complex dispersion and levels of food in an acarine predator-prey interaction. *Hilgardia* 34:305–30.
16.   Hughes, R. D., Gilbert, N. 1968. A model of an aphid population—a general statement. *J. Anim. Ecol.* 40:525–34.

17. Hutchinson, G. E. 1970. Ianula: an account of the history and development of the Lago di Monterosi, Latium, Italy. *Trans. Am. Phil. Soc.* 60:1–178.

18. Larkin, P. A. 1971. Simulation studies of the Adams River Sockeye Salmon (*Oncarhynchus nerka*). *J. Fish. Res. Bd. Can.* 28:1493–1502.

19. Le Cren, E. D., Kipling, C., McCormack, J. C. 1972. Windermere: effects of exploitation and eutrophication on the salmonid community. *J. Fish. Res. Bd. Can.* 29:819–32.

20. Lewontin. R. C. 1969. The meaning of stability. *Diversity and Stability of Ecological Systems, Brookhaven Symp. Biol.* 22:13–24.

21. Loucks, O. L. 1970. Evolution of diversity, efficiency and community stability. *Am. Zool.* 10:17–25.

22. MacArthur, R. 1955. Fluctuations of animal populations and a measure of community stability. *Ecology* 36:533–6.

23. May, R. M, 1971. Stability in multi-species community models. *Math Biosci.* 12:59–79.

24. May, R. M. 1972. Limit cycles in predator-prey communities. *Science* 177:900–2.

25. May, R. M. 1972. Will a large complex system be stable? *Nature* 238:413–14.

26. Minorsky, N. 1962. *Nonlinear Oscillations.* Princeton, N.J.: Van Nostrand.

27. Morris, R. F. 1963. The dynamics of epidemic spruce budworm populations. *Mem. Entomol. Soc. Can.* 31:1–332.

28. Neave, F. 1953. Principles affecting the size of pink and chum salmon populations in British Columbia. *J. Fish. Res. Bd. Can.* 9:450–91.

29. Nicholson, A. J., Bailey, V. A. 1935. The balance of animal populations— Part I. *Proc. Zool. Soc. London* 1935: 551–98.

30. Ricker, W. E. 1954. Stock and recruitment. *J. Fish. Res. Bd. Can.* 11:559–623.

31. Ricker, W. E. 1963. Big effects from small causes: two examples from fish population dynamics. *J. Fish. Res. Bd. Can.* 20:257–84.

32. Roff, D. A. 1973. Spatial heterogeneity and the persistence of populations. *J. Theor. Pop. Biol.* In press.

33. Rosenzweig, M. L., MacArthur, R. H. 1963. Graphical representation and stability condition of predator-prey interactions. *Am. Natur.* 97:209–23.

34. Rosenzweig, M. L. 1971. Paradox of enrichment: destabilization of exploitation ecosystems in ecological time. *Science* 171:385–7.

35. Rosenzweig, M. L. 1972. Stability of enriched aquatic ecosystems. *Science* 175:564–5.

36. Slobodkin, L. B. 1964. The strategy of evolution. *Am. Sci.* 52:342–57.

37. Smith, S. H. 1968. Species succession and fishery exploitation in the Great Lakes. *J. Fish. Res. Bd. Can.* 25:667–93.

38. Steele, J. H. 1971. Factors controlling marine ecosystems. *The Changing Chemistry of the Oceans*, ed, D. Dryssen, D. Jaquer, 209–21. Nobel Symposium 20, New York: Wiley.

39. Walters, C. J. 1971. Systems ecology: the systems approach and mathematical models in ecology. *Fundamentals of Ecology*, ed. E. P. Odum. Philadelphia: Saunders. 3rd ed.

40. Wangersky, P. J., Cunningham, W. J. 1957. Time lag in prey-predator population models. *Ecology* 38:136–9.

41. Watt, K. E. F. 1968. A computer approach to analysis of data on weather, population fluctuations, and disease. *Biometeorology, 1967 Biology Colloquium*, ed. W. P. Lowry. Corvallis, Oregon: Oregon State Univ. Press.

42. Watt, K. E. F. 1960. The effect of population density on fecundity in insects. *Can. Entomol.* 92:674–95.

43. Wellington, W. G. 1952. Air mass climatology of Ontario north of Lake Huron and Lake Superior before outbreaks of the spruce budworm and the forest tree caterpillar. *Can. Jour. Zool.* 30:114–27.

44. Wellington, W. G. 1964. Qualitative changes in populations in unstable environments. *Can. Entomol.* 96:436–51.

45. Wellington, W. G. 1965. The use of cloud patterns to outline areas with different climates during population studies. *Can. Entomol.* 97:617–31.

Engineering Resilience versus Ecological Resilience

C. S. HOLLING

Pp. 31–44 in *Engineering Within Ecological Constraints*, Edited by Peter C. Schulze National Academy Of Engineering, National Academy Press, Washington, DC. 1996.

Reprinted with permission from the National Academies Press, Copyright 1996, National Academy of Sciences.

Ecosystem Structure and Function

ECOLOGICAL SCIENCE has been shaped largely by the biological sciences. Environmental science, on the other hand, has been shaped largely by the physical sciences and engineering. With the beginning of interdisciplinary efforts between the two fields, some of the fundamental differences between them are generating conflicts caused more by misunderstanding of basic concepts than by any difference in social purposes or methods. Those differences are most vivid in that part of ecology called ecosystem science, for it is there that it is obvious that both the biota and the physical environment interact such that not only does the environment shape the biota but the biota transforms the environment.

The accumulated body of empirical evidence concerning natural, disturbed, and managed ecosystems identifies key features of ecosystem structure and function (Holling et al., 1995) that probably are not included in many engineers' image of ecology:

- Ecological change is not continuous and gradual; Rather it is episodic, with slow accumulation of natural capital such as biomass or nutrients, punctuated by sudden releases and reorganization of that capital as the result of internal or external natural processes or of man-imposed catastrophes. Rare events, such as hurricanes, or the arrival of invading species, can unpredictably shape structure at critical times or at locations of increased vulnerability. The results of these rare events can persist for long periods. Therein lies one of the sources of new options that biological diversity provides. Irreversible or slowly reversible states exist–that is, once the system flips into such a state, only explicit management intervention can return its previous self-sustaining state, and even then success is not assured (Walker, 1981). *Critical processes function at radically different rates covering several orders of magnitude, and these rates cluster around a few dominant frequencies.*

- Spatial attributes are not uniform or scale invariant. Rather, productivity and textures are patchy and discontinuous at all scales from the leaf to the individual, the vegetation patch, the landscape, and the planet. There are several different ranges of scales each with different attributes of patchiness and texture (Holling, 1992). *Therefore scaling up from small to large cannot be a process of simple linear addition; nonlinear processes organize the shift from one range of scales to another. Not only do the large and slow variables control small and fast ones, the latter occasionally "revolt" to affect the former.*

- Ecosystems do not have single equilibria with functions controlled to remain near them. Rather, destabilizing forces far from equilibria, multiple equilibria, and disappearance of equilibria define functionally different states, and movement between states maintains structure and diversity. *On the one hand, destabilizing forces are important in maintaining diversity, resilience, and opportunity. On the other hand, stabilizing forces are important in maintaining productivity and biogeochemical cycles, and even when these features are perturbed, they recover rather rapidly if the stability domain is not exceeded* (e.g., recovery of lakes from eutrophication or acidification, Schindler, 1990; Schindler et al., 1991).

- Policies and management that apply fixed rules for achieving constant yields (such as constant carrying capacity of cattle or wildlife or constant sustainable yield of fish, wood, or water), independent of scale, lead to systems that gradually lose resilience and suddenly break down in the face of disturbances that previously could be absorbed (Holling, 1986). *Ecosystems are moving targets, with multiple potential futures that are uncertain and unpredictable. Therefore management has to be flexible, adaptive, and experimental at scales compatible with the scales of critical ecosystem functions* (Walters, 1986).

The features described above are the consequence of the stability properties of natural systems. In the ecological literature, these properties have been given focus through debates on the meaning and reality of the resilience of ecosystems. For that reason, and because the same debate seems to be emerging in economics, I will review the concepts to provide a foundation for understanding.

The Two Faces of Resilience

Resilience of a system has been defined in two different ways in the ecological literature. These differences in definition reflect which of two different aspects of stability are emphasized. I first emphasized the consequences of those different aspects for ecological systems to draw attention to the paradoxes between efficiency and persistence, or between constancy and change, or between predictability and unpredictability (Holling, 1973). One definition focuses on efficiency, constancy, and predictability—all attributes at the core of engineers' desires for fail-safe design. The other focuses on persistence, change, and unpredictability—all attributes embraced and celebrated by biologists with an evolutionary perspective and by those who search for safe-fail designs.

The first definition, and the more traditional, concentrates on stability near an equilibrium steady state, where resistance to disturbance and speed of return to the equilibrium are used to measure the property (O'Neill et al., 1986; Pimm, 1984; Tilman and Downing, 1994). That view provides one of the foundations for economic theory as well and may be termed *engineering resilience*.

The second definition emphasizes conditions far from any equilibrium steady state, where instabilities can flip a system into another regime of behavior—that is, to another stability domain (Holling, 1973). In this case the measurement of resilience is the magnitude of disturbance that can be absorbed before the system changes its structure by changing the variables and processes that control behavior. We shall call this view *ecological resilience* (Walker et al., 1969).

The same differences have also begun to emerge in economics with the identification of multistable states for competing technologies because of increasing returns to scale (Arthur, 1990). Thus, increasingly it seems that effective and sustainable development of technology, resources, and ecosystems requires ways to deal not only with near-equilibrium efficiency but with the reality of more than one equilibrium. If there is more than one equilibrium, in which direction should the finger on the invisible hand of Adam Smith point? If there is more than one objective function, where does the engineer search for optimal designs?

These two aspects of a system's stability have very different consequences for evaluating, understanding, and managing complexity and change. I argue that designing with ecosystems requires an emphasis on the second definition of resilience, that is, the amount of disturbance that can be sustained before a change in system control and structure occurs—ecological resilience. I do so because that interplay between stabilizing and destabilizing properties is at the heart of present issues of development and the environment—global change, biodiversity loss, ecosystem restoration, and sustainable development.

The two contrasting aspects of stability—essentially one that focuses on maintaining *efficiency* of function (engineering resilience) and one that focuses on maintaining *existence* of function (ecological resilience)—are so fundamental that they can become alternative paradigms whose devotees reflect traditions of a discipline or of an attitude more than of a reality of nature.

Those who emphasize the near-equilibrium definition of engineering resilience, for example, draw predominantly from traditions of deductive mathematical theory (Pimm, 1984) where simplified, untouched ecological systems are imagined, or from traditions of engineering, where the motive is to design systems with a single operating objective (DeAngelis, 1980; O'Neill et al., 1986; Waide and Webster, 1976). On the one hand, that makes the mathematics more tractable, and on the other, it accommodates the engineer's goal to develop optimal designs. There is an implicit assumption of global stability, that is, that only one equilibrium steady state exists, or, if other operating states exist, they should be avoided (Figure 1) by applying safeguards.

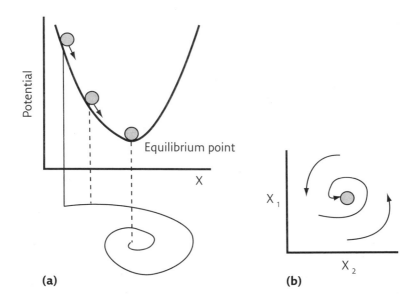

FIGURE 1: *Two views of a single, globally stable equilibrium. (a) Provides a mechanical ball and topography analogy. (b) Provides an abstract state space view of a point's movement toward the stable equilibrium, with x_1 and x_2 defining, for example, population densities of predator and prey, or of two competitors. This is an example of engineering resilience. It is measured by the resistance of the ball to disturbances away from the equilibrium point and the speed of return to it.*

Those who emphasize the stability domain definition of resilience (ecological resilience), on the other hand, come from traditions of applied mathematics and applied resource ecology at the scale of ecosystems. Examples include the dynamics and management of freshwater systems (Fiering, 1982), of forests (Holling et al., 1977), of fisheries (Walters, 1986), of semiarid grasslands (Walker et al., 1969) and of interacting populations in nature (Dublin et al., 1990; Sinclair et al., 1990). Because these studies are rooted in inductive rather than deductive theory formation and in experience with the impacts of large-scale management disturbances, the reality of flips from one operating state to another cannot be avoided. Moreover, it becomes obvious that the variability of critical variables forms and maintains the stability landscape (Figure 2).

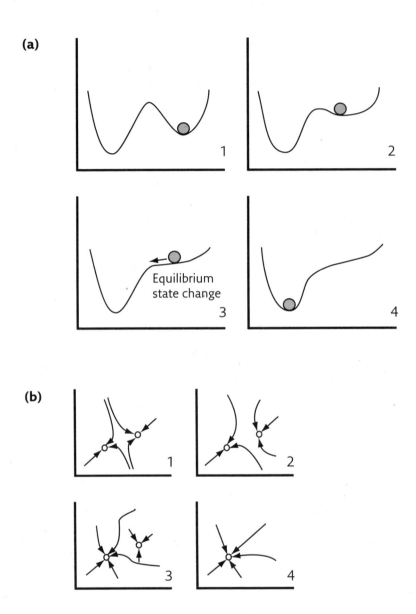

FIGURE 2: *Topographic analogy and state space views of evolving nature.*
The system modifies its own possible states as it changes over time from 1 to 4.
In this example, as time progresses, a progressively smaller perturbation is needed
to change the equilibrium state of the system from one domain to the other, until
the system spontaneously changes state. (a) Ball and topography analogy.
(b) Equivalent state space representation.

Managing For Engineering Resilience

Management and resource exploitation can overload waters with nutrients, turn forests into grasslands, trigger collapses in fisheries, and transform savannas into shrub-dominated semideserts. One example, described by Walker et al. (1969) concerns grazing of semiarid grasslands. Under natural conditions in east and south Africa, the grasslands were periodically pulsed by episodes of intense grazing by various species of large herbivores. Directly as a result, a dynamic balance was maintained between two groups of grasses. One group contains species able to withstand grazing pressure and drought because of their deep roots. The other contains species that are more efficient in turning the sun's energy into plant material, are more attractive to grazers, but are more susceptible to drought because of the concentration of biomass above ground in photosynthetically active foliage.

The latter, productive but drought-sensitive grasses, have a competitive edge between bouts of grazing so long as drought does not occur. But, because of pressure from pulses of intense grazing, that competitive edge for a time shifts to the drought-resistant group of species. As a result of these shifts in competitive advantage, a diversity of grass species serves a set of interrelated functions—productivity on the one hand and drought protection on the other.

When such grasslands are converted to cattle ranching, however, the cattle have been typically stocked at a sustained, moderate level, so that grazing shifts from the natural pattern of intense pulses separated by periods of recovery, to a more modest but persistent impact. Natural variability is replaced by constancy of production. The result is that, in the absence of intense grazing, the productive but drought-sensitive grasses consistently have advantage over the drought-resistant species and the soil- and water-holding capacity they protect. The land becomes more productive in the short-term, but the species assemblage narrows to emphasize one functional type. Droughts can no longer be sustained and the system can suddenly flip to become dominated and controlled by woody shrubs. That is, ecological resilience is reduced. It is an example of what Schindler (1990, 1993) has demonstrated experimentally in lakes as the effect of a reduction of species diversity when those species are part of a critical ecosystem function.

There are many examples of managed ecosystems that share this same feature of gradual loss of functional diversity with an attendant loss of resilience followed by a shift into an irreversible state, such as occurs in agriculture and in forest, fish, and grasslands management (as summarized in Holling, 1986). In each case the cause is reduction of natural variability of the critical structuring variables such as plants, insect pests, forest fires, fish populations, or grazing pressure to achieve a social, economic, or engineering objective. The result is that the ecosystem evolves to become more spatially uniform, less functionally diverse, and more sensitive to disturbances that otherwise could have been absorbed. That is, ecological resilience decreases even though engineering resilience might be great. Short-term success in stabilizing production leads to long-term surprise.

Moreover, such changes can be essentially irreversible because of accompanying changes in soils, hydrology, disturbance processes, and species complexes that regulate or control ecological structure and dynamics. Control of ecosystem function shifts from one set of interacting physical and biological processes to a different set (Holling et al., 1995).

In the examples of resource management that I have explored in depth, not only do ecosystems become less resilient when they are managed with the goal of achieving constancy of production, but the management agencies, in their chive for efficiency, also become more myopic, the relevant industries become more dependent and static, and the public loses trust (Gunderson et al., 1995). This seems to define an ultimate pathology that typically can lead to a crisis triggered by unexpected external events, sometimes followed by a reformation of policy. I first saw the form of this pathology emerging in the early stages of testing and developing theories, methods, and case study examples of adaptive environmental assessment and management. Those cases and their diagnoses were summarized in Holling (1986).

Those cases involved a number of different examples of forest development, fisheries exploitation, semiarid grazing systems, and disease management in crops and people. We have greatly expanded and deepened the case studies and tests since then, adding examples that are presented in a new book that explores both the dynamics of ecosystems and the dynamics of the institutions that attempt to manage them (Gunderson et al., 1995). Two of the original examples continue to provide insights.

In those two examples, the initial diagnoses of the pathology as I saw it in the early 1970s were as follows:

- Successful suppression of spruce budworm populations during the 1950s and 1960s in eastern Canada, using insecticide, certainly preserved the pulp and paper industry in the short-term by significantly reducing defoliation by the insect so that tree mortality was delayed. This encouraged expansion of pulp mills but left the forest, and hence the economy, more vulnerable to an outbreak that would cause more intense and more extensive tree mortality than had ever been experienced before. That is, the short-term success of spraying led to moderate levels of infestation and partially protected foliage that became more homogeneous over larger areas, demanding ever more vigilance and control.
- Effective protection and enhancement of salmon spawning through use of fish hatcheries on the west coast of North America quickly led to more predictable and larger catches by both sport and commercial fishermen. That triggered increased fishing pressure and investment in both sectors, pressure that caused more and more of the less productive natural stocks to become locally extinct.

  That left the fishing industry precariously dependent on a few artificially enhanced stocks, whose productivity began declining in a system where larger-scale physical oceanic changes contributed to unexpected impacts on the distribution and abundance of fish.

In both those cases, however, by the 1980s I began to realize that the phase of a growing pathology was transient and could be broken by a spasmodic readjustment, an adaptive lurch of learning that created new opportunity. It is that creation of something fundamentally novel that gives an evolutionary character to development of a region that might make sustainable development an achievable reality rather than an oxymoron.

The heart of these two different views of resilience lies in assumptions regarding whether multistable states exist. If it is assumed that only one stable state exists *or can be designed to so exist*, then the only possible definitions for, and measures of, resilience are near-equilibrium ones—such as characteristic return time. While that is certainly consistent with the engineer's desire to make things work, not to make things that break down or suddenly shift their behavior. But nature is different.

There are different stability domains in nature, and variation in critical variables tests the limits of those domains. Thus, a near-equilibrium focus seems myopic and attention shifts to determining the constructive role of instability in maintaining diversity and persistence and to designs of management that maintain ecosystem function in the face of unexpected disturbances. Such designs would maintain or expand ecological resilience. It is those ecosystem functions and ecological resilience that provide the ecological "services" that invisibly provide the foundations for sustaining economic activity.

Managing For Ecological Resilience

There is a puzzle in these examples and this analysis. It implies that efficient control and management of renewable resources in an engineering sense leads initially to success in managing a target variable for sustained production of food or fiber but ultimately to a pathology of less resilient and more vulnerable ecosystems, more rigid and unresponsive management agencies, and more dependent societies. But there seems to be something inherently wrong with that conclusion, implying, as it does, that the only solution is humanity's radical return to being "children of nature." The puzzle needs to be clarified to test its significance and generality.

The above conclusion is based on two critical points. One is that reducing the variability of critical variables within ecosystems inevitably leads to reduced resilience and increased vulnerability. The second is that there is, in principle, no different way for agencies and people to manage and benefit from resource development. Both points are explored in more detail in a new book on barriers and bridges to ecosystem and institutional renewal (Gunderson et al., 1995), so here I will deal only with highlights.

Puzzles can sometimes be solved by searching for counterexamples. Oddly, nature itself provides such counterexamples of tightly regulated yet sustainable systems in the many examples of physiological homeostasis. Consider temperature regulation of endotherms (warm-blooded animals). The internal body temperature of endotherms is not only tightly regulated within a narrow band, but among present-day birds and mammals, the average temperature is perilously close to lethal. Moreover, the cost of achieving that regulation requires ten times the energy for metabolism that is required by ectotherms (cold-blooded animals). That would seem to be a recipe for not only disaster but a very inefficient one at that. And yet evolution somehow led to the extraordinary success of the animals having such an adaptation—the birds and mammals.

To test the generality of the variability-loss/resilience-loss hypothesis, I have been collecting data from the physiological literature on the viable temperature range within the bodies of organisms exposed to different classes of variability. I have organized the data into three groups ranging from terrestrial ectotherms, which are exposed to the greatest variability of temperature from unbuffered ambient conditions, to aquatic ectotherms, which are exposed to an intermediate level of variability because of the moderating attributes of water, to endotherms, which regulate temperature within a narrow band. The viable range of internal body temperature decreases from about 40 degrees centigrade for the most variable group to about 30 degrees for the intermediate, to 20 degrees for the tightly regulated endotherms. Therefore resilience, in this case the range of internal temperatures that separates life from death, clearly does contract as variability in internal temperature is reduced, just as in the resource management cases. I conclude, therefore, that reduction of variability of living systems, from organisms to ecosystems, inevitably leads to loss of resilience in that part of the system being regulated.

But that seems to leave an even starker paradox for management; seemingly successful control inevitably leads to collapse. But, in fact, endothermy does persist and flourish. It therefore serves as a revealing metaphor for sustainable development. This metaphor contains two features that were not evident in my earlier descriptions of examples of resource management.

First, the kind of regulation is different. Five different mechanisms, from evaporative cooling to metabolic heat generation, control the temperature of endotherms. Each mechanism is not notably efficient by itself. Each operates over a somewhat different but overlapping range of conditions and with different efficiencies of response. It is this overlapping "soft" redundancy that seems to characterize biological regulation of all kinds. It is not notably efficient or elegant in the engineering sense. But it is robust and continually sensitive to changes in internal body temperature. That is quite unlike the examples of rigid regulation by management where goals of operational efficiency gradually isolated the regulating agency from the things it was regulating.

Examples of similar regulation of ecosystem dynamics in nature include the set of herbivorous antelope species that structure the vegetation of the savannas of East Africa at intermediate scales from meters to kilometers (Walker et al., 1969) or the suite of 35 species of insectivorous birds that, through their predation on insect larvae, set the timing for outbreaks of spruce budworm in the forests of eastern Canada (Holling, 1988). In these examples, each species performs its actions somewhat differently from others, and each responds differently to external variability because of differences in habitat preference and the scales of choice for its resources (Holling, 1992). As an example, some species of insectivorous birds exert modest predation pressure over a broad range of prey densities, whereas others exert strong pressure over narrow ranges of density and still others function between those extremes. The densities at which the predation impact is maximal also differ between species. Competition occurs among these species such that the aggregate predation effect is inefficient when predators are abundant and prey scarce and efficient when the reverse is true. As a consequence, the result of their joint action is an overlapping set of reinforcing influences that are less like the redundancy of engineered devices and more like portfolio diversity strategies of investors. The risks and benefits are spread widely to retain overall consistency in performance independent of wide fluctuations in the individual species. That is at the heart of the role of functional diversity in maintaining the resilience of ecosystem structure and function.

We chose the term *functional diversity* to describe this process, following the terms suggested by Schindler (1990) and by Holling et al. (1995). Such diversity provides great robustness to the process and, as a consequence, great resilience to the system behavior.

The second feature of nature's way of tightly regulating variability that is different from traditional management is the tendency to function near the edge of instabilities, not far away from them. That is where information and opportunity are the greatest. Again endothermy provides an example. Endothermy is a true innovation that explosively released opportunity for the organisms that evolved the ability to regulate their body temperature. Maintaining high body temperature, just short of death, allows the greatest range of external activity for an animal. Speed and stamina increase and activity can be maintained at both high and low external temperatures. A range of habitats forbidden to an ectotherm is open to an endotherm. The evolutionary consequence of temperature regulation was to open opportunity suddenly for dramatic organizational change and the adaptive radiation of new life forms. Variability is therefore not eliminated. It is reduced in one place and transferred from the animal's internal environment to its external environment as a consequence of allowing continual probes by the whole animal for opportunity and change. Hence the price of reducing internal resilience, maintaining high metabolic levels, and operating close to an edge of instability is more than offset by that creation of evolutionary opportunity. Nature's policy of ecological resilience, if we can call it that, seems far from those of traditional engineering safeguards or economic efficiency, where operating near an equilibrium far from an instability defines engineering resilience.

But ascribing that designation to engineering is to stereotype the field with only one face of its activities, just as ecological resilience represents only one face of ecology. At least some aspects of ecologically resilient control are equally familiar to the control engineer, for operation at the edge of instability is characteristic of designs for high-performance aircraft. Oddly, the result is opportunity. Effective control of internal dynamics at the edge of instability generates external options. Operating at the edge of instability generates immediate signals of changing opportunity.

That surely is at the heart of sustainable development—the release of human opportunity. It requires flexible, diverse, and redundant regulation, early signals of error built into incentives for corrective action, and continuous experimental probing of the changes in the external world. Those are the features of adaptive environmental and resource management. Those are the features missing in the descriptions I presented of traditional, piecemeal, exploitive resource management and its ultimate pathology.

Conclusion

There are indeed strong suggestions that management and institutional regimes can be designed to preserve or expand resilience of systems as well as provide developmental opportunity. It is a central issue that only now is beginning to be the focus of serious scholarship and practice. Of the cases I know well, management of the forests of New Brunswick seems most clearly to demonstrate the cycles of crisis and learning and the hesitant emergence of a more sustainable path.

In the New Brunswick example, one major crisis and several minor ones have occurred since the early 1950s. During this period, the new technologies of airplanes and pesticides developed in World War II were adapted for spraying operations and their use was progressively refined to achieve high mortality of insects while reducing environmental side effects. These procedures for pesticide control of budworm synergized with other technological developments in tree harvesting, pulp production chemistry, and mill construction and resulted in large investments in pulp production. Minor crises occurred when effects on human health were linked to the pesticides. Key pieces of integrated understanding of the natural system were achieved by the teams of Morris (1963) and the modelers of the 1970s (Clark et al., 1979). The brittleness that developed (defined by a loss of ecological resilience together with an increase in institutional efforts to control information and action) reflected the complacent belief among agency staffs that budworm damage was controlled in an efficient and cost-effective manner and that there was plenty of wood available for harvest. In reality, the costs of using pesticides were rapidly increasing because of increases in oil prices and because of modification of pesticide application in response to public pressure. In addition, available stocks of harvestable trees were decreasing because of past harvests and because more and more mature stands over larger areas were gradually deteriorating from the pressure of moderate but persistent budworm defoliation. The major crises occurred during the late 1970s when a forest inventory report finally indicated that there would not be sufficient stock to support the current mills, thereby confirming an earlier prediction of the models. This led to a new law that restructured the licensing and forest management policies and freed the innovative capacity of local industries within a regional set of goals and constraints. A sequence of adaptive responses among the actors began to develop regional forest policy in a way that now engages local industrial, environmental, and recreational goals.

The examples of growing pathology are caused by the very success of achieving near equilibrium behavior and control of a single target variable independently of the larger ecosystem, economic, and social interactions. When that orientation or goal is abandoned, it happens suddenly, in response to perceived or real crises. The scale of the issues becomes redefined more broadly from a local to a regional setting and from short-term to long-term. The scientific understanding of the natural system becomes more integrated, and the issues themselves are not posed in response to needs to maximize constancy or productivity of yield, but to ones of designing interrelations between people and resources that are sustainable in the face of surprises and the unexpected. If there is such a thing as sustainable development, then that is it. The key features are integration of knowledge at a range of scales, engagement of the public in exploring alternative potential futures, adaptive designs that acknowledge and test the unknown, and involvement of citizens in monitoring and understanding outcomes. That is possible only in situations where ecological resilience and public trust have not been degraded. If they have, as in many situations, then the initial goal has to be the restoration of both resilience and trust.

References

Arthur, B. 1990. Positive feedback in the economy. Scientific American 262:92–99.

Clark, W. C., D. D. Jones, and C. S. Holling. 1979. Lessons for ecological policy design: A case study of ecosystem management. Ecological Modeling 7:1–53.

DeAngelis, D. L. 1980. Energy flow, nutrient cycling and ecosystem resilience. Ecology 61:764–771.

Dublin, H. T., A. R. E. Sinclair, and J. MeGlade. 1990. Elephants and fire as causes of multiple stable states in the Serengeti-Mara woodlands. Journal of Animal Ecology 59:1147–1164.

Fiering, M. B. 1982. Alternative indices of resilience. Water Resources Research 18:33–39.

Gunderson, L. H., C. S. Holling, and S. Light. 1995. Barriers and Bridges to Renewal of Ecosystems and Institutions. New York: Columbia University Press.

Holling, C. S. 1973. Resilience and stability of ecological systems. Annual Review of Ecology and Systematics 4:1–23.

Holling, C. S. 1986. Resilience of ecosystems; local surprise and global change. Pp. 292–317 in Sustainable Development of the Biosphere, W. C. Clark and R. E. Munn, eds. Cambridge, England: Cambridge University Press.

Holling, C. S. 1988. Temperate forest insect outbreaks, tropical deforestation and migratory birds. Memoirs of the Entomological Society of Canada 146:21–32.

Holling, C. S. 1992. Cross-scale morphology, geometry and dynamics of ecosystems. Ecological Monographs 62(4):447–502.

Holling, C. S., D. D. Jones, and W. C. Clark. 1977. Ecological policy design: A case study of forest and pest management. IIASA CP-77-6:13–90 in Proceedings of a Conference on Pest Management, October 1976, G. A. Norton and C. S. Holling, eds. Laxenburg, Austria.

Holling, C. S., D. W. Schindler, B. Walker, and J. Roughgarden. 1995. Biodiversity in the functioning of ecosystems: An ecological primer and synthesis. In Biodiversity Loss: Ecological and Economic Issues, C. Perrings, K. G. Mäler, C. Folke, C. S. Holling, and B. O. Jansson, eds. Cambridge, England: Cambridge University Press.

Morris, R. F. 1963. The dynamics of epidemic spruce budworm populations. Memoirs of the Entomological Society of Canada 21:332.

O'Neill, R. V., D. L. DeAngelis, J. B. Waide, and T. F. H. Allen. 1986. A Hierarchical Concept of Ecosystems. Princeton, N.J.: Princeton University Press.

Pimm, S. L. 1984. The complexity and stability of ecosystems. Nature 307:321–326.

Schindler, D. W. 1990. Experimental perturbations of whole lakes as tests of hypotheses concerning ecosystem structure and function. Proceedings of 1987 Crafoord Symposium. Oikos 57:25–41.

Schindler, D. W. 1993. Linking species and communities to ecosystem management. Proceedings of the 5th Cary Conference, May 1993.

Schindler, D. W., T. M. Frost, K. H. Mills, P. S. S. Chang, I. J. Davis, F. L. Findlay, D. F. Malley, J. A. Shearer, M. A. Turner, P. J. Garrison, C. J. Watras, K. Webster, J. M. Gunn, P. L. Brezonik, and W. A. Swenson. 1991. Freshwater acidification, reversibility and recovery: Comparisons of experimental and atmospherically-acidified lakes. Volume 97B: 193–226 in Acidic Deposition: Its Nature and Impacts, F. T. Last and R. Watling, eds. Proceedings of the Royal Society of Edinburgh.

Sinclair, A. R. E., P. D. Olsen, and T. D. Redhead. 1990. Can predators regulate small mammal populations? Evidence from house mouse outbreaks in Australia. Oikos 59:382–392.

Tilman, D., and J. A. Downing. 1994. Biodiversity and stability in grasslands. Nature 367:363–365.

Waide, J. B., and J. R. Webster. 1976. Engineering systems analysis: Applicability to ecosystems. Volume IV, pp. 329–371 in Systems Analysis and Simulation in Ecology, B. C. Patten, ed. New York: Academic Press.

Walker, B. H. 1981. Is succession a viable concept in African savanna ecosystems? Pp. 431–447 in Forest Succession: Concepts and Application, D. C. West, H. H. Shugart, and D. B. Botkin, eds. New York: Springer-Verlag.

Walker, B. H., D. Ludwig, C. S. Holling, and R. M. Peterman. 1969. Stability of semi-arid savanna grazing systems. Ecology 69:473–498.

Walters, C. J. 1986. Adaptive Management of Renewable Resources. New York: McGraw Hill.

The Resilience of Terrestrial Ecosystems

Local Surprise and Global Change

C. S. HOLLING

Pages 292–320 in *Sustainable development of the biosphere*:
By W. C. Clark and R. E. Munn (eds). Cambridge University Press,
Cambridge, UK.

Source: Clark WC, Munn RE (eds) (1986) *Sustainable Development of the Biosphere*. Cambridge University Press, Cambridge, UK.

ADEQUATE EXPLANATIONS of long-term global changes in the biosphere often require an understanding of how ecological systems function and of how they respond to human activities at local levels.

Outlined in this chapter is one possible approach to the essential task of linking physical, biological, and social phenomena across a wide range of spatial and temporal scales. It focuses on the dynamics of ecological systems, including processes responsible for both increasing organization and for occasional disruption. Special attention is given to the prevalence of discontinuous change in ecological systems, and to its origins in specific nonlinear processes interacting on multiple time and space scales. This ecological scale of analysis is linked "upward" to the global scale of biogeochemical relationships and the "Gaia" hypothesis (see Chapters 7,

8, and 9 in *Clark and Munn* volume), and "downward" to the local scale of human activities and institutions (see Chapters 3, 11, and 14 in *Clark and Munn* volume).

Introduction

Considerable understanding has been accumulated during the last decade of the way the world "ticks" in its various parts, and progress has been made in recognizing what those parts are and the need to inter-relate them. Only after such developments, and therefore only recently, could we begin to address global ecological questions effectively. There are four key questions. How do the Earth's land, sea, and atmosphere interact through biological, chemical, and physical processes? How do ecosystems function and behave to absorb, buffer, or generate change? How does the development of man's economic activities, particularly in industry and agriculture, perturb the global system? How do people—as individuals, institutions, and societies—adapt to change at different scales? In this chapter I respond to the second question by exploring the way ecosystems function and behave, but with the other three problems in mind.

During much of this century global change has been slow, although some important cumulative effects have occurred. The gradual expansion, on a global scale, of economic and agricultural development is well represented by regular increases in atmospheric carbon dioxide (CO_2) of approximately one part per million per year [1]. This increase is attributable to a 2–4% annual increase in burning fossil fuels and, in part, to deforestation. The climate during this century has been benign relative to other periods. Marine fisheries stocks, although typically variable, were largely steady during the period 1920–1970, at least in relation to the apparent sharp shifts that occur among such species as Pacific sardine or Atlantic herring every 50 to 100 years [2]. Problems of environmental pollution have increased in geographical scale from the highly local to the size of air basins or watersheds, but slowly enough that the effects have been largely ameliorated [3]. The deterioration of Lake Baikal in the USSR has been slowed and that of Lake Erie in North America has been stopped. Fish have returned to the River Thames and the extreme smogs of London are now only memories. Atmosphere, oceans, and land, coupled through biological, chemical, and physical forces have apparently been able to absorb the global changes of this century.

But now qualitative change [4], as distinct from gradual quantitative change, seems possible. Man's industrial and agricultural activities have speeded up many terrestrial and atmospheric processes, expanded them globally, and homogenized them. Four qualitative changes are suggested. First, such changes are being considered as ecological, not simply environmental. For example, pollution can no longer be viewed as inertly burdening the atmosphere. Rather, its impacts on vegetation can accelerate the consequences by impairing the regulatory processes that are mediated by vegetation. Second, the intensity of the impact of man's activities and their acceleration of the time dynamics of natural processes can influence the coupling of long-term regulatory phenomena that link atmosphere, oceans, and land. Third, some of these qualitative changes are likely to be irreversible in principle. So long as the change is local, it can be reversed because there are alternative sources both of genetic variability and species and for the renewal of air and water. But this becomes less and less an option as the change becomes more homogeneous with increasing scale, from local to continental to hemispheric, and then to global. Finally, with the option of reversibility reduced, increasing emphasis will be placed on adapting to the inevitable. But individual, institutional, and social adaptation each have their own time dynamics and histories. There has been little experience in translating the remarkable adaptive responses of individuals to local changes [5] into responses to international and global ones.

In order to analyze ecosystem function and behavior in such a way that global changes can be related to local events and action, I consider four topics. The first is a conceptual framework that can help focus treatment of the contrasts between global and local behavior and between continuous and discontinuous behavior. Since the framework describes different perceptions of regulation and stability, it provides the necessary background for the second topic: the particular causal relations and processes within ecosystems, the influence of external variation on them, and their behavior in time and space. The third topic synthesizes our present understanding of the structure and behavior of ecosystems in a way that has considerable generality, and organizational power. The fourth connects that understanding to our knowledge of global phenomena and of local perception and action.

The Conceptual Framework: Gaia and Surprise

In this chapter I discuss ecosystems, but first the relationships between ecosystems and two other key aspects of the global puzzle must be established; namely, with global biogeophysical events, and with societal perception and management. In the former case, some image is needed of the way the global systems in the atmosphere, oceans, and land interact. That image is provided by the Gaia hypothesis [6]. And to relate our understanding of the behavior of local ecosystems to the way societies perceive and manage those systems the concept of surprise is needed: Gaia and surprise are dealt with in turn.

Gaia

Gaia is the "global biochemical homeostasis" hypothesis, proposed by Lovelock and Margulis [6,7], that life on Earth controls atmospheric conditions optimal for the contemporary biosphere. The Gaia hypothesis presumes homeostatic regulation at a global level. An example is the maintenance of 21% oxygen in the air, a composition representing the highest possible level to maximize aerobic metabolism, but just short of the level that would make Earth's vegetation inflammable. The residence time of atmospheric oxygen is of the order of thousands of years, a time scale that renders methane (CH_4) production by anaerobic organisms an important regulator of oxygen concentration. The mechanism proposed includes the burying of a small amount of the carbonaceous material of living matter each year and the production of CH_4, which reacts with oxygen, thereby providing a negative feedback loop in the system of oxygen control. Similarly, linked biological and geological feedback mechanisms have been proposed for the regulation of global temperature. The regulation is mediated by control of CO_2 in the atmosphere at concentrations that have compensated for increasing solar radiation over geological time [8].

Even though the Gaia hypothesis is speculative, at least there is more and more evidence for the dynamic role of living systems in determining the composition of many chemicals in the air, soil, and water [9]. And at a smaller geographical scale, as discussed later, there are many ecosystem processes that cybernetically regulate conditions for life.

There are three reasons why I use the Gaia hypothesis as one of my two organizing themes. First, by being rooted in questions of regulation and stability through identifiable biological, chemical, and physical processes, it gives a direction for relevant scientific research—for disproof of the hypothesis if nothing else. Second, this is the only concept I know that can, in principle, provide a global rationale for giving priority to rehabilitation, protection of ecosystems, and land use management. If CH_4 production, for example, provides an essential negative feedback control for ozone (O_3) concentration, then the recent 1–2% annual increase in CH_4 content is important and priority should be given to considering major changes in its primary sources—i.e., wetlands, biomass burning, and ecosystems containing ruminants and termites [10]. Finally, an examination of global change concerns not only science but also policy and politics. In a polarized society where certitude is lacking, Gaia has some potential for bridging extremes by providing a framework for understanding and action.

Surprise

Just as Gaia is global, the second organizing theme of surprise is, necessarily, local. Surprise concerns both the natural system and the people who seek to understand causes, to expect behaviors, and to achieve some defined purpose by action. Surprises occur when causes turn out to be sharply different than was conceived, when behaviors are profoundly unexpected, and when action produces a result opposite to that intended —in short, when perceived reality departs *qualitatively* from expectation.

Expectations develop from two interacting sources: from the metaphors and concepts we evolve to provide order and understanding and from the events we perceive and remember. Experience shapes concepts; concepts, being incomplete, eventually produce surprise; and surprise accumulates to force the development or those concepts. This sequence is qualitative and discontinuous. The longer one view is held beyond its time, the greater the surprise and the resultant adjustment. Just such a sequence of three distinct viewpoints, metaphors, or myths has dominated perceptions of ecological causation, behavior and management [11].

Equilibrium-Centered View: Nature Constant: This viewpoint emphasizes not only constancy in time, but also spatial homogeneity and linear causation. A familiar image is that of a landscape with a bowl-shaped valley within which a ball moves in a way determined by its own acceleration and direction and by the forces exerted by the bowl and gravity. If the bowl was infinitely large, or events beyond its rim meaningless, this would be an example of global stability. Such a viewpoint directs attention to the equilibrium and near-equilibrium conditions. It leads to equilibrium theories and to empirical measures of constancy that emphasize averaging variability in time and "graininess" in space. It represents the policy world of a benign nature where trials and mistakes of any scale can be made with recovery assured once the disturbance is removed. Since there are no penalties or size, only benefits to increasing scale, this viewpoint leads to notions of large and homogeneous economic developments that affect other biophysical systems, but are not affected by them.

Multiple Equilibria States: Nature Engineered and Nature Resilient: This second viewpoint is a dynamic one that emphasizes the existence of more than one stable state. In one variant the instability is seen as maintaining the resilience of ecological systems [12]. It emphasizes variability, spatial heterogeneity, and nonlinear causation. A useful image is that of a landscape of hills and valleys with the ball journeying among them, in part because of internal processes and in part because exogenous events can flip the ball from one stability domain to another. This viewpoint emphasizes the qualitative properties of important ecological processes that determine the existence of stable regions and of boundaries separating them. Continuous behavior is expected over defined periods that end with sharp changes induced by internal dynamics or by exogenous events, at times large, at times small.

The length of the period of continuous behavior often determines the magnitude of the subsequent change and affects policy recommendations. For example, one would argue from an equilibrium-centered viewpoint that climate warming due to the accumulation of "greenhouse" gases will proceed slowly enough for ecological and social processes to adapt of their own accord. Efforts to facilitate adjustment are unnecessary because existing crop types, for example, are likely to develop and be well adapted to prevailing conditions. However, the second viewpoint of dynamic, nonlinear nature suggests just the opposite: that slow changes of the type expected might be so successfully absorbed and ignored that a sharp, discontinuous change becomes inevitable.

Similarly, spatial graininess, which is small relative to the range of movement of an organism, is presumed to be averaged out in the equilibrium-centered approach [18]. The nonlinear viewpoint, however, presents the possibility that small-scale events cascade upward, as has been described for climatic behavior [14]. But for ecological systems, Steele [15,16] notes that widely ranging animals feed on small-scale spatial variability. For example, if fish could not discover and remain in plankton patches they could not exist.

This second viewpoint can produce two variants of policy. One assumes that the landscape is fixed or that sufficient knowledge is available to keep it fixed. It is a view of nature engineered to keep variables (the ball) away from dangerous neighboring domains. It occurs in the responsible tradition of engineering for safety, of fixed environmental and health standards, and of nuclear safeguards.

The alternative variant sees that key features of the landscape are maintained by the journeys of the ball, by variability itself testing and maintaining the configuration. This is resilient nature in which the experience of instability is used to maintain the structure and general patterns of behavior. It is assumed in the design that there is insufficient knowledge to control the landscape and hence one attempts to retain variability while producing economic and social benefit [12, 17]. In such cases variables are allowed to exceed flexible limits so long as natural and designed recovery mechanisms are encouraged. Designs have been proposed for example, for dealing with pollution [18], environmental hazards [5], water resources [19], and pest management [20].

Organizational Change: Nature Evolving: The final viewpoint is one of evolutionary change. Later a number of examples are presented to demonstrate that successful efforts to constrain natural variability lead to self-simplification and so to fragility of the ecosystem. A variety of genetic, competitive, and behavioral processes maintain the values of parameters that define the system. If the natural variability changes, the values shift: the landscape of hills and valleys begins to alter. Stability domains shrink, key variables become more homogeneous (e.g., species composition, age structure, spatial distribution), and perturbations that previously could be absorbed no longer can be.

The resultant surprises can be pathological if continuing control requires ever-increasing vigilance and cost. But if control is internal and self-regulated, i.e., homeostatic, then the possibility opens for organizational change because the benefits of being embedded in a larger ecological or social system significantly exceed the costs of local control.

An example from biological evolution is the remarkably constant internal temperature maintained by endothermic (warm-blooded) animals in the presence of large changes in external temperature. A large metabolic load is required to maintain a constant temperature. As expected, the range of internal temperatures that sustains life becomes narrower than for {cold-blooded} ectotherms. Moreover, the typical endotherm body temperature of around 37° C is close to the upper lethal temperature for most living protoplasm. It does not represent a "policy" of keeping well away from a dangerous threshold.

The evolutionary significance of this internal temperature regulation is that maintenance of the highest body temperature, short of death, allows the greatest range of external activity for an animal [21]. Speed and stamina increase and activity can be maintained at high and low external temperatures, rather than forcing aestivation or hibernation. There is hence an enhanced capability to explore environments and conditions that otherwise would preclude life. The evolutionary consequence of such temperature regulation was the suddenly available opportunity for dramatic organizational change and explosive radiation of adaptive life forms. Hence the reduction of internal resilience as a consequence of effective self-regulation was more than offset by the opportunities offered by other external settings.

Hence the study of evolution requires not only concepts of function but also concepts of organization—of the way elements are connected within subsystems and the way subsystems are embedded in larger systems. Food webs and the trophic relations that represent them are an example and have long been a part of ecology. Recently some revealing empirical analyses have demonstrated remarkable regularities in such ecosystem structures [22], with food webs of communities in fluctuating environments having a more constrained trophic structure than those in constant environments [23].

These and related developments, connected in turn to hierarchy theory [24] on the one hand, and the stability and resilience concepts described earlier, on the other, are starting to provide the framework required for comprehending organizational evolution [25]. Although not as well developed as equilibrium, engineering, and resilience concepts, such developments are an essential part of any effort to understand, guide, or adapt to global change.

These views of nature represent the different concepts people have of the way natural systems behave, are regulated, and should be managed. Surprise can occur when the real world is found to behave in a sharply different way from that conceived. The perception can be ignored, resisted or acknowledged depending on how extreme the departure is and depending on how flexible and adaptable the observer is. Although observer and system are interlinked, I do not explore the psychology and dynamics of individual, institutional, and social adaptation in this chapter, though this is ultimately necessary if we want to understand and design sustainable systems. But in the next section I examine a number of ecological systems to determine which of the views of nature most closely matches reality.

Dynamics of Ecosystems

Resilience and Stability

This chapter relies heavily on the distinction between resilience and stability. Since that distinction was first emphasized [12] a significant literature has developed to test its reality in nature, to expand the theory, and to apply this to management and design. Much of what follows is drawn from the literature. The distinction relies on definitions that recognize the existence of different stability structures of the kind described in the previous section. There are four main points. First, there can be more than one stability region or domain, i.e., multiequilibrium structures are possible. Second, the behavior is discontinuous when variables (i.e., elements of an ecosystem) move from one domain to another because they become attracted to a different equilibrium condition. Third, the precise

kind of equilibrium—steady state or stable oscillation—is less important than the fact of equilibrium. Finally, parameters of the system that
define the existence, shape, and size of stability domains depend on a
balance of forces that may shift if variability patterns in space and time
change. In particular, reduced variability through management or other
activities is likely to lead to smaller stability regions whose contraction
can lead to sharp changes because the stability boundary crosses the variables, rather than the reverse.

This leads to the following definitions. Stability *(sensu stricto)* is the
propensity of a system to attain or retain an equilibrium condition of
steady state or stable oscillation. Systems of high stability resist any
departure from that condition and, if perturbed, return rapidly to it with
the least fluctuation. It is a classic equilibrium-centered definition.

Resilience, on the other hand, is the ability of a system to maintain
its structure and patterns of behavior in the face of disturbance. The size
of the stability domain of residence, the strength of the repulsive forces
at the boundary, and the resistance of the domain to contraction are all
distinct measures of resilience.

Stability, as here defined, emphasizes equilibrium, low variability,
and resistance to and absorption of change. In sharp contrast, resilience
emphasizes the boundary of a stability domain and events far from equilibrium, high variability, and adaptation to change.

However, one school of ecology so strongly emphasizes linear interactions and steady state properties [26, 27, 28] that resilience is treated
in the opposite way to that described above. It is defined as "how fast
the variables return towards their equilibrium following a perturbation
[28]" and is measured by the characteristic return times. In terms of the
definitions used in this chapter, this concerns only one facet of stability
and has nothing to do with the qualitative distinctions that I believe are
important.

In addition to the growing number of tests and demonstrations of the
key features of resilience, there have been two major expansions of theory and example. One is Levin's excellent analysis and review of patterns
in ecological communities [29]. Levin first placed experimental, functional, and behavioral descriptions within a formal mathematical frame.
More important, he made two qualitative additions. One was to explore
spatial patterns of multistable systems by analyzing the consequences of

diffusion. The second was to make a sharp distinction between variables associated with different speeds or rates of activity, partly to facilitate analysis but more to stress the consequences of coupling subsystems whose cycles are of different lengths. The second major expansion was that of Allen and Starr who extended the analyses of ecosystem patterns for a wide range of examples [25]. Most significantly, they embedded theory, measurement, and modeling relevant to resilience and stability into a hierarchical framework. More than any recent development, this framework provides a means of studying community structure and of treating evolution or organizational change.

Ecosystem Scale

These three developments in analysis—of multistable systems [12], of spatial diffusion [29], and of hierarchies [25]—concern the coupling of nonlinear subsystems of different scales in time and space. They are fundamental to understanding how predictable change is, whether or not historical accidents are important, and how to achieve a balance between anticipatory design and adaptive design. Clark [30] has provided a useful classification of the relevant scales for a wide range of geophysical, ecological and social phenomena. The scales range from square centimeters to global and from minutes to thousands of years. In the present analysis I concentrate on ecological systems covering scales from a few square meters to a few thousand square kilometers and from a few years to a few hundred years.

These scales represent ecosystems, which are defined here as communities of organisms in which internal interactions between the organisms determine behavior more than do external biological events. External abiotic events do have a major impact on ecosystems, but are mediated through strong biological interactions within the ecosystems. It is through such external links that ecosystems become part of the global system. Hence, the spatial scale is determined by the dispersal distance of the most mobile of the key biological variables. The structure of eastern North American spruce–fir forests, for example, is profoundly affected by the spruce budworm, which periodically kills large areas of balsam fir. The modal distance of dispersal of adult budworms is of the order of 50 km [20], but movements are known to extend up to 200 km. The relevant spatial area over which internal events dominate can therefore cover a good part of east–central North America. And the minimum area for analysis has to be of the order of 70 000 km².

Similarly, the time span of up to a few hundred years is set by the longest-lived (slowest-acting) key biological variables. In the case of the spruce–fir forests the trees are the slowest variables with a rotation age of about 70 years. Any effective analysis therefore must consider a time span that is a small multiple of that—of the order of 200 years.

Eugene Odum [31], more than anyone else, has emphasized that such ecosystems are legitimate units of investigation, having properties of production, respiration, and exchange that are regulated by biological, chemical, and physical processes. Hence they represent distinct sub-systems of the biogeochemical cycles of the Earth. They are open, since they receive energy from the sun and material and energy from larger cycles. In regulating and cycling this material through biotic and abiotic processes, outputs are discharged to larger cycles. Ecosystems hence are Gaia writ small.

Succession

One dominant theme of ecosystem study has been succession—the way complexes of plants develop after a disturbance. Clements' scheme of succession has played an important role in guiding study and theory [32]. He emphasized that succession leads to a climax community of a self-replicating assemblage of plants. The species comprising the assemblage are determined by precipitation and temperature. Plant colonization and growth are seen as proceeding to the stable climax. Initial colonization is by pioneer species that can grow rapidly and withstand physical extremes. They so ameliorate these conditions as to allow entry of less robust but more competitive species. These species in turn inhibit the pioneers but set the stage for their own replacement by still more effective competitors. Throughout this process, biomass accumulates, regulation of biological, chemical, and physical processes becomes tighter, and variability is reduced until the stable climax condition is reached and maintained. This scheme represents a powerful equilibrium-centered view in which disturbances by fire, storm, or pest are treated as exogenous (and somehow inappropriate) intrusions into a natural order. Clements gave an analogy to an organism and its ability to repair damage [32].

During the past 15 years this view has been significantly modified by a wide range of studies of ecosystems—some dominated by disturbance, some not—and by experimental manipulation of defined ecosystem units such as the classic Hubbard Brook Watershed Study [33]. Before describing those developments, however, it is useful to relate this view of succession to another powerful equilibrium-centered notion: that of r and K strategies.

MacArthur and Wilson [34] proposed this classification to distinguish between organisms selected for efficiency of food harvest in crowded environments (K-selected) and those selected simply to maximize returns without constraint (r-selected). The designations come from the terms of the logistic equation, where K defines the saturation density (stable equilibrium population) and r the instantaneous rate of increase. MacArthur [35] pointed out the contrast between "opportunist" species in unpredictable environments (r-strategists) and "equilibrium" species in predictable ones (K-strategists). Pianka [36] and Southwood et al. [37] have emphasized that these represent extremes of a continuum, but that a variety of life histories and biological and behavioral features correlate with the two strategies. Briefly, r-strategists have a high reproductive potential, short life, high dispersal properties, small size, and resistance to extremes. They are the pioneers of newly disturbed areas or the fugitive species that ever occupy transient habitats. K-strategists have lower reproductive potential, longer life, lower dispersal rates, large size, and effective competitive abilities. They represent, therefore, the climax species of Clements or those that occupy stable, long-lasting habitats.

There clearly are communities that have developed a climax maintained through plant-by-plant replacement in the manner proposed by Clements. Lorimer [38], for example, examined the history of pre-settlement forests in northeastern Maine, USA, and found that the time interval between severe disturbances was much longer than that needed to obtain a climax, all-age structure. Other examples are presented in an extensive review of forest succession edited by West et al. [39]. But the Clementsian view of succession as analogous to the recovery of an organism from injury, with an ordered and obligatory sequence of replacements of one species by another, is oversimplified and limiting for several reasons.

First, many communities are subjected to regular or irregular disturbances severe enough to kill established plants over areas of a few square meters (the size of a tree) to several thousand square kilometers. Traditionally these disturbances—fires, landslides, storms, floods, disease, insect pests, and herbivore grazing—have usually been viewed as external to the system. But when they occur at a frequency related to the life span of the longest-lived species, the plants themselves can become increasingly adapted to the disturbance and so make the event an internally triggered and maintained phenomenon. This is particularly well recognized for fire. Mutch [40], for example, demonstrated that vegetation of fire-adapted species was significantly more combustible than that of related species in communities not subject to fires. Similarly, Biswell [41] describes the twig development and proliferation of chaparral species that significantly increase the inflammability of plants that are 15 years and older. This coincides with a typical burn cycle of similar duration. Such "accidents designed to happen" are more common than is usually recognized and further examples involving agents other than fire are described later.

As a consequence, there are many examples of what I imagine consistent Clementsian ecologists would be forced to see as self-inflicted wounds to the ecosystem "organism". Such disturbances have a wide range of periodicities set by the dynamics of the slowest variable [42]. Fire frequency in the Pacific Northwest of North America, for example, occurs every 400–500 years, and this period is related to the potentially 100-year life span of Douglas fir [43]. Eastern white pine forests experience a fire periodicity of 100–300 years in presettlement times [44, 45], while cyclic changes of 200-year periodicity are proposed for elephant populations; this period is due to the recovery time required for the tasty (to elephants) and long-lived baobab trees of East Africa [46]. The fire controlled period throughout much of the boreal forest of northern North America was between 51 and 120 years [47, 48]. Chaparral in California [49] is adapted to a more frequent cycle of 10–50 years.

The second significant departure from Clements' notions is even more fundamental. Some disturbances can carry the ecosystem into a different stability configuration or domain. At times this happens after a long period of exploitation has apparently reduced the resilience of the ecosystem. For example, fishing in the Great Lakes has been argued to have set the stage for a radical change in fish communities from a system dominated by a large species to one of smaller species [50], the overall biomass remaining constant. The resilience was reduced to a level where small stresses from the physical environment, from man, or from biological invasion triggered the new configuration. In a similar way, shifts of savannas from mixed grass-shrub systems to ones dominated by shrubs are often triggered by a modest drought after being conditioned by an extended period of cattle grazing [51]. In other instances, the magnitude of the triggering event is so great or the resilience of the ecosystem so naturally low that new configurations emerge quite independent of previous management. Bormann and Likens [33, p. 189] present one such example of a burn in a spruce–hardwood forest on thin soils that transformed part of the ecosystem into a bare-rock-shrub system. Hence there is not just one climax state; there can be more than one.

Third, species that are important late in the sequence can be present together with pioneer species at the initiation of old field succession [52] or forest succession [43,48]. The resultant successional sequence is hence much more in the form of a competitive hierarchy as described by Horn [42]. Early in a sequence, the opportunist species grow rapidly, dominate for a short time, but ultimately cannot withstand crowding and competition from other more persistent species. Marks [53] presents a particularly clear demonstration of this sequence and of the opportunist role of pin cherry in reestablishing disturbed hardwood systems in New Hampshire, USA. Late in the sequence pin cherry trees are almost totally absent, having been squeezed out by more competitive trees like beech and red maple. After disturbance, however, seeds long dormant in the soil germinate, pin cherry trees flourish, and begin to be eliminated again after about 20 years.

Finally, invasion of species after disturbance as well as during succession is highly probable, particularly in tropical lowlands [54]. This, combined with the competitive hierarchical relations mentioned above, makes the tropical forest highly individual in character and very diverse. So many species are capable of filling a particular niche that succession is better described by life history traits and tree geometry. There is, moreover, considerable advantage in dealing with succession in terms of such properties, since they determine successional status [55]—whether in tropical, temperate, or arctic regions.

That is why the idea of r-selected and K-selected species was introduced earlier. Each strategy is associated with distinctive traits that, in exaggerated form, contrast the exploitative and opportunistic species that dominate early in the succession with the consolidating or conservative species that dominate later through competition. Moreover, the terms can be used to refer to two principal ecosystem functions: exploitation and conservation. Early in succession biotic and abiotic exploitative processes dominate. These lead to the organization and binding of nutrients, rapid accumulation of biomass, and modification of the environment. Eventually, conservative forces begin to dominate, with competition being the most important aspect. This leads to increased organization through trophic and competitive connections, to reduced variability, and, if not interrupted, to reduced diversity.

Ecosystems, however, are also systems of discontinuous change. In addition to the successional processes leading to increasing order there are periods of disorganization. The examples mentioned earlier were of large-scale disruptions that affect extensive areas. But change of this kind also occurs in the ecosystems that most closely achieve a climax condition. Individual trees senesce, creating local gaps. The only difference is that the spatial scale is small and the disruptions are not necessarily synchronous. A complete picture of the dynamics of ecosystems therefore requires additional functions to those of exploitation and conservation. Such functions relate more to the generation of change and the introduction of disorder.

Forces of Change

To identify these functions and their effects, I initially analyze a small number of examples of ecosystems that demonstrate pronounced change and that have been examined in detail. They can be classified as follows: forest insect pests, forest fires, grazing of semiarid savannas, fisheries, and human disease. Many of the examples also represent systems that have been subjected to management. In a sense, the management activities can be viewed as diagnostic, for they introduced external changes that helped expose some of the internal workings of the natural, unmanaged system. In addition, a number of the management approaches were very much dominated by the goal of achieving constancy through externally imposed regulations. Hence the implicit hypothesis was an equilibrium-centered one and the experiences in managing forests, fish, and other organisms can be viewed as weak tests of that hypothesis.

To give an impression of the consequences I consider the following examples:

(1) Successful suppression of spruce budworm populations in eastern Canada using insecticides certainly preserved the pulp and paper industry and employment in the short term by partially protecting the forest. But this policy has left the forest and the economy more vulnerable to an outbreak over a larger area and of an intensity not experienced before [20].

(2) Suppression of forest fire has been remarkably successful in reducing the probability of fire in the national parks of the USA. But the consequence has been the accumulation of fuel to produce fires of an extent and cost never experienced before [56].

(3) Semiarid savanna ecosystems have been turned into productive cattle-grazing systems in the Sahel zone of Africa, southern and east Africa, the southern USA, northern India, and Australia. But because of changes in grass composition, an irreversible switch to woody vegetation is common and the systems become highly susceptible to collapse, often triggered by drought [51].

(4) Effective protection and enhancement of salmon spawning on the west coast of North America are leading to a more predictable success. But because this triggers increased fishing and investment pressure, less productive stocks become extinct, leaving the fishing industry precariously dependent on a few enhanced stocks that are vulnerable to collapse [57].

(5) Malaria eradication programs in Brazil, Egypt, Italy, and Greece have been brilliant examples of sophisticated understanding combined with a style of implementation that has all the character of a military campaign. But in other areas of the world, where malaria was neither marginal nor at low endemic levels, transient success has led to human populations with little immunity, and mosquito vectors resistant to DDT. As a consequence, during the past five years some

countries have reported a 30- to 40-fold increase in malaria cases compared with 1969–1970, signaling a danger not only to the health of the population, but also to overall socioeconomic development.

In each of these examples, the policy successfully reduced the probability of an event that was perceived as socially or economically undesirable. Each was successful in its immediate objective. Each produced a system with qualitatively different properties. All of these examples, and others that fall into the five classes, represent "natural" systems that are coupled to management institutions and to the society that experiences the success or endures the failure of management. Here I focus principally on the natural system.

Despite the large number of variables in each example, the essential causal structure and behavior can be represented by interaction among three sets of variables. These represent three qualitatively different speeds, or rates of activity, corresponding to rates of growth, generation times, and life spans (Table 1). It becomes possible, as a consequence, to proceed in two directions: to achieve a qualitative understanding of the natural system and to achieve a detailed policy design. The first objective draws upon the theory of differential equations [58]. The second draws upon more recent developments in simulation modeling and optimization [20]. Both are very much connected to hierarchy theory [25].

In many of the examples, both objectives have been pursued. Here I concentrate more on the efforts to achieve a qualitative understanding in order to define research agendas. The other objective, concentrating on detail, is also useful but more in terms of defining operational management agendas. Since distinctively different speeds can be identified for the variables, four steps of analysis are possible [58]:

(1) Analyze the long-term behavior of the fast variable, while holding the slow variables constant.
(2) Define the response of the slow variables when the fast ones are held fixed.
(3) Analyze the long-term behavior of the slow variables, with the fast variables held at their corresponding equilibria.

TABLE 1 : *Key variables and speeds in five classes of ecosystems.*

	The variables			
The system	*Fast*	*Intermediate*	*Slow*	*Key references*
Forest insect pest	Insect	Foliage	Trees	[58, 59]
Forest fire	Intensity	Fuel	Trees	[60]
Savanna	Grasses	Shrubs	Herbivores	[51]
Fishery	Phytoplankton	Zooplankton	Fish	[16]
Human disease	Disease organism	Vector and susceptibles	Human population	[61, 62]

(4) Combine the preceding steps to identify needs for extra coupling, so that, when added, the behavior of the full system is described.

I now summarize the procedure, emphasizing the main conclusions. The "fast" dynamics are determined by the way key processes affect change in the fast variable at different fixed values of the slow variables. An example is shown in Figure 1. The important point is that a long history of experimental analysis of ecological processes has led to generalization of the qualitative form of system response, the condition for each distinct form, and the features that determine where impact is greatest (see Holling and Buckingham [63] for predation and competition and Peters [64] for a variety of ecological processes related to body size). Hence, this knowledge can be applied to understanding behavior at a more aggregate level of the hierarchy. Equally important, it frees the analysis from the need for detailed quantification, setting the stage for research designs that are both more economical and more appropriate.

In continuing this qualitative emphasis, attention is then focused on conditions for increase and conditions for decrease of the fast variable. The boundary between the two represents either transient or potentially stable equilibria. The conditions for these equilibria can be organized to

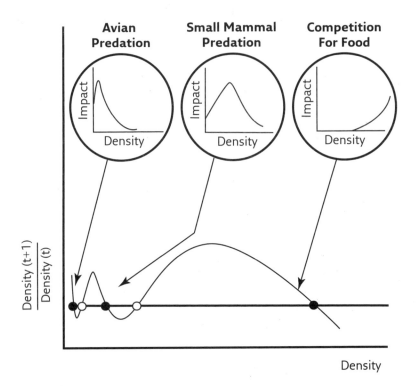

FIGURE 1: *A stylized recruitment curve for jack-pine sawfly at a fixed level of slow and driving variables, showing the contributions of three of the key processes. The horizontal line represents the conditions where the population density of the next generation is the same as that of the present generation. Intersections of the recruitment curve with this line indicate potential equilibria, some potentially stable (closed circles), some unstable (open circles).*

show the set of all equilibria, i.e., the zero isoclines for the fast variable. Four examples are shown in Figure 2: for spruce budworm (*a*), jack-pine sawfly (*b*), forest fire (*c*), and savanna grazing (*d*).

These equilibrium structures show that there are a number of stability states controlled by the slow variable. There are many other examples given in the literature: for 15 other forest insect pests [59], for other grazing systems [62,65], for fish [17,66], and for human host–parasite systems [62].

Two main points emerge at this stage of the analysis. First, discontinuous change occurs because there are multiple stable states. As the

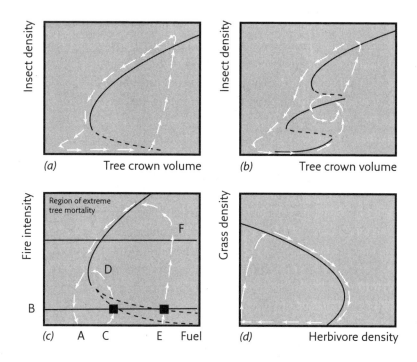

FIGURE 2: *Zero-isocline surfaces showing the equilibrium values of the fast variable at different fixed levels of the slow variables. The full lines represent stable surfaces and the broken lines unstable ones. Typical trajectories are shown by the arrows. (a) Spruce budworm and balsam fir; (b) jack-pine and sawfly; (c) forest fire and fuel; (d) savanna grass and herbivore grazing.*

slow variables change (tree growth, fuel accumulation, herbivore population increase), different equilibria suddenly appear, and when other equilibria disappear, the system is suddenly impelled into rapid change after periods of gradual change. The basic timing of these events is set by the dynamics of the slow variable.

Second, external stochastic events can lead to highly repetitive consequences. All of the surfaces shown in Figure 2 would be better represented as fuzzy probability bands to reflect "white noise" variability in weather conditions. But this modification typically changes the precise timing of events by a trivial amount. For long periods the systems are in a refractory state and the triggering event is totally or strongly inhibited.

In the insect pest cases [Figures 2(*a*) and (*b*)], for example, a variety of predators, chiefly birds, introduce such a strong "predator pit" that insect populations are either becoming extinct or being kept at very low densities. Similarly, the reflexively folded set of unstable equilibria for fire [Figure 2(*c*)] can turn stochastic ignition events, such as lightning strikes, into highly predictable outbreaks of fire. If the surface is low [dotted line in Figure 2(*c*)], then the average ignition intensity of B triggers a fire at C which consumes the fuel and hence extinguishes itself at A. This is similar to the cycles of ground fires experienced prior to fire management in the mixed-conifer forests of the Sierra Nevada in western USA [56,60]. In several areas they occurred with a remarkably consistent interval of seven to eight years, and helped maintain conditions for tree regeneration and nutrient cycling. In addition, these light fires killed only some of the young white fir, thereby introducing and maintaining gaps in the forest canopy and, in essence, producing natural fire breaks. However, if the undersurface is raised because of increased moisture or effective fire control practices [broken line in Figure 2(*c*)], more fuel must accumulate before an average ignition event triggers a fire [point E, Figure 2(*c*)]. This results in a longer period before a fire, but also in a more intense fire, corresponding to a natural long-period fire cycle of the kind mentioned earlier or to the unexpected failure of a fire control policy.

Because of these properties, pulses of disturbance should not be seen as exogenous events. Insect outbreaks, forest fires, overgrazing, sudden changes in fish populations, and outbreaks of disease are determined by identifiable processes affecting the fast variable, whose impacts are modified by the magnitudes of the slow variables. As a consequence, changes in the slow variables eventually result in a condition where a sharp disturbance is inevitable.

A fuller definition of those properties requires two more steps. First, the equilibrium structure set by the fast variable is affected by both intermediate-speed and slow variables. A three-dimensional representation of the zero isocline can then be shown. An example for spruce budworm is shown in Figure 3. Finally, zero-isocline surfaces are constructed for both the intermediate and slow variables in order to explore the intersections between them. A more formal and rigorous treatment can be found in Ludwig et al. [58]. Figure 4 shows, as an example, the iso-

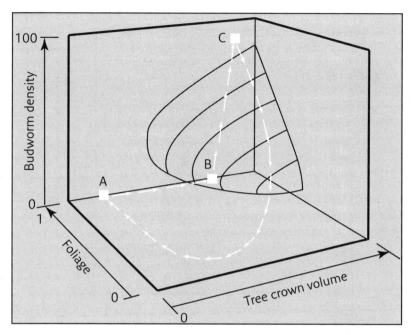

FIGURE 3: *Zero-isocline surface for budworm as a function of foliage and tree crown volume. The trajectory shows a typical unmanaged outbreak sequence.*

cline surface for tree crown volume laid over that for budworm. Where these two surfaces intersect (line AB) represent the only places where a stable equilibrium for both budworm and trees might be possible. But this can only be realized if the stable portion of the zero isocline for foliage, the intermediate-speed variable, also intersected the line AB. The surface for foliage is folded something like that of the budworm, with a stable surface and an unstable reflexed one. For values of foliage area below this unstable surface, foliage production cannot match natural foliage depletion, so that the foliage eventually disappears. Although it cannot be clearly shown in Figure 4, it happens that the foliage zero-isocline surface lies under the major portion of the budworm surface. As a consequence, there is no stable intersection.

Thus the unmanaged budworm system is in a state of continuous and fundamental disequilibrium. If one variable is on a stable zero isocline, the others are usually not on theirs. If two of the variables happen to be simultaneously on their stable isoclines, the third one is never on its

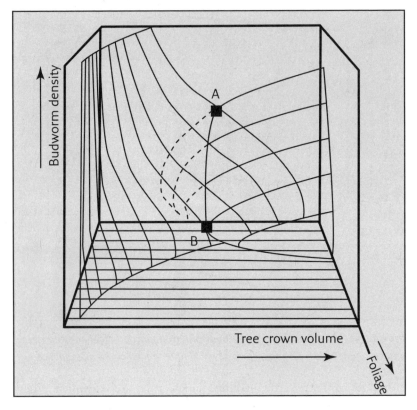

FIGURE 4: *Overlay of the zero-isocline surface for tree crown volume (a measure of forest age) against the budworm surface of Figure 2. The line AB is where the two surfaces intersect.*

stable isocline. It is a system under continual dynamic change, always chasing ever-receding equilibria. But control is never completely lost because of the existence of the independent single- or two-variable equilibrium states.

Only an analysis involving three variables would expose this behavior. In such cases, equilibrium-centered concepts, definitions, and measurements that require the existence of at least one non-zero equilibrium (termed here a system equilibrium) are simply irrelevant. There is no system equilibrium.

With these examples I can now add to the review of the main conclusions by summarizing the role of the three distinct speeds of variables in producing cyclic variations of different periods. Under different conditions each of these speeds can dominate the dynamics.

In the spruce–budworm system the period is set by the slowest variable, the tree, and the other variables interact with it. The same is true of unmanaged savanna-grazing systems, where intensive periods of overgrazing lead to depletion of above-ground vegetation. This is followed by emigration or high mortality of ungulates allowing early recovery of perennial grasses with underground storage. Sometime later, annual grasses begin to dominate through competition with the former. Many forest fire systems are similarly controlled, in that the slowest variable sets the cycle and the fast and intermediate variables follow.

In all these examples the variability produces diversity as a consequence of a cyclic shifting of the competitive advantage between species. Balsam fir can outgrow spruce, and would do so except that budworm preferentially attacks balsam and suddenly shifts the balance [67]. The high photosynthetic efficiency of annual grasses places them at an advantage over those perennial grasses that invest a considerable part of their biomass in underground storage. But intense overgrazing tips the balance the other way, so that both types are refused [51]. And Loucks' analysis of long-period forest fire cycles makes a similar point regarding maintenance of species diversity [44]. In all these cases the cycle length is set by the slowest variables and other variables are driven according to that cycle. In every example, the high variability encourages species diversity and spatial heterogeneity.

In a second class of cases the basic timing is set by the intermediate-speed variable. The slowest variable is largely disengaged because a stable oscillation develops around a single system equilibrium in which this variable persists. An example is the ground fire cycle described earlier for the mixed-conifer forests of the Sierra Nevada [Figure 2(c), cycle C–A]. A number of forest insect systems show this pattern [59]. For example, the European larch–budmoth system in Alpine regions of Switzerland [68] shows a remarkably persistent cycle with an 8- to 10-year period that has persisted for centuries. Both insect and foliage follow this cycle but there is little tree mortality.

Finally, there are patterns in which the fast variable dominates and the intermediate and slow variables are little affected. A particularly interesting example is the jack-pine-sawfly system, [69] and Figure 2(b).

At intermediate tree ages, predation by small mammals establishes a "predator pit" at moderate population densities of sawfly. At these densities, the small-mammal predation causes enough mortality to allow the sawfly and its parasites to establish a high-speed stable limit cycle of 3–4 years. Foliage is little affected and, as a consequence, neither is tree mortality. Such oscillations can persist for some years, but eventually the system is shifted to a different pattern by a change in climate or forest stand. Other forest insect systems show this pattern, as does endemic malaria where vectoral capacity is high [61].

In summary, the key features of this analysis of the forces of change lead to the following observations:

(1) There can be a number of locally stable equilibria and stability domains around these equilibria.

(2) Jumps between the stability domains can be triggered by exogenous events, and the size of these domains is a measure of the sensitivity to such events.

(3) The stability domains themselves expand, contract, and disappear in response to changes in slow variables. These changes are internally determined by processes that link variables and, quite independently of exogenous events, force the system to move between domains.

(4) Besides exogenous stochastic events, different classes of variability and of temporal and spatial behavior emerge from the form of equilibrium surfaces and the manner in which they interact. There can be conditions of low equilibrium with little variability. There can be stable-limit cyclic oscillations of various amplitudes and periods. And there can be dynamic disequilibrium in which there is no global equilibrium condition and the system moves in a catastrophic manner between stability domains, occasionally residing in extinction regions. There also exists the possibility of "chaotic" behavior.

The one overall conclusion is that discontinuous change is an internal property of each system. For long periods change is gradual and discontinuous behavior is inhibited. Conditions are eventually reached, however, when a jump event becomes increasingly likely and ultimately inevitable.

Synthesis of Ecosystem Dynamics

Ecosystem Functions

It was mentioned in an earlier section that there are two aggregate functions that determine ecosystem succession: an exploitation function (related to the notion of r-strategies) that dominates early and a conservation function (related to K-strategies) that dominates late in the succession. The conclusion of the preceding analysis of forces of change is that there is a third major ecosystem function. The increasing strength of connection between variables in the maturing ecosystem eventually leads to an abrupt change. In a sense, key structural parts of the system become "accidents waiting to happen."

When the timing is set by the slowest variable, the forces of change can lead to intense, widespread mortality. When the timing is set by the faster variables the changes are less intense and the spatial impact, while synchronous over large areas, is more patchy. But even in those instances, individuals constituting the slow variables eventually senesce and die. The difference is that the impact is local and is not synchronous over space.

There is both a destructive feature to such changes and a creative one. Organisms are destroyed, but this is because of their very success in competing with other organisms and in appropriating and accumulating the prime resources of energy, space, and nutrients. The accumulated resources, normally bound tightly and unavailable, are suddenly released by the forces of change. Such forces therefore permit creative renewal of the system. I call this third ecosystem function "creative destruction," a term borrowed from Schumpeter's economic theory [70].

Although the change is triggered by such a function, the bound energy, nutrients, and biomass that accumulated during the succession are not immediately available. There is therefore a fourth and final ecosystem function. One facet of that function is the mobilization of this stored capital through processes of decomposition that lead to mineralization of nutrients and release of energy into the soil. The other facet includes biological, chemical, and physical processes that retain these released nutrients, minimizing losses from leaching. This fourth function is one of ecosystem renewal.

These processes result in a pulse of available nutrients after disturbance. In many instances surprisingly little is lost from the ecosystem through leaching. In other instances so much is lost that algal blooms may be triggered in receiving waters [33, 71]. The kinds of retention mechanisms are not well understood because of the difficulty of studying soil dynamics at an ecosystem scale. But experimental manipulation of whole watersheds through harvesting, removal of structural organic material from the soil surface, and herbicidal inhibition of vegetative regrowth has begun to allow some of the mechanisms to be identified [33, 72]. They include colloidal behavior of soil, rapid uptake by the remaining vegetation whose growth is accelerated by the disturbance, and low rates of nitrification that keep inorganic nitrogen in ammonia pools rather than as the more soluble nitrates [73]. In addition a recent experiment demonstrated that rapid uptake of nutrients by microbes during decomposition is a major process preventing nitrogen losses from areas of harvested forests [72].

Such processes of release and retention after disturbance define the renewal function. Hill [74] emphasizes their importance in reestablishing the cycle of change and hence in determining the resilience of ecosystems. Of particular importance are the processes of retention. When savannas become dominated by woody shrubs, it is because of the loss of water retention capacity of perennial plants and soil. Similarly, intensified burning of upland vegetation in Great Britain has caused the vegetation to shift irreversibly from forest cover to extensive blanket bogs [75]. On sites where soils are poor, rainfall is high, and temperatures are low, the result has been loss of nutrients and reduced transpiration and rainfall interception, leading to waterlogged soils, reduced microbial decomposition, and the development of peat. The original tree species, such as oak, cannot regenerate because of wetness, acidity, and nutrient deficiency. In a similar vein, tropical rain forests may have a low resilience to large-scale disturbance. Many of the tree species have large seeds with short dormancy periods. These features facilitate rapid germination and regrowth of vegetation in small disturbed areas, but make it impossible to recolonize extensive areas of cleared land [76]. Partly as

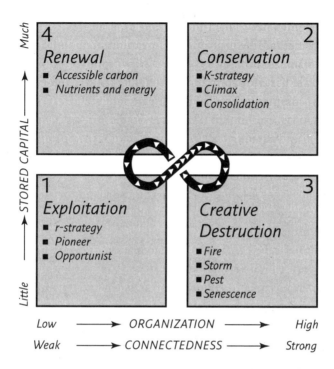

FIGURE 5: *The four ecosystem functions and their relationship to the amount of stored capital and the degree of connectedness. The arrowheads show an ecosystem cycle. The interval between arrowheads indicates speed, i.e., a short interval means slow change, a long interval rapid change.*

a result, extensive land clearance in the Amazon basin has led to permanent transformation of tropical forest areas into scrub savanna [77].

The full dynamic behavior of ecosystems at an aggregate level can therefore be represented by the sequential interaction of four ecosystem functions: exploitation, conservation, creative destruction, and renewal (Figure 5). The progression of events is such that these functions dominate at different times: from exploitation, 1, slowly to conservation, 2, rapidly to creative destruction, 3, rapidly to renewal, 4, and rapidly back to exploitation. Moreover, this is a process of slowly increasing organization or connectedness (1 to 2) accompanied by gradual accumulation of capital. Stability initially increases, but the system becomes so overconnected that rapid change is triggered (3 to 4). The stored capital is then released and the degree of resilience is determined by the balance between the processes of mobilization and of retention. Two properties are being controlled: the degree of organization and the amount of capital accumulation and retention. The speed and amplitude of this cycle, as indicated earlier, are determined by whether the fast, intermediate, or slow variable dominates the timing.

These patterns in time have consequences for patterns in space. Rapidly cycling systems generate ecosystems that are patchy. Tropical ecosystems are an example. Slowly cycling systems produce higher amplitude, discontinuous change that tends to occur as a wave moving across space. In the case of uncontrolled budworm outbreaks, for example, the wave takes about 10 years to sweep across the province of New Brunswick, Canada.

The factors determining the size distribution of the areas of disturbance, however, are not well understood. If the considerable understanding of time dynamics could be connected to an equal understanding of spatial patch dynamics, then questions of global change could be better anticipated and better dealt with by local policies.

Levin [129, 78] has developed an effective framework for analysis and description of patches. There are two parts:

(1) Patch size and age distributions as related to birth and death rates of patches.
(2) The response of species to the regeneration opportunities existing in patches of different sizes and ages.

The "fast" and "slow" designations are part of the analysis, as well as diffusion and extinction rates. It is, therefore, completely compatible with the analysis presented here, and has begun to be applied to forest systems [79] together with Mandelbrot's theory of fractals to relate extinction laws and relative patchiness [80]. As a consequence Mandelbrot proposes a descriptive measure of patchiness and succession that is scale independent and has considerable value for any effort to measure patch dynamics and disturbance.

Complexity, Resilience, and Stability
This synthesis helps clarify the relationship between complexity and stability. It was long argued that more species and more interactions

in communities conferred more stability, the intuitive notion being that the more pathways that were available for movement of energy and nutrients, the less would be the effect of removal of one. However, May's analysis of randomly connected networks showed that increased diversity, in general, lowered stability [81]. This means that ecosystems are not randomly connected. The issue has been significantly clarified by Allen and Starr [25] and the treatment here provides further support of May's observation.

First, measures of stability referred to typically did not distinguish between stability and resilience—systems with low stability can often demonstrate high resilience. Second, ecosystems have a hierarchical structure, and for this reason it has been possible to capture the essential discontinuous behaviors with three sets of variables operating at different speeds. Other species and variables are dramatically affected by that structure and the resultant behavior, but do not directly contribute to it. Hence the relevant measures of species diversity, which is one measure of complexity, should not involve all species, but only those contributing to the physical structure and dynamics.

The significant measure of complexity, therefore, concerns the degree of connectedness within ecosystems. As Allen and Starr demonstrate, the higher the connectedness, i.e., the complexity, the lower the probability of stability. They present examples both from theory and from the empirical literature to demonstrate the point. A system can also become so underconnected that critical parts go their own way independent of each other—resilience disappears. Tropical farming based upon monocultures and extensive land clearance is highly overconnected, particularly through pest loads [82]. Hence stability and ultimately resilience are lost. Introduction of patches through traditional shifting of agriculture or through the breaking of connections by interplanting different cultivars produces a farming pattern more akin to the highly patchy, less connected natural system.

Third, the pattern of connectedness and the resultant balance between stability and resilience are a consequence of the pattern of external variability that the system has experienced. Systems such as those in the tropics, which have developed in conditions of constant temperature and precipitation, therefore demonstrate high stability but low resilience. They are very sensitive to disturbances induced by man. On the other hand, temperate systems exposed to high climatic variability typically show low stability and high resilience, and are robust to disturbance by man.

The present analysis adds an important fourth ingredient. Hierarchies are not static in the kinds or strengths of connections. The degree of connectedness changes as the ecosystem is driven by the four ecosystem functions. Succession introduces more connectedness, and hence increasing likelihood of instability. An overconnected condition develops, triggering a discontinuous change. The connectedness is sharply reduced thereby, to be followed by reorganization and renewal. The destabilizing effect produced by overconnectedness generates variability, which in turn encourages the development and maintenance of processes conferring resilience, particularly during the period of low connectedness and recovery. Collapse of resilience, or escape to a different stability domain, can occur, however, if the system becomes too underconnected during the destabilized phase of this cycle. This can happen if processes of mobilization are not balanced by processes of retention. Since those processes occur dominantly in soil, any exploration of global change must place a high priority on developing a better and more extensive understanding of soil dynamics in relation to the cycles driven by the four ecosystem functions.

Connections

The previous section addressed the question posed in the introduction concerning the capacity of ecosystems to absorb, buffer, or generate change. It concentrated on the processes and functions that lead to cycles of ecosystem growth, disruption, and renewal. The periods and amplitudes of those cycles are defined by qualitatively distinct speeds of a small number of key variables. Their ability to maintain structure and patterns of behavior in the face of disturbance, i.e., their resilience, is determined by the renewal function whose properties are, in part, maintained by pulses of disturbance.

The timing and spatial extent of the pulses emerge from the interaction between external events and an internally generated rhythm of stability/instability. Industrial societies are changing the spatial and temporal patterns of those external events. Spatial impacts are more homogeneous; temporal patterns are accelerated. An understanding of

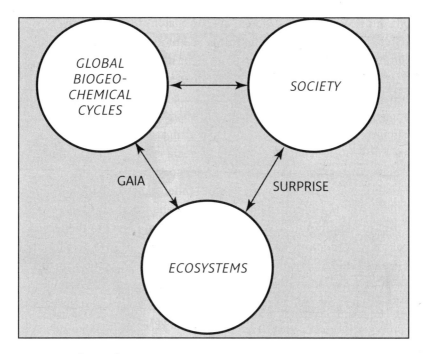

FIGURE 6: *Connections*

impacts of global change therefore requires a framework to connect the understanding developed here for ecosystem dynamics to that developed for global biogeochemical changes on the one hand and societal developments on the other hand. There are transfers of energy, material, and information among all three, as suggested in Figure 6. Biosphere studies now concentrate on changes in the amount, speed, and scale of those transfers and what should or should not be done about them. The Gaia hypothesis, as indicated earlier, provides a focus for discussing the interaction between ecosystems and global biogeochemical cycles. Surprise provides a focus for discussing the interaction between ecosystems and society.

Gaia

The spatial and temporal patterns generated by the four ecosystem functions form the qualitative structure of ecosystems. A small number of variables and species are fundamental to determining that structure. And the resultant architecture of an ecosystem offers a variety of niches which are occupied by different species that are affected by the ecosystem structure, but contribute little to it. But while contributing little to the structure, they could contribute significantly to exogenous biogeochemical cycles. This could be determined by drawing on the extensive literature that identifies the major components of geochemical exchange and regulation in plants, animals, and soils. The scheme presented here can provide a way to organize this knowledge so that the interactions of specific ecosystems with external biogeochemical cycles and their possible regulatory roles can be better understood.

Moreover, if the key processes of homeostatic regulation in atmospheric cycles could be demonstrated, such an approach could also provide a way to identify the ecosystems that contribute most to the feedback control. In order to do so, the qualitative analysis outlined here for a few systems could be expanded into a comparative study of the structure of ecosystem dynamics in each of the life zones defined by Holdridge [83] or by Soviet geographers (as reviewed in Grigor'yev [84]) on the basis of climate data.

Such a study would provide a descriptive classification for determining ecosystem responses to global environmental changes. The responses that are most critical are the qualitative patterns of behavior. These patterns are determined by the fast, intermediate, and slow variables during ecosystem growth and disruption and by the mobilization/retention processes during renewal. They also emerge from the way the resultant internal dynamics modify climatic variability. The latter is determined by global atmospheric and oceanic processes, which in turn set the variability in the physical environment.

Steele has reviewed the temporal behavior of physical variables in the ocean and atmosphere [85]. If the well defined periods of days and seasons are removed, the underlying trend for physical conditions in the oceans is for variance to increase as a function of period. The increase is close to the square of the period and occurs at all time (and space) scales. This "red noise" is in contrast to "white" noise where variance is independent of scale (see Dickinson, Chapter 9, pp. 257–260, this volume).

For periods up to about 50 years, physical variation in the atmosphere, unlike in the oceans, is close to being white noise. Thereafter variation seems to follow a red spectrum, suggesting a coupling of atmospheric and oceanic processes. As described earlier, many terrestrial ecosystem cycles have a period from a few decades to one or two hundred years, driven by the slowest variables. Even if this similarity in cycle periods is a coincidence rather than due to adaptation, changes in the external forcing frequency induced by man's activities could be transmitted and transformed by the existing response times of ecosystems in unexpected ways. Now that the qualitative dynamics of a number of ecosystems are beginning to be better understood, a fruitful area of research can be developed to demonstrate, by example, how changes in the frequency pattern of external forcing can affect ecosystem stability and resilience.

Surprise

Man's efforts to manage ecosystems can be viewed as weak experiments testing a general hypothesis of stability/resilience. In many of the examples discussed earlier, the management goal was to reduce the *variability* of a target variable by applying external controls. Crudely, it represented an equilibrium-centered view of constant nature. All the cases examined were successful in achieving their short-term objectives, but as a consequence of that success, each system evolved into a qualitatively different one.

The evolution took place in three areas. First, the social and economic environment changed. More pulp mills were built to exploit the protected spruce–balsam forests; more recreational demand was developed in the parks protected from fire; more efficient and extensive fisheries were developed to exploit salmon; more land was used for cattle ranches on the savannas; and more development was possible in those areas protected from malaria.

Second, the management agencies began to evolve. Effective agencies were formed to spray insects, fight fires, operate fish hatcheries, encourage cattle ranching, and reduce mosquito populations. And the objective of these agencies naturally shifted from the original socioeconomic objective to one that emphasized operational efficiency: better and better aircraft, navigation, and delivery systems to distribute insecticide; better and better ways to detect fires and control them promptly.

These changes in the socioeconomic environments and in the management institutions were generally perceived and were rightly applauded. But evolution occurred in a third area—the biophysical—whose consequences were not generally perceived.

Because of the initial success in reducing the variability of the target variable, features of the biophysical environment which were implicitly viewed as constants began to change to produce a system that was structurally different and more fragile. Reduction of budworm populations to sustained moderate levels led to accumulation and persistence of foliage over larger and larger areas. Any relaxation of vigilance could then lead to an outbreak in a place where it could spread over enormous areas. Reduction of fire frequency led to accumulation of fuel and the closing of forest crowns so that what were once modest ground fires affecting limited areas and causing minor tree mortality became catastrophic fires covering large areas and causing massive tree mortality. And similarly, increased numbers of salmon led to increases in size and efficiency of fishing fleets and extinction of many native stocks; maintenance of moderate numbers of cattle led to changes in grass composition toward species more vulnerable to drought and errors of management; persistent reduction in mosquitoes led to gradual increases in the number of people susceptible to malaria, and to mosquitoes resistant to insecticide.

In short, the biophysical environment became more fragile and more dependent on vigilance and error-free management at a time when greater dependencies had developed in the socioeconomic and institutional environment. The ecosystems simplified into less resilient ones as a consequence of man's success in reducing variability. In these cases, connectedness increased because of spatial homogenization of key variables: foliage for budworm, fuel and canopy structure for fires, efficient but vulnerable grasses for savannas, numbers of stocks and ages of fish, and the number of people susceptible to malaria.

The hypothesis of constant nature encountered the surprising reality of resilient nature. If control falters, the magnitude and extent of the resultant disruptive phase can be great enough to overwhelm the renewal process.

Just as ecosystems have their own inherent response times, so do societal, economic and institutional systems. How long an inappropriate policy is successful depends on how slowly the ecosystem evolves to the point when the increasing fragility is perceived as a surprise and potential crisis.

The response to such surprises is alarm, denial, or adaptation and is similarly related to the response times of different groups in society and of the management institutions. For example, forest fire policy in the national parks of the western USA has recently changed radically to reinstate fire as the natural "manager" of the forests. This rather dramatic adaptation was not made easily and rested upon the existence of an alternative policy and of technologies to implement it, on a climate of understanding, and on costs that were relatively modest compared with my other examples. But it might be equally important that the critical variables of fuel and forest tree composition changed at the slowest rate of all the examples. It was some sixty years before the change became critical. I argue that the relevant time unit of change for a management institution is of the order of 20 to 30 years, the turnover rate of employees. As a consequence, by the time the problem became critical there was a new generation of experts and policy advisors who would be more willing to recognize failures of their predecessors than of their own. In addition, the slowness of change allowed the accumulation of knowledge of the processes involved and the communication of that growing understanding to a wide range of actors.

In contrast, the changes in the budworm/forest systems proceeded faster. Insecticide spraying began on a large scale in the mid-1950s with conditions of vulnerability building to a critical point by the early 1970s. In this case, the 15–20 year-period was insufficient to accumulate and, most important, disseminate an understanding of the problem. Alternative policies or technologies were not developed and the parents of the original policies were still central actors and defenders of the past. Adaptive change has been an agonizing process and is only now showing signs of occurring [86].

There are insufficient examples to make these remarks anything more than speculation: but they do identify a research priority to determine the time dynamics that lead to increasing dependencies of societies on policies that have succeeded in the past, to examine increasing rigidities of management institutions, and to increase sensitivity to surprise. The research effort should be based on case studies that cover as wide a spectrum of man's activities as possible—economic, technological, and behavioral.

TABLE 2: *Possible analogies between ecosystem function and functions or typologies proposed for other systems.*

Subject	Function or typology			
Ecosystem	Exploitation (*r*)	Conservation (*K*)	Creative destruction	Renewal
Economics [e.g., 70]	Innovation, Market, Entrepreneur	Monopolism, Hierarchy, Saturation, Social rigidity	Creative destruction	Invention
Technology [e.g., Brooks 87]	Innovation	Technological monoculture, Technological stalemate	Participatory paralysis	Expert knowledge
Institutions [e.g., 88, 89]	Entrepreneurial market	Caste, Bureaucracy	Sect	Ineffectual
Psychology [e.g., Jung as in 90]	Sensation	Thinking	Intuition	Feeling

Such a comparative study requires collaboration among a number of disciplines. But it is essential to involve practical experience in business, government, and international organizations as well. It is only possible now because so many place priority on understanding change. Equally important, frameworks for understanding change can be found in economics, technology, institutional behavior, and psychology that provide some possible connections to the framework presented here for ecosystems. Examples are suggested in Table 2.

The analogies suggested by Table 2 might simply represent common ways for people to order their ignorance. But there are strong hints, at least from analysis of institutional organizations from the perspectives of cultural anthropology [88, 89] and of technological developments [87], that functions similar to the four ecosystem functions operate in societal settings, although the results can be very different. Some comparative studies already exist that have both predictive and descriptive power. An example is Thompson's analysis of the very different decisions that were made in the UK and California concerning the siting of liquefied gas plants [89]. And regarding technological development, consider Brooks' argument (Chapter 11, this volume) and this quote [87, p. 253]:

> One reason for this situation is that, as a particular technology matures, it tends to become more homogeneous and less innovative and adaptive. Its very success tends to freeze it into a mould dictated by the fear of departing from a successful formula, and by massive commitment to capital investments, marketing structures and supporting bureaucracies. During the early stages of a new technology many options and choices are possible, and there are typically many small competing units, each supporting a different variation of the basic technology, and each striving to dominate the field. Gradually one variation begins to win, and the economies of scale in marketing and production then begin to give it a greater and greater competitive edge over rival options. The technical options worth considering become narrower and narrower; research tends to be directed increasingly at marginal product improvements or product differentiation, and the broader consequences of application tend to be taken more and more for granted. Elsewhere I have spoken of this as a "technological monoculture". What happens is that through its very success a new technology and its supporting systems constitute a more and more self-contained social system, unable to adapt to the changes necessitated by its success.

He later adds [87, p 256]:

> The paralysis of the decision process by excessive participation will eventually result in a movement to hand the process back to elites with only broad accountability for results according to then current social values. Eventually effects on certain social expectations will become sufficiently serious so that distrust of the experts will revive and there will be a new wave of demands for participation until the frustration of more diffuse social interests will again result in reversion to experts.

This analysis and speculation is completely in harmony with the cyclic processes described here for ecosystems.

Such analogues at the least suggest that a formal comparative study of different cases could help provide an empirical basis to classify the timing of key phases of societal response to the unexpected: in detecting surprise, in understanding the source and cause of surprise, in communicating that understanding, and in responding to surprise. Such a classification can help introduce a better balance between prediction, anticipation, and adaptation to the known, the uncertain, and the unknown features of our changing world.

Recommendations

Ecosystems have a natural rhythm of change the amplitude and frequency of which is determined by the development of internal processes and structures in a response to past external variabilities. These rhythms alternate periods of increasing organization and stasis with periods of reorganization and renewal. They determine the degree of productivity and resilience of ecosystems.

Modern technological man affects these patterns and their causes in two ways. First, traditional resource-management institutions constrain the rhythms by restricting them temporally and homogenizing them spatially. Internal biophysical relationships then change, leading to systems of increasing fragility, i.e., to a reduced resilience. Moreover, modern man and his institutions operate with a different historical rhythm that can mask indications of slowly increasing fragility and can inhibit effective adaptive responses, resulting in the increased likelihood of internally generated surprises, i.e., crises. Second, the increasing extent and intensity of modern industrial and agricultural activities have modified and accelerated many global atmospheric processes, thereby changing the external variability experienced by ecosystems. This imposes another set of adaptive pressures on ecosystems when they are already subject to local ones. As a consequence, locally generated surprises can be more frequently affected by global phenomena, and in turn can affect these global phenomena in a web of global ecological interdependencies.

We now have detailed examples and analyses of ecological patterns, largely from northern temperate regions, that demonstrate the role of variables of different rates of action and reveal the importance of functions that trigger change and renewal in maintaining resilience. The resultant synthesis indicates that there is now less of a priority to develop predictive tools than to design systems with enough flexibility to allow recovery and renewal in the face of unexpected events—in short, there needs to be a better balance established between anticipation, monitoring, and adaptation [91].

The design effort would be facilitated by research to test and expand the conclusions in three ways. First, the ecosystem synthesis should be extended to further examples of four critical ecosystems: arctic, arid, humid tropical, and marine, since each has patterns and structures different from northern temperate ecosystems. Second, the analyses of time responses and rhythms of change described here should be extended more explicitly to the links between natural/societal systems, particularly regarding the history of economic, technological, and resource development. Third, there is a need and opportunity to develop a set of well replicated mesoscale experiments in order to reduce the ambiguity of problems occurring because of local surprise and global interconnection. These are given more specifically in the following sections.

A Comparative Study of Resilience and Ecosystem Recovery
Purpose: to define early warning signals of pathologically destructive change and to design self-renewing resource systems.

Data are required to extend the description of time patterns to allow comparison between northern temperate, arctic, arid, humid tropical, and marine systems. Processes that trigger change and facilitate renewal should be identified and classified in terms of their effects on stability, productivity, and resilience. The former requires information as to the role of slow variables in triggering pulses of disturbance. The latter particularly requires a study of soil processes, the balance between nutrient mobilization and retention, their sensitivity to disturbance, and their rates of recovery after small- to large-scale disturbances (e.g., from natural patch formations to man-made. land clearance and drainage).

A Comparative Study of Sources and Responses to Surprise in
Natural–Social Systems, Particularly Economic Technological,
and Resource Development
Purpose: to define conditions that determine how much to invest in action (decide policy and act now), anticipation (delay and find out more), or adaptation (forget the immediate problem and invest in innovation).

The generation of sharp change, its detection, and adaptation of policy responses depend on the interaction between the response times of the managed (natural) system, of the institutions managing the systems, and of the economic and social dependencies that develop.

It now seems possible to classify resource, ecological, and environmental problems not only in terms of uncertainty of their consequences, but also in terms of uncertainty of societal response. Those requiring priority attention are not necessarily those that have the greatest impact, but those likely to generate a pathological policy response. The analysis presented here for ecosystems could be usefully applied to interactions between three components.

One of these components concerns the organization and time dynamics of management institutions. Focus and direction can be given by combining the analysis of surprise with the experience and orientation that has matured in hazards research studies [5] and in institutional analyses from the perspective of cultural anthropology [88, 89]. The second component concerns the geophysicochemistry of the atmosphere and oceans that increasingly connects regional economic development with global ecological interdependency through the ecosystems. Focus and direction can be given by the Gaia hypothesis of Lovelock [6] and the system dynamic studies of Steele [16] which view the atmosphere, oceans, and living systems as an interacting, self-regulated whole. The third and final component is society itself, particularly the historical patterns of economic and technological development that reveal how attitudes are formed, technological monocultures developed, and innovations either inhibited or enhanced. Focus and direction can be given by combining an understanding of ecosystem surprise with historical analyses of change, such as those of McNeill [92].

International Mesoscale Experiments

Purpose: to develop a set of internationally replicated experiments involving areas from a few square kilometers to a few thousand that can test alternative hypotheses developed to explain particular impacts of man's activities and to determine remedial policies.

Our understanding of the structure and behavior of ecosystems, and of how exploitation and pollution affect them, comes from a synthesis of knowledge of ecological, behavioral, physiological, and genetic processes. Much of that knowledge has been developed from the solid tradition of experimental, quantitative, and reductionist science which can now be generalized and synthesized to propose quantitative structures, qualitative behaviors, and qualitative consequences of impacts. Although synthetic, they essentially represent hypotheses because the arguments are based on studies that could be accommodated in the laboratory or in a few hectares. Ecosystems (as well as people's responses to them) operate on scales of a few square kilometers to several thousand square kilometers. That is where our knowledge and experience is the weakest.

The experiments would be designed to clarify alternative explanations of and policies for problems that emerge from the extension and intensification of industrial and agricultural development. Rather than discussing or investigating these endlessly, it should now be possible to design experiments that distinguish between alternatives. It is essential to concentrate on experiments in which the tests are qualitative in nature, the duration short (less than 5 years by drawing on fast/slow definitions of variables), the spatial scale in the "meso" range, and the policy consequences international. International replication and collaboration then becomes part of the design, which could ultimately contribute to institutional solutions as well as to scientific and policy understanding [93].

Notes and References

[1]   See, for example, McElroy (Chapter 7) and Dickinson (Chapter 9).

[2]   Steele, J. H. and Henderson, E. W. (1984), Modeling long-term fluctuations in fish stocks, *Science*, 224, 985–987.

[3]   Clark, W. C. and Holling, C. S. (1985), Sustainable development of the biosphere: human activities and global change, in T. Malone and J. Roederer (Eds), *Global Change*, pp. 474–490, Proceedings of a symposium sponsored by the ICSU in Ottawa, Canada (Cambridge University Press, Cambridge).

[4]   Qualitative change, in the sense used here, is structural change; that is, changes in the character of relationships between variables and in the stability of parameters. Such changes challenge traditional approaches to development, as well as to control, which is in itself a qualitative change in the combined ecological–social system.

[5]   Burton, I., Kates, R. W., and White, G. F. (1977), *The Environment as Hazard* (Oxford University Press, New York).

[6]   Lovelock, J. E. (1979), *Gaia: A New Look at Life on Earth* (Oxford University Press, Oxford, UK).

[7]   Lovelock, J. E. and Margulis, L. (1974), Atmospheric homeostasis by and for the biosphere: the Gaia hypothesis, *Tellus*, 26, 1–10.

[8]   Lovelock, J. E. and Whitfield, M. (1982), Life span of the biosphere, *Nature*, 296, 561–563.

[9]   Bolin, B. and Cook, R. B. (1983), The *Major Biogeochemical Cycles and Their Interactions* (John Wiley, Chichester, UK).

[10] Crutzen, P. J. (1983), Atmospheric interactions — homogeneous gas reactions of C, N and S containing compounds, in Bolin and Cook, note [9], pp. 67–114.

[11] Holling, C. S. (1977), Myths of ecology and energy, in *Proceedings of the Symposium on Future Strategies for Energy Development*, Oak Ridge, TN, 20–21 October, 1976, pp. 36–49 (Oak Ridge Associated Universities, TN). Republished in L. C. Ruedisili and M. W. Firebaugh (Eds) (1978), *Perspectives on Energy: Issues, Ideas and Environmental Dilemmas* (Oxford University Press, New York).

[12] Holling, C. S. (1973), Resilience and stability of ecological systems, *Annual Review of Ecology and Systematics*, 4, 1–23.

[13] Levins, R. (1968), *Evolution in Changing Environments* (Princeton University Press, Princeton, NJ).

[14] Lorenz, E. N. (1964), The problem of deducing climate from the governing equations, *Tellus*, 16, 1–11.

[15] Steele, J. H. (1974), Spatial heterogeneity and population stability, *Nature*, 248, 83.

[16] Steele, J. H. (1974), *Structure of Marine Ecosystems* (Harvard University Press, Cambridge, MA).

[17] Peterman, R., Clark, W. C., and Holling, C. S. (1979), The dynamics of resilience: shifting stability domains in fish and insect systems, in R. M. Anderson, B. D. Turner, and L. R. Taylor (Eds), *Population Dynamics*, pp. 321–341 (Blackwell Scientific, Oxford, UK).

[18] Fiering, M. B. and Holling, C. S. (1974), Management and standards for perturbed ecosystems, *AgroEcosystems*, 1, 301–321.

[19] Fiering, M. B. (1982), A screening model to quantify resilience, *Water Resources Research*, 18, 27–32; and Fiering, M. B. (1982), Alternative indices of resilience, *Water Resources Research*, 18, 33–39.

[20] Clark, W. C., Jones, D. D., and Holling, C. S. (1979), Lessons for ecological policy design: a case study of ecosystem management, *Ecological Modelling*, 7, 1–53.

[21] Bennett, A. F. and Ruben, J. A. (1979), Endothermy and activity in vertebrates, *Science*, 206, 649–654.

[22] Briand, F. and Cohen, J. E. (1984), Community food webs have scale-invariant structure, *Nature*, 307, 264–267.

[23] Cohen, J. E. and Briand, F. (1984), Trophic links of community food webs, *Proceedings of the National Academy of Sciences*, 81, 4105.

[24] Simon, H. A. (1973), The organization of complex systems, in H. H. Pattee (Ed), *Hierarchy Theory*, pp. 1–28 (George Braziller Inc., New York).

[25] Allen, T. F. H. and Starr, T. B. (1982), *Hierarchy, Perspectives for Ecological Complexity* (University of Chicago Press, Chicago, IL, and London).

[26] Patten, B. C. (1975), Ecosystem linearization: an evolutionary design problem, *American Naturalist*, 109, 529–539.

[27] Webster, J. R., Waide, J. B., and Patten, B. C. (1975), Nutrient recycling and the stability of ecosystems, in *Mineral Cycling in Southeastern Ecosystems*, ERDA Symposium Series, CON-740-513, pp. 1–27.

[28] Pimm, S. L. (1984), The complexity and stability of ecosystems, *Nature*, 307, 321–326.

[29] Levin, S. A. (1978), Pattern formation in ecological communities, in Steele, J. A. (Ed), *Spatial Patterns in Plankton Communities*, pp. 433–470 (Plenum Press, New York).

[30] Clark, W. C. (1985), Scales of climate impacts, *Climatic Change*, 7(1), 5–27.

[31] Odum, E. P. (1971), *Fundamentals of Ecology* (W. B. Saunders, Philadelphia, PA).

[32] Clements, F. E. (1916), Plant succession: an analysis of the development of vegetation, *Carnegie Institution of Washington Publication*, 242, 1–512.

[33] Bormann, F.H. and Likens, G. E. (1981), *Patterns and Processes in a Forested Ecosystem* (Springer, New York).

[34] MacArthur, R. H. and Wilson, E. O. (1967), *The Theory of Island Biogeography* (Princeton University Press, Princeton, NJ).

[35] MacArthur, R. H. (1960), On the relative abundance of species, *American Naturalist*, 94, 25–36.

[36] Pianka, E. R. (1970), On r- and K-selection, *American Naturalist*, 104, 592–597.

[37] Southwood, T. R. E., May, R. M., Hassell, M. P., and Conway, G. R. (1974), Ecological strategies and population parameters, *American Naturalist*, 108, 791–804.

[38] Lorimer, C. G. (1977), The presettlement forest and natural disturbance cycle of northeastern Maine, *Ecology*, 58, 139–148.

[39] West, D. C., Shugart, H. H., and Botkin, D. B. (1981), *Forest Succession. Concepts and Application* (Springer, New York).

[40] Mutch, R. W. (1970), Wildland fires and ecosystems—a hypothesis, *Ecology*, 51, 1046–1051.

[41] Biswell, H. H. (1974), Effects of fire on chaparral, in T. T. Kozlowski and C. E. Ahlgren (Eds), *Fire and Ecosystems*, pp. 321–364 (Academic Press, New York).

[42] Horn, H. S. (1976), Succession, in R. M. May (Ed), *Theoretical Ecology. Principles and Application*, pp. 187–204 (Blackwell Scientific, Oxford, UK).

[43] Franklin, J. F. and Hemstrom, M. A. (1981), Aspects of succession in the coniferous forests of the Pacific Northwest, in West et al., note [39], pp. 212–229.

[44] Loucks, O. L. (1970), Evolution of diversity, efficiency, and community stability, *American Zoologist*, 10, 17–25.

[45] Heinselman, M. L. (1973), Fire in the virgin forests of the Boundary Waters Canoe Area, Minnesota, *Quarterly Research*, 3, 329–382.

[46] Caughley, G. (1976), The elephant problem—an alternative hypothesis, *East African Wildlife Journal*, 14, 265–283.

[47] Rowe, J. S. and Scotter, G. W. (1973), Fire in the boreal forest, *Quarterly Research*, 3, 444–464.

[48] Heinselman. M. L. (1981), Fire and succession in the conifer forests of northern North America, in West et al., note [39], pp. 374–405.

[49] Hanes, T. L. (1971), Succession after fire in the chaparral of southern California, *Ecological Monographs*, 41, 27–52.

[50] Regier, H. A. (1973), The sequence of exploitation of stocks in multi-species fisheries in the Laurentian Great Lakes, *Journal of the Fisheries Research Board, Canada*, 30, 1992–1999.

[51] Walker, B. H., Ludwig, D., Holling, C. S., and Peterman, R. M. (1981), Stability of semi-arid savanna grazing systems, *Journal of Ecology*, 69, 473–498.

[52] Pickett, S. T. A. (1982), Population patterns through twenty years of oldfield succession, *Vegetatio*, 49, 45–59.

[53] Marks, P. L. (1974), The role of pin cherry (*Prunus pennsylvanica* L.) in the maintenance of stability in northern hardwood ecosystems, *Ecological Monographs*, 44, 73–88.

[54] Goméz-Pompa, A. and Vázquez-Yanes, C. (1981), Successional studies of a rain forest in Mexico, in West et al., note [39], pp. 246–266.

[55] Horn, H. S. (1981), Some causes of variety in patterns of secondary succession, in West et al., note [39], pp. 24–35.

[56] Kilgore, B. M. (1976), Fire management in the national parks: an overview, in *Proceedings of Tall Timbers Fire Ecology Conference*, Vol. 14, pp. 45–57 (Florida State University Research Council, Tallahassee, FL).   .

[57] Larkin, P. A. (1979), Maybe you can't get there from here: history of research in relation to management of Pacific salmon, *Journal of the Fisheries Research Board, Canada*, 36, 98–106.

[58] Ludwig, D., Jones, D. D., and Holling, C. S. (1978), Qualitative analysis of insect outbreak systems: the spruce budworm and the forest, *Journal of Animal Ecology*, 44, 315–332.

[59] McNamee, P. J., McLeod, J. M., and Holling, C. S. (1981), The structure and behavior of defoliating insect/forest systems, *Research on Population Ecology*, 23, 280–298.

[60] Holling, C. S. (1980), Forest insects, forest fires and resilience, in H. Mooney, J. M. Bonnicksen, N. L. Christensen, J. E. Latan, and W. A. Reiners (Eds) *Fire Regimes and Ecosystem Properties*, USDA Forest Service General Technical Report, pp. 20–26 (USDA Forest Service, Washington, DC).

[61] Macdonald, G. (1973), *Dynamics of Tropical Disease*, (Oxford University Press, London. UK).

[62] May, R. M. (1977), Thresholds and breakpoints in ecosystems with a multiplicity of stable states, *Nature*, 269, 471–477.

[63] Holling, C. S. and Buckingham, S. (1976), A behavioral model of predator-prey functional responses, *Behavioral Science*, 3, 185–195.

[64] Peters, R. H. (1983), The *Ecological Implication of Body Size* (Cambridge University Press, Cambridge, UK).

[65] Noy-Meir, I. (1982), Stability of plant–herbivore models and possible applications to savannah, in B. J. Hundey and B. H. Walker (Eds), *Ecology of Tropical Savannahs* (Springer, Heidelberg, FRG).

[66] Clark, C. W. (1976), *Mathematical Bioeconomics* (Wiley, New York).

[67] Baskerville, G. L. (1976), Spruce budworm: super silviculturist, *Forestry Chronicle*, 51, 138–140.

[68] Baltensweiler, W., Benz, G., Boven, P., and Delucchi, V. (1977), Dynamics of larch budmoth populations, *Annual Review of Entymology*, 22, 79–100.

[69] McLeod, J. M. (1979), Discontinuous stability in a sawfly life system and its relevance to pest management strategies, *Current Topics in Forest Entomology*, General Technical Report WO-8 (USDA Forest Service, Washington, DC).

[70] Schumpeter, J. A. (1950), *Capitalism, Socialism and Democracy* (Harper, New York); and Elliott, J. E. (1980), Marx and Schumpeter on capitalism's creative destruction: a comparative restatement, *Quarterly Journal of Economics*, 95, 46–58.

[71] Vitousek, P. M. and White, P. S. (1981), Process studies in succession, in West et al., note [39].

[72] Vitousek, P. M. and Matson, P. A. (1984), Mechanisms of nitrogen retention in forest ecosystems: a field experiment, *Science*, 225, 51–52.

[73] Marks, P. L. and Bormann, F. H. (1972), Revegetation following forest cutting: mechanisms for return to steady-state nutrient cycling, *Science*, 176, 914–915.

[74] Hill, A. R. (1975), Ecosystem stability in relation to stresses caused by human activities, *Canadian Geographer*, 19, 206–220.

[75] Moore, P. D. (1982), Fire: catastrophic or creative force? *Impact of Science on Society*, 92, 5–14.

[76] Goméz-Pompa, A., Vázquez-Yanes, C., and Guevara, S. (1972), The tropical rain forest: a non-renewable resource, *Science*, 177, 762–765.

[77] Denevan, W. M. (1973), Development and imminent demise of the Amazon rain forest, *Professional Geographer*, 25, 130–135.

[78] Levin, S. A. and Paine, R. T. (1974), Disturbance, patch formation and community structure, *Proceedings of the National Academy of Sciences*, 71, 2744–2747; and Paine, R. T. and Levin, S. A. (1981), Intertidal landscapes: disturbance and the dynamics of pattern, *Ecological Monographs*, 51, 145–178.

[79] Hastings, H. M., Pekelney, R., Monticcolo, R., van Kannon, D., and Del Monte, D. (1982), Time scales, persistence and patchiness, *BioSystems*, 15, 281–289.

[80] Mandelbrot, B. B. (1977), *Fractals: Form, Chance and Dimension* (Freeman, San Francisco, CA).

[81] May, R. M. (1971), Stability in multi-species community models, *Mathematical Biosciences*, 12, 59–79.

[82] Janzen, D. H. (1983), Tropical agroecosystems, in P. H. Abelson (Ed), *Food: Politics, Economics, Nutrition and Research* (AAAS, Washington, DC).

[83] Holdridge, L. R. (1947), Determination of world plant formations from simple climate data, *Science*, 105, 367–308.

[84] Grigor'yev, A. Z. (1958), The heat and moisture regime and geographic zonality, *Third Congress of the Geographical Society of the USSR*, pp. 3–16.

[85] Steele, J. H. (1985), A comparison of terrestrial and marine ecological systems, *Nature*, 313, 355–358.

[86] Baskerville, G. L. (1983), *Good Forest Management, a Commitment to Action* (Dept. Natural Resources, New Brunswick, Canada).

[87] Brooks, H. (1973), The state of the art: Technology assessment as a process, *Social Sciences Journal*, 25, 247–256.

[88] Douglas, M. (1978), *Cultural Bias*. Occasional Paper for the Royal Anthropological Institute No. 35 (Royal Anthropological Institute, London).

[89] Thompson, M. (1983), A cultural bias for comparison, in H. C. Kunreuther and J. Linnerooth (Eds), *Risk Analysis and Decision Processes: The Siting of Liquified Energy Gas Facilities in Four Countries* (Springer, Berlin).

[90] Mann, H., Siegler, M., and Osmond, H. (1970), The many worlds of time, *Journal of Analytical Psychology*, 13, 35–56.

[91] Walters, C. J. and Hilborn, R. (1978), Ecological optimization and adaptive management, *Annual Review of Ecology and Systematics*, 9, 157–188.

[92] McNeill, W. H. (1982), *The Pursuit of Power* (University of Chicago Press, Chicago, IL).

[93] An earlier version of this chapter was published in Malone and Roederer, note [3].

Commentary

F. di Castri

From my own perception, the chief aim of Holling's chapter is to provide from the start a kind of conceptual framework for the proposed program, the *Sustainable Development of the Biosphere*. There have been and will be other international endeavors such as this: worth mentioning are UNESCO's *Man and the Biosphere Programme* (MAB), which started in 1971 and is ongoing, and ICSU's *International Geosphere–Biosphere Programme* (IGBP)—also called *Global Change*—for which a feasibility study is being undertaken. For none of these has an essay of conceptualization—comparable to this chapter—been attempted; probably there were good reasons for disregarding this aspect, not the least being consideration of the difficulties involved in achieving a sensible agreement on such conceptual issues, especially when large and heterogeneous groups of countries and scientists have to work together.

In addition, the MAB Program has almost unavoidably shifted toward a loose coordination of very heterogeneous packages of national projects; a number of which certainly provide enlightening empirical insights for a theorization of some man–ecosystem interactions. Nevertheless, the local and specifically problem-oriented nature of the best MAB field projects prevents an approach—through MAB—to global concerns.

On the other hand, IGBP initially had almost exclusively global views. It is now increasingly recognized within IGBP that a large number of explanations for the functioning of the biosphere should come from local and regional studies—chiefly of a biological nature—and that the notion of spatial and temporal scale is the key for the success of the overall program. However, it is unlikely that a program of fundamental science, like IGBP, can become really involved in local societal issues.

There is, therefore, a niche for the kind of concerns addressed in Holling's chapter, where a conciliation is proposed between two outcomes, local surprise (of a biological but mostly societal type) and global change (considered chiefly from a biogeochemical point of view). There would be merit if Holling's views were also discussed within MAB and IGBP, as these two programs are seeking new avenues of research and a better defi-nition of their framework.

Concerning specifically Holling's chapter, one can be tempted to react in two oppos-ing ways: either to accept everything by intuition and sympathy—being attracted by the cohesion of the argument and the stimulating challenge of the ideas—or to block reject these speculations. In fact, several statements are not supported by scientific evidence (but this is to be expected in such an article), some hypotheses are not testable in an experimental and quantitative way (as the author recognizes), and one cannot easily visualize how they could be proved or disproved; there is also much use (and perhaps some abuse) of analogical reasoning (with all the charm and the risk involved in analo-gies). I suspect that whichever position one is inclined to—acceptance or rejection—is a matter of personal feeling and behavior rather than of a difference in scientific back-ground. From my viewpoint, I will try to establish which are topics where a consensus might emerge, as well as which are the most controversial aspects. I focus my comments on the three pillars of Holling's conceptual building: local surprise, the four ecosystem functions, and the Gaia hypothesis.

As regards local surprise in relation to societal and institutional behavior, while admittedly similar concepts might be put forward using different terms—e.g., perception—I believe that Holling's presentation has definite advantages, even from a pedagogical viewpoint *vis-a-vis* both scientists and policy-makers. I cannot agree more that in the contemporary context—both socioeconomic and scientific—management institutions (and political establishments) should not desperately seek prediction and control, but rather should settle their organization and their policy on the acceptance of, and the adaptation to, surprise effects. Flexibility should be the *sine qua non* condition for their decisions. By the way, little prediction can be provided at present by the key science of ecology, and I suspect that the same is true for the key disciplines of economy and sociology.

In relation to the Gaia hypothesis, it seems to be at present one of the few workable concepts when addressing research on global biospheric problems, even if the end result be the disproof of it. I imagine that Gaia is also the underlying hypothesis of IGBP. In addition, the use of Gaia and of local surprise to enlighten the interfaces between biospheric and more traditional ecological issues, on the one hand, and between eco-logical and societal issues on the other, has the advantage of focusing on *processes* of a scientific and decision-making nature rather than on ill-defined interdisciplinary link-ages. Interdisciplinarity, when it is considered and implemented as an end in itself—and not as a tool for addressing new complex problems—leads too often to ster-ile research and to verbose descriptions and nonexplanatory results.

Almost paradoxically, in view of Holling's background, the most controversial points may well refer to the ecosystem concept as defined by Holling's four functions — exploitation, conservation, creative destruction, and renewal. To be kindly sarcastic, and admittedly somewhat unfair, one would be tempted to think that Holling's conceptual construction is too harmonious and too beautiful to accurately equate with reality. The supporting examples given by Holling are well presented, but I am doubtful whether they should be generalized to such an extent, and anyway they may lead to different interpretations. Furthermore, I wonder whether these views on ecosystem functions can convince anyone who is not already converted or preadapted to such ideas. From my viewpoint, I believe that Holling's views on this matter represent a very stimulating framework for discussion, but I am somewhat skeptical of the possibility of his hypotheses being proved (or disproved). In fact, too many elements evoked to support these four ecosystem functions are either nonmeasurable or interpreted at present in ways too controversial for good scientific communication. No one is more aware than Holling of the many (and sometimes opposite) interpretations of the terms stability and resilience [1]; but also the adaptive strategies of species represent more a reference concept than a quantifiable parameter (and the place of so-called r- and K-species in an ecological succession is not so clearly defined or so fixed as Figure 5 may suggest).

As a matter of fact, I believe that the whole ecosystem concept is in crisis at present, at least if the ecosystem is taken as a kind of well-defined supraorganism, and not as a useful methodological tool to study problems of interactions and of system behavior as regards different ecological multispecies units. A too extreme and exclusive view of ecosystem properties can alienate from ecology some of the disciplines, such as ecophysiology, population biology, and genetics, that are essential for the explanation of most of the ecological processes (and this is already happening in a number of countries [2]).

The other key aspect of the theoretical framework proposed by Holling deals with *connections*, as exemplified in Figure 6. Ecosystems clearly represent the crossroads of all the systems, being intermediate between biosphere, and society. Some geographers and economists could again object to the use of ecosystems; they may argue that the exchanges of energy and products exceed the boundaries of any given ecosystem, and may prefer other units such as "human use systems" or "resource systems." Nevertheless, even if these systems help to establish interfaces with societies, they cannot facilitate linkages with global biospheric issues; furthermore, as regards ecosystems, these possible objections may imply simply a different interpretation of the definition and hierarchical scale of ecosystems. I am personally more worried, as regards connections, about the excessive use of analogies. I need evidence to be convinced that ecosystems are just "Gaia writ small"; changing of scale implies usually the emerging (and the disappearing) of new functions and properties. On the other hand, analogies, such as those presented in Table 2 between ecosystems and other systems of a societal nature *sensu lato*, can perhaps improve, the understanding between disciplines and their different approaches, but do not serve operational purposes for research and may be dangerous and misleading if they are taken too strictly.

After all, it is amazing to see to what extent a science in crisis, such as ecology—in crisis for several reasons, not least the confusion between ecological and environmental problems [2]—is capable of "exporting" so many concepts, through an analogical process, to other sciences that probably face a comparable crisis, such as geography, economics, or sociology. I wonder whether ecology itself is not shifting at present from a biological science to a sociological one, with all the epistemological and, methodological implications involved in this move.

Finally, in relation to Holling's recommendations, they all represent essential avenues for research. However, on the basis of previous experience in international programs, such as the *International Biological Programme* and the *Man and the Biosphere Programme* [3], I suspect that it is seldom realized how high are the costs, the manpower involvement, and the intrinsic difficulties represented by these proposals. Another obstacle, which must be overcome for the development of ecology, is our lack of a theoretical basis for the comparison of ecosystems. We do not know yet how to evaluate the degree of extrapolatability and predictability of our research from one to another ecosystem, or from one to another period of time. This is true even within the same ecosystem type, e.g., comparisons between temperate forests of the northern and southern hemispheres; the difficulties in comparing more complex and less studied ecosystems, such as those of the humid tropics, or of more heterogeneous systems, such as the Mediterranean-type ones, increase exponentially. When different types of ecosystems are compared, for instance terrestrial and aquatic ones, or forests and grasslands, it becomes almost impossible to escape from the pitfall of overgeneralization, with little regard for reality.

This remark is not intended to underestimate the importance of a comparative ecology or to discourage students from undertaking this kind of research. On the contrary, I feel that these comparisons represent now the main challenge for ecology. It is, nevertheless, essential to fully understand the limits and the methodological background for such comparisons. If they are placed in the framework of strictly climatic and old-fashioned classifications of life zones, such as those of Holdridge and Grigor'yev, as suggested here, I do not hesitate in predicting many "surprise effects" when the results are compared; and these surprises may well derive from the lack of understanding or knowledge of the diverse origins of these ecosystems, on the weight of the past, including both geological and historical factors (and man's impact in fashioning ecosystem patterns is much older than usually considered). A strong injection into such research of evolutionary ecology (I mean evolution in the strict sense of biological evolution) and of explanatory biogeography—with special emphasis on the invasion of alien species - would be likely to reduce considerably these surprises.

I regret that, because of space limitations, I cannot refer to many other enlightening points of this chapter, such as continuity and discontinuity in ecological processes, linear and nonlinear interactions, etc. As with most of Holling's work one is captured by the argument and—in agreement or disagreement—cannot resist the temptation of being involved in the discussion; and this is precisely the goal that a chapter of this nature should achieve.

Notes and References (Comm.)

[1]   As quoted by Holling, Pimm, S. L. (1984), The complexity and stability of ecosystems, *Nature*, 307, 321–326.

[2]   di Castri, F. (1984), *L'écologie. Les défis d'une science en temps de crise*. Rapport au Ministre de l'lndustrie et de la Recherche (La Documentation Françise, Paris).

[3]   di Castri, F. (1985), Twenty years of international programmes on ecosystems and the biosphere: an overview of achievements, shortcomings and possible new perspectives, in T. F. Malone and J. R. Roederer (Eds), *Global Change* (Cambridge University Press. Cambridge, UK).   .

Regime Shifts, Resilience, and Biodiversity in Ecosystem Management

CARL FOLKE, STEVE CARPENTER, BRIAN WALKER, MARTEN SCHEFFER,

THOMAS ELMQVIST, LANCE GUNDERSON, AND C. S. HOLLING

THOMAS ELMQVIST AND CARL FOLKE, Department of Systems Ecology, Stockholm University.

STEVE CARPENTER AND CARL FOLKE, Beijer International Institute of Ecological Economics, Royal Swedish Academy of Sciences, Stockholm, Sweden.

STEVE CARPENTER, Center for Limnology, University of Wisconsin, Madison, Wisconsin 53706; email: srcarpen@wisc.edu.

BRIAN WALKER, Sustainable Ecosystems, CSIRO, Canberra, ACT, 2601, Australia; email: Brian.Walker@csiro.au.

MARTEN SCHEFFER, Aquatic Ecology and Water Quality Management Group, Wageningen Agricultural University, Wageningen, The Netherlands; email: Marten.Scheffer@wur.nl.

LANCE GUNDERSON, Department of Environmental Studies, Emory University, Atlanta, Georgia 30322; email: lgunder@emory.edu.

C. S. HOLLING, 16871 Sturgis Circle, Cedar Key, Florida 32625; email: holling@zoo.ufl.edu.

Key Words: *alternate states, regime shifts, response diversity, complex adaptive systems, ecosystem services.*

Annual Review Ecology Evolution Systematics 2004. 35:557–81
Originally Published by *Annual Reviews.*

Abstract

WE REVIEW THE EVIDENCE of regime shifts in terrestrial and aquatic environments in relation to resilience of complex adaptive ecosystems and the functional roles of biological diversity in this context. The evidence reveals that the likelihood of regime shifts may increase when humans reduce resilience by such actions as removing response diversity, removing whole functional groups of species, or removing whole trophic levels; impacting on ecosystems via emissions of waste and pollutants and climate change; and altering the magnitude, frequency, and duration of disturbance regimes. The combined and often synergistic effects of those pressures can make ecosystems more vulnerable to changes that previously could be absorbed. As a consequence, ecosystems may suddenly shift from desired to less desired states in their capacity to generate ecosystem services. Active adaptive management and governance of resilience will be required to sustain desired ecosystem states and transform degraded ecosystems into fundamentally new and more desirable configurations.

Introduction

Humanity strongly influences biogeochemical, hydrological, and ecological processes, from local to global scales. We currently face more variable environments with greater uncertainty about how ecosystems will respond to inevitable increases in levels of human use (Steffen et al. 2004). At the same time, we seem to challenge the capacity of desired ecosystem states to cope with events and disturbances (Jackson et al. 2001, Paine et al. 1998). The combination of these two trends calls for a change from the existing paradigm of command-and-control for stabilized "optimal" production to one based on managing resilience in uncertain environments to secure essential ecosystem services (Holling & Meffe 1996, Ludwig et al. 2001). The old way of thinking implicitly assumes a stable and infinitely resilient environment, a global steady state. The new perspective recognizes that resilience can be and has been eroded and that the self-repairing capacity of ecosystems should no longer be taken for granted (Folke 2003, Gunderson 2000). The challenge in this new situation is to actively strengthen the capacity of ecosystems to support social and economic development. It implies trying to sustain desirable pathways and ecosystem states in the face of continuous change (Folke et al. 2002, Gunderson & Holling 2002).

Holling (1973), in his seminal paper, defined ecosystem resilience as the magnitude of disturbance that a system can experience before it shifts into a different state (stability domain) with different controls on structure and function and distinguished ecosystem resilience from engineering resilience. Engineering resilience is a measure of the rate at which a system approaches steady state after a perturbation, that is, the speed of return to equilibrium, which is also measured as the inverse of return time. Holling (1996) pointed out that engineering resilience is a less appropriate measure in ecosystems that have multiple stable states or are driven toward multiple stable states by human activities (Nyström et al. 2000, Scheffer et al. 2001).

Here, we define resilience as the capacity of a system to absorb disturbance and reorganize while undergoing change so as to retain essentially the same function, structure, identity, and feedbacks (Walker et al. 2004). The ability for reorganization and renewal of a desired ecosystem state after disturbance and change will strongly depend on the influences from states and dynamics at scales above and below (Peterson et al. 1998). Such cross-scale aspects of resilience are captured in the notion of a panarchy, a set of dynamic systems nested across scales (Gunderson & Holling 2002). Hence, resilience reflects the degree to which a complex adaptive system is capable of self-organization (versus lack of organization or organization forced by external factors) and the degree to which the system can build and increase the capacity for learning and adaptation (Carpenter et al. 2001b, Levin 1999).

Several studies have illustrated that ecological systems and the services that they generate can be transformed by human action into less productive or otherwise less desired states. The existence of such regime shifts (or phase shifts) is an area of active research. Regime shifts imply shifts in ecosystem services and consequent impacts on human societ-ies. The theoretical basis for regime shifts has been des-cribed by Beisner et al. (2003), Carpenter (2003), Ludwig et al. (1997), Scheffer & Carpenter (2003), and Scheffer et al. (2001).

Here, we review the evidence of regime shifts in terrestrial and aquatic ecosystems in relation to resilience and discuss its implications for the generation of ecosystem services and societal development. Regime shifts in ecosystems are increasingly common as a consequence of human activ-ities that erode resilience, for example, through resource exploitation, pollution, land-use change, possible climatic impact and altered distur-bance regimes. We also review the functional role of biological diversity in relation to regime shifts and ecosystem resilience. In particular, we focus on the role of biodiversity in the renewal and reorganization of eco-systems after disturbance—what has been referred to as the back-loop of the adaptive cycle of ecosystem development (Holling 1986). In this con-text, the insurance value of biodiversity becomes significant. It helps sustain desired states of dynamic ecosystem regimes in the face of uncertainty and surprise (Elmqvist et al. 2003). Strategies for transforming degraded ecosystems into new and improved configurations are also discussed.

Regime Shifts and Dynamics of Resilience in Ecosystems

Ecosystems are complex, adaptive systems that are characterized by historical dependency, nonlinear dynamics, threshold effects, multiple basins of attraction, and limited predictability (Levin 1999). Increasing evidence suggests that ecosystems often do not respond to gradual change in a smooth way (Gunderson & Pritchard 2002). Threshold effects with regime shifts from one basin of attraction to another have been doc-umented for a range of ecosystems (see Thresholds Database on the Web site www.resalliance.org). Passing a threshold marks a sudden change in feedbacks in the ecosystem, such that the trajectory of the system changes direction—toward a different attractor. In some cases, cross-ing the threshold brings about a sudden, sharp, and dramatic change in the responding state variables, for example, the shift from clear to tur-bid water in lake systems (Carpenter 2003). In other cases, although the dynamics of the system have "flipped" from one attractor to another, the transition in the state variables is more gradual, such as the change from a grassy to a shrub-dominated rangeland (Walker & Meyers 2004). In

Table 1, we provide examples of documented shifts between alternate states and expand on some of them in the text.

Temperate Lakes

Lake phosphorus cycles exhibit multiple regimes, each stabilized by a distinctive set of feedbacks. Generally, two regimes have attracted the most interest, although deeper analyses have revealed even greater dynamic complexities (Carpenter 2003, Scheffer 1997). The two regimes of most concern to people who use the lakes are the clear-water and turbid-water regimes. In the clear-water regime, phosphorus inputs, phytoplankton biomass, and recycling of phosphorus from sediments are relatively low. In the turbid-water regime, these same variables are relatively high. The turbid-water regime provides lower ecosystem services because of abundant toxic cyanobacteria, anoxic events, and fish kills (Smith 1998).

In the clear-water regime of shallow lakes (lakes that do not stratify thermally), extensive beds of higher aquatic plants (macrophytes) stabilize sediments and reduce recycling of phosphorus to phytoplankton (Jeppesen et al. 1998, Scheffer 1997). Macrophyte beds may be lost because of shading by phytoplankton when high phosphorus inputs stimulate phytoplankton growth. Bottom-feeding (benthivorous) fishes that increase in abundance with nutrient enrichment damage the macrophytes and cause turbidity by resuspending sediment. Once the macrophytes are lost, sediments are more easily resuspended by waves, and rapid recycling of phosphorus from sediments maintains the turbid regime. Reversion to the clear-water state requires reduction of phosphorus inputs, but even if phosphorus inputs are reduced, the turbid state may remain resilient. With sufficiently low levels of phosphorus the ecosystem can possibly be perturbed to the clear-water state by temporarily removing fish, which allows macrophytes to recover, stabilizes the sediments, and reduces phosphorus cycling, thereby consolidating the clear-water state.

A different mechanism operates in deep (thermally stratified) lakes, although the clear-water and turbid-water states are similar (Carpenter 2003). Interactions of iron with oxygen are the key (Caraco et al. 1991, Nürnberg 1995). In the clear-water regime, rates of phytoplankton production, sedimentation, and oxygen consumption in deep water are low. Consequently the deep water remains oxygenated most of the time, and iron remains in the oxidized state, which binds phosphorus in insoluble

TABLE 1: *Documented shifts between states in different kinds of ecosystems*

Ecosystem type	Alternate state 1	Alternate state 2	References
Freshwater systems			
Temperate lakes	Clear water	Turbid eutrophied water	Carpenter 2003
	Game fish abundant	Game fish absent	Post et al. 2002, Walters & Kitchell 2001, Carpenter 2003
Tropical lakes	Submerged vegetation	Floating plants	Scheffer et al. 2003
Shallow lakes	Benthic vegetation	Blue-green algae	Blindow et al. 1993, Scheffer et al. 1993, Scheffer 1997, Jackson 2003
Wetlands	Sawgrass communities	Cattail communities	Davis 1989, Gunderson 2001
	Salt marsh vegetation	Saline soils	Srivastava & Jefferies 1995
Marine systems			
Coral reefs	Hard coral dominance	Macroalgae dominance	Knowlton 1992, Done 1992, Hughes 1994, McCook 1999
	Hard coral dominance	Sea urchin barren	Glynn 1988, Eakin 1996
Kelp forests	Kelp dominance	Sea urchin dominance	Steneck et al. 2002, Konar & Estes 2003
	Sea urchin dominance	Crab dominance	Steneck et al. 2002
Shallow lagoons	Seagrass beds	Phytoplankton blooms	Gunderson 2001, Newman et al. 1998
Coastal seas	Submerged vegetation	Filamentous algae	Jansson & Jansson 2002, Worm et al. 1999
Benthic foodwebs	Rock lobster predation	Whelk predation	Barkai & McQuaid 1988
Ocean foodwebs	Fish stock abundant	Fish stock depleted	Steele 1998, Walters & Kitchell 2001, de Roos & Persson 2002

TABLE 1: *Continued*

Ecosystem type	Alternate state 1	Alternate state 2	References
Forest systems			
Temperate forests	Spruce-fir dominance	Aspen-birch dominance	Holling 1978
	Pine dominance	Hardwood dominance	Peterson 2002
	Hardwood-hemlock	Aspen-birch	Frelich & Reich 1999
	Birch-spruce succession	Pine dominance	Danell et al. 2003
Tropical forests	Rain forest	Grassland	Trenbath et al. 1989
	Woodland	Grassland	Dublin et al. 1990
	Native crab consumers	Invasive ants	O'Dowd et al. 2003
Savanna and grassland			
Grassland	Perennial grasses	Desert	Wang & Eltahir 2000, Foley et al. 2003, van de Koppel et al. 1997
Savanna	Native vegetation	Invasive species	Vitousek et al. 1987
	Tall shrub, perennial grasses	Low shrub, bare soil	Bisigato & Bertiller 1997
	Grass dominated	Shrub dominated	Anderies et al. 2003, Brown et al. 1997
Arctic, sub-Arctic systems			
Steppe/tundra	Grass dominated	Moss dominated	Zimov et al. 1995
	Tundra	Boreal forest	Bonan et al. 1992, Higgins et al. 2002

forms. When phosphorus inputs are high, rates of phytoplankton production, sedimentation, and oxygen consumption increase. The deep
water is anoxic part if not all of the time, iron is in the reduced state, and
phosphorus dissolves into the water. Recycling of phosphorus from sediments makes the turbid-water state resilient. The regime can be shifted
back to clear water by extreme reductions of phosphorus input or by various manipulations that decrease recycling of phosphorus (Carpenter et
al. 1999, Cooke et al. 1993).

Similar mechanisms may have operated during massive oceanic events
in the remote past. Episodes of large-scale phosphorus release in the oceans
may represent a regime shift in which high phytoplankton production
forms a strong positive feedback with phosphorus recycling from deep
waters or sediments (Algeo & Scheckler 1998, Van Capellen & Ingall 1994).

Tropical Lakes

Experiments, field data, and models suggest that a situation with extensive free-floating plant cover and a state characterized by submerged
plants tend to be alternate regimes (Scheffer et al. 2003). Dense mats of
floating plants have an adverse effect on freshwater ecosystems because
they create anoxic conditions, which strongly reduce animal biomass
and diversity. Floating plants are superior competitors for light and carbon. Submerged plants are better competitors for nutrients and may
prevent expansion of free-floating plants through a reduction of available nutrients in the water column. As a result, over a range of conditions,
the lake can exist in either a floating-plant-dominated state or a sub-
merged-plant dominated state. Both states are resilient, but nutrient
enrichment reduces the resilience of the submerged plant state. A single
drastic harvest of floating plants can induce a permanent shift to an
alternate state dominated by rooted submerged growth forms if the
nutrient loading is not too high (Scheffer et al. 2003).

Wetlands, Estuaries, and Coastal Seas

In the Everglades, the freshwater marshes have shifted from wetlands
dominated by sawgrass to cattail marshes because of nutrient enrichment. The soil phosphorous content defines the alternate states, and
several types of disturbances (fires, drought, or freezes) can trigger a
switch between these states (Gunderson 2001).

In Florida Bay, the system has flipped from a clear-water, seagrass-dominated state to one of murky water, algae blooms, and recurrently stirred-up sediments. Hypotheses that have been proposed to explain this shift include change in hurricane frequency, reduced freshwater flow entering the Bay, higher nutrient concentrations, removal of large grazers such as sea turtles and manatees, sea-level rise, and construction activities that restrict circulation in the Bay (Gunderson 2001).

The Baltic Sea is eutrophied and overfished (Elmgren 2001), and a shift has occurred in the coastal subsystem from submerged vegetation dominated by perennial fucoids to filamentous and foliose annual algae with lower levels of diversity (Kautsky et al. 1986). Jansson & Jansson (2002) propose that these conditions represent alternate states. Grazers on turf algae, such as gastropods, contribute to the maintenance of the fucoid-dominated state, but nutrient influx overrides grazing control and shifts the system into a less desired state (Worm et al. 1999). In addition to shading by filamentous algae (Berger et al. 2003), increased sedimentation caused by excessive phytoplankton production hinders recruitment and settlement of fucoids plants (Eriksson & Johansson 2003), which may lock the coastal subsystem into the undesired state.

Coral Reefs

The current shift of coral reefs into dominance by fleshy algae was long preceded by diminishing stocks of fishes and increased nutrient and sediment runoff from land (Jackson et al. 2001). The grazing of algae by fish species and other grazers contributes to the resilience of the hard-coral-dominated reef by, for example, keeping the substrate open for recolonization of coral larvae after disturbances such as hurricanes (Nyström et al. 2000).

In the Caribbean, overfishing of herbivores (dominated by fishes on reefs) led to expansion of sea urchin populations as the key grazers on invading algae. Thereby, the coral-dominated state was maintained, albeit at low resilience. The high densities of the sea urchin populations may have contributed to their eventual demise when a disease outbreak spread throughout the Caribbean and reduced their numbers by two orders of magnitude, which precipitated the shift to the algal-dominated state that persists today (Hughes 1994, Knowlton 1992).

In other areas, high densities of grazing echinoids erode the reef matrix and, if unchecked, have the capacity to destroy reefs, as documented in the Galapagos Islands and elsewhere in the East Pacific (Eakin 1996, Glynn 1988). Loss of macrofauna, reduced fish stocks, replacement of herbivorous fishes by a single species of echinoid, overgrazing by food-limited sea urchins, high levels of erosion by echinoids, and reduced coral recruitment make coral reefs vulnerable to change and subject to regime shifts (Bellwood et al. 2004, Done 1992, Hughes et al. 2003). A significant ecological restructuring of reefs towards "weedy" generalist species of low trophic levels that are adapted to variable environments is underway (Knowlton 2001, McClanahan 2002). Bellwood et al. (2004) describe six different reefs transitions towards less desired states as a consequence of human-induced erosion of resilience (Figure 1).

Kelp Forests

Remarkably sharp boundaries are found between kelp forests and neighboring "deforested" areas (Konar & Estes 2003). Also, remarkable switches have occurred between kelp dominance and sea-urchin dominance over time (Steneck et al. 2002). Experiments reveal that sea urchins can control kelp growth in the open areas, but the sweeping of kelp foliage over the rocks in the border region prevents migration of sea urchins into the kelp stands. Thus, kelp stands can withstand herbivory by combining their flexible morphology with the energy of wave-generated surge (Konar & Estes 2003). Kelp forests can recover if the numbers of sea urchins are reduced by an external factor. Proposed mechanisms include increased predation by sea otters, cycles of diseases, and introduction of sea urchin fishery. The fishing down of coastal food webs has in some locations led to the return of kelp forests devoid of vertebrate apex predators. Large predatory crabs have filled this void in areas of the western North Atlantic (Steneck et al. 2002).

Pelagic Marine Fisheries

Pelagic marine fish stocks sometimes exhibit sharp changes consistent with regime shifts (Steele 1996, 1998). Large changes in fish stocks have occurred ever since the introduction of fisheries (Jackson et al. 2001). Similar rapid and massive changes have occurred in freshwater ecosystems subject to sport fishing (Post et al. 2002). Cascading changes

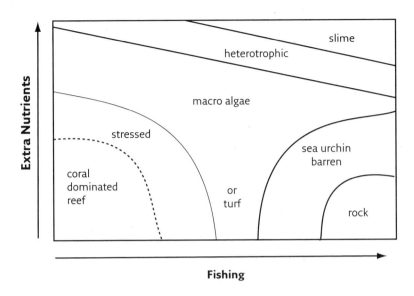

FIGURE 1: *Effects of eutrophication and fishing and observed shifts between states in coral reefs (modified from Bellwood et al. 2004).*

are often related to size-structured predation (de Roos & Persson 2002). Larger individuals of one species eat smaller individuals of other species. If larger individuals of one species become rare for some reason, that species' recruitment can be eliminated by predation from the other species and perhaps lead to severe decline of the population of the first species. Walters & Kitchell (2001) call this dynamic "cultivation/depensation." If adults of the former species are abundant, they create favorable conditions for their own offspring by reducing the abundance of the latter species. If adults of the former species are overfished, expansion of latter species may permanently prevent reestablishment of the former species.

Savannas
Marked fluctuations in grass and woody plant biomass are a characteristic feature of savannas, because of their highly variable rainfall, and primary productivity varies up to tenfold from one year to the next (Kelly & Walker 1976). Herbivores cannot respond fast enough to track these fluctuations, and the accumulation of grass during wet periods means periodic accumulation of fuel and, therefore, fires. The net effect of fires

has been to maintain savanna rangelands in more open, grassy states than would be achieved without fires (Scholes & Walker 1993).

The interaction of fire, herbivory, and variable rainfall has resulted in a grass-shrub-livestock system that exhibits regime shifts between an open, grassy state and a dense, woody state, particularly where humans have altered the pattern and intensity of grazing (Anderies et al. 2002). Establishment of shrub seedlings occurs in wet periods when the seedlings can get their roots below the grass-rooting zone to survive the first dry season. A vigorous grass layer for the first few years strongly suppresses established seedlings, but once established, grasses have little effect on woody plant growth. Fire has little effect on grasses because it occurs at the end of the dry season when grasses are dormant, but it has a severe effect on woody plants by killing many and reducing others to ground level.

The change from a grassy to a woody state comes about through a combination of sustained grazing pressure and lack of fire. Periods of drought with high stock numbers bring about the death of perennial grasses and lead to reduced grass cover. When followed by good rains this reduced grass cover, in turn, leads to a profusion of woody seedlings. If, at this point, all livestock were removed, enough grass growth would still occur to enable an effective fire and keep the system in a grassy state. However, if grazing pressure is sustained a point is reached in the increasing woody:grass biomass ratio after which, even if all livestock are removed, the competitive effect of the woody plants is such that it prevents the build up of sufficient grass fuel to sustain a fire. The system then stays in the woody state until the shrubs or trees reach full size and, through competition among them, begin to die. The vegetation then opens up for the reintroduction of grass and fire. This process can take 30 or 40 years.

The flip in the rangeland occurs when the resilience of the grassy state has been exceeded—that is, when the amount of change in the ratio of grass:woody vegetation needed to push the system into the woody state falls within the range of the annual fluctuations of this ratio (because of fluctuations in rainfall and grazing pressure). Once this situation is reached, the conditions needed to flip the system (e.g., a period of low rainfall) will inevitably follow.

Forests

The boreal forests of North America experience distinctive outbreaks of the spruce budworm, with 30 to 45 years and occasionally 60 to 100 years between outbreaks. This defoliating insect destroys large areas of mature softwood forests, principally spruce and fir. Once the softwood forest is mature enough to provide adequate food and habitat for the budworm, and if a period of warm dry weather occurs, budworm numbers can increase sufficiently to exceed the predation rate and trigger an outbreak. A local outbreak can spread over thousands of square kilometers and eventually collapse after 7 to 16 years. Programs of spraying insecticide to control spruce budworm outbreaks exacerbated the conditions for outbreaks over even more extensive areas (Holling 1978). After a defoliation event, aspen and birch often dominate the regenerating forest, but over a period of 20 to 40 years, selective browsing by moose can shift this forest back to a state dominated by conifers (Ludwig et al. 1978).

Browsing that causes change in dominance between tree species that have different effects on ecosystem functions can lead to dramatic effects in forest ecosystems. For example, the gradual reduction of willows by ungulates on the Alaskan floodplain makes room for nitrogen-fixing alders that increase soil fertility and cause overall vegetation change. In the mountain range of Scandinavia, birches dominate young stands, followed by Norway spruce in the forest succession. If the birches are heavily browsed by ungulates, spruce does not get shelter and fails. Instead, pines may establish and become dominant, which causes long-term changes in soil fertility (Danell et al. 2003). Forestry and hunting policies affect and shape those trajectories.

Shifts in forest cover, associated with management of fire regimes, in the well-drained soils of the southeastern United States reflect alternate states (Peterson 2002). A pine-dominated savanna, with grasses, palms, or shrubs in the understory, historically covered the region and was the result of frequent ground fires. Hardwood shrubs would invade during fire-free periods but their dominance was inhibited by frequent burning. Because of fire suppression and fragmentation of the landscape, fire frequency decreased and led to either mixed pine-hardwood forests or hardwood forests. Once a canopy of hardwoods is established, the site becomes less flammable and precludes pine regeneration (Peterson 2002).

Regime Shifts and Irreversibility

In some cases, regime shifts may be largely irreversible. Loss of trees in cloud forests is one example. In some areas, the forests were established under a wetter rainfall regime thousands of years previously. Necessary moisture is supplied through condensation of water from clouds intercepted by the canopy. If the trees are cut, this water input stops and the resulting conditions can be too dry for recovery of the forest (Wilson & Agnew 1992).

A continental-scale example of an irreversible shift seems to have occurred in Australia, where overhunting and use of fire by humans some 30,000 to 40,000 years ago removed large marsupial herbivores. Without large herbivores to prevent fire and fragment vegetation, an ecosystem of fire and fire-dominated plants expanded and irreversibly switched the ecosystem from a more productive state, dependent on rapid nutrient cycling, to a less productive state, with slower nutrient cycling (Flannery 1994). Similarly, extinction of megafauna at the end of the Pleistocene in Siberia, possibly through improvement in hunting technology, may have triggered an irreversible shift from steppe grassland to tundra. The resulting increase in mosses led to cooler soils, less decomposition, and greater carbon sequestration in peat (Zimov et al. 1995).

Vulnerability Through Human-Induced Loss of Resilience

As illustrated by the foregoing examples, undesired shifts between ecosystem states are caused by the combination of the magnitudes of external forces and the internal resilience of the system. As resilience declines, the ecosystem becomes vulnerable, and progressively smaller external events can cause shifts. Human actions have increased the likelihood of undesired regime shifts. In Figure 2, we summarize shifts into less desired states as a consequence of human-induced loss of resilience.

Humans have, over historical time but with increased intensity after the industrial revolution, reduced the capacity of ecosystems to cope with change through a combination of top-down (e.g., overexploitation of top predators) and bottom-up impacts (e.g., excess nutrient influx), as well as through alterations of disturbance regimes including climatic change (e.g., prevention of fire in grasslands and forest or increased bleaching of coral reefs because of global warming) (Nyström et al. 2000, Paine et al.

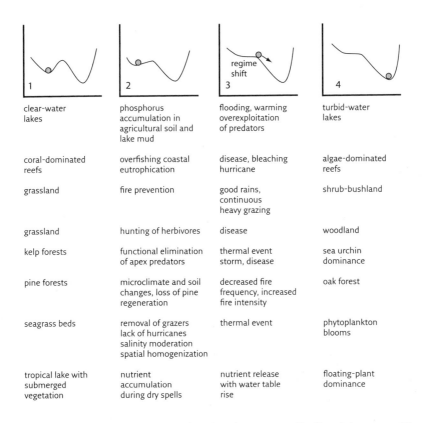

clear-water lakes	phosphorus accumulation in agricultural soil and lake mud	flooding, warming overexploitation of predators	turbid-water lakes
coral-dominated reefs	overfishing coastal eutrophication	disease, bleaching hurricane	algae-dominated reefs
grassland	fire prevention	good rains, continuous heavy grazing	shrub-bushland
grassland	hunting of herbivores	disease	woodland
kelp forests	functional elimination of apex predators	thermal event storm, disease	sea urchin dominance
pine forests	microclimate and soil changes, loss of pine regeneration	decreased fire frequency, increased fire intensity	oak forest
seagrass beds	removal of grazers lack of hurricanes salinity moderation spatial homogenization	thermal event	phytoplankton blooms
tropical lake with submerged vegetation	nutrient accumulation during dry spells	nutrient release with water table rise	floating-plant dominance

FIGURE 2: *Alternate states in a diversity of ecosystems (1, 4) and the causes (2) and triggers (3) behind loss of resilience and regime shifts. For more examples, see Thresholds Database on the Web site www.resalliance.org.*

1998, Worm et al. 2002). The result of those combined impacts tends to be leaking, simplified, and "weedy" ecosystems characterized by unpredictability and surprise in their capacity to generate ecosystem services.

The likelihood that an ecological system will remain within a desired state is related to slowly changing variables that determine the boundaries beyond which disturbances may push the system into another state (Scheffer & Carpenter 2003). Consequently, efforts to reduce the risk of undesired shifts between ecosystem states should address the gradual changes that affect resilience rather than focus all effort into trying to control disturbance and fluctuations. The slowly changing variables include such things as land use, nutrient stocks, soil properties, freshwater dynamics, and biomass of long-lived organisms (Gunderson & Pritchard 2002). In the following sections we focus on biological diversity as a slowly changing variable and its significance in ecosystem resilience.

Trophic Cascades

Loss of top predators can increase the vulnerability of aquatic ecosystems to eutrophication by excessive nutrient input (Carpenter 2003). In manipulations of whole-lake ecosystems, removal of top predators allows primary producers to respond much more strongly to experimental inputs of nutrients (Carpenter et al. 2001a). The mechanism is a trophic cascade — in the absence of top predators, planktivorous fishes become abundant, and grazing zooplankton is suppressed. When nutrients are added, phytoplankton grow unconstrained by grazers.

Over human history, removal of top predators from near-shore marine ecosystems through fishing may have increased the vulnerability of the ecosystems to coastal nutrient inputs and paved the way for impacts such as eutrophication, algal blooms, disease outbreaks, and species introductions in coastal areas (Jackson et al. 2001). In the Black Sea, for example, overfishing and eutrophication changed the ecosystem from one dominated by piscivorous fishes (bluefish, bonito) and dolphins to one dominated by jellyfish, small-bodied planktivorous fishes, and phytoplankton (Daskalov 2002, Zaitsev & Mamaev 1997). In the Baltic Sea, removal of top predators such as seals (through pollution and hunting), fishing pressure, and influx of excessive nutrients have caused widespread eutrophication and oxygen deficiency in deeper waters that have wiped out important food chains over nearly 100,000 km^2 of sea bottom (Elmgren 1989).

Trophic cascades occur in a diversity of ecosystem, and examples are known from terrestrial systems (Pace et al. 1999). Trophic cascades are becoming another signature of the vast and growing human footprint (Terborgh et al. 2001). Cascading provides nonlinear and often surprising changes in ecosystem dynamics and may lead to regime shifts (Diaz & Cabido 2001). Trophic cascades seem to be less likely under conditions of high diversity or extensive omnivory in food webs (Pace et al. 1999).

Biodiversity and Resilience Dynamics

The diversity of functional groups in a dynamic ecosystem undergoing change, the diversity within species and populations, and the diversity of species in functional groups appear to be critical for resilience and the generation of ecosystem services (Chapin et al. 1997, Luck et al. 2003). Two aspects of diversity are distinguished: functional-group diversity and functional-response diversity.

Functional-Group Diversity

Functional groups of species in a system refers to groups of organisms that pollinate, graze, predate, fix nitrogen, spread seeds, decompose, generate soils, modify water flows, open up patches for reorganization, and contribute to the colonization of such patches. The persistence of functional groups contributes to the performance of ecosystems and the services that they generate. Loss of a major functional group, such as apex predators, other consumers, or benthic filter-feeders, may, as previously discussed, cause drastic alterations in ecosystem functioning (Chapin et al. 1997, Duffy 2002, Jackson et al. 2001).

However, in systems that lack a specific functional group, the addition of just one species may dramatically change the structure and functioning of ecosystems (Chapin et al. 2000). In Hawaii, the introduced nitrogen-fixing tree *Myrica faya* has dramatically changed the structure and function in ecosystems where no native nitrogen-fixing species had been present. Once established, *M. faya* can increase nitrogen inputs up to five times, thereby facilitating establishment of other exotic species (Vitousek et al. 1987). Studies of coastal environments and reefs suggest that more diverse ecosystems are less sensitive to invasion of exotic species (Knowlton 2001, Stachowicz et al. 1999).

Functional-Response Diversity

Recently, Naeem & Wright (2003) argued that functional-response traits should be considered and distinguished from functional-effect traits when analyzing biodiversity effects on ecosystem functioning. Variability in responses of species within functional groups to environmental change is critical to ecosystem resilience (Chapin et al. 1997, Norberg et al. 2001). Elmqvist et al. (2003) call this property *response diversity*, and it is defined as the diversity of responses to environmental change among species that contribute to the same ecosystem function. For example, in lake systems, animal plankton species with higher tolerance to low pH sustain the grazing function on phytoplankton during acid conditions

(Frost et al. 1995). In semiarid rangelands, resilience of production to grazing pressure is achieved by maintaining a high number of apparently less important and less common, or apparently "redundant," species from the perspective of those who want to maximize production, each with different capacities to respond to different combinations of rainfall and grazing pressures. The species replace each other over time, which ensures maintenance of rangeland function over a range of environmental conditions (Walker et al. 1999).

The role of genetic and population diversity for response diversity is illustrated through sockeye salmon production in the rivers and lakes of Bristol Bay, Alaska. Several hundred discrete spawning populations display diverse life-history characteristics and local adaptations to the variation in spawning and rearing habitats. Geographic regions and life-history strategies that were minor producers during a certain climatic regime have been the major producers during other climatic regimes, which allowed the aggregate of the populations to sustain its productivity in fluctuating freshwater and marine environments. The response diversity of the fish stocks has been critical in sustaining their resilience to environmental change. Such management is in stark contrast to the common focus on only the most productive runs at a certain moment in time (Hilborn et al. 2003).

Springer et al. (2003) propose that the decimation of the great whales since the World War II caused their foremost natural predator, killer whales, to begin feeding more intensively on seals, sea lions, and sea otters, which are the major predators on sea urchins, thereby possibly contributing to shifts from kelp-dominated to sea urchin–dominated coastal areas. In areas where sea-urchin predator diversity is low (e.g., in western North Atlantic and Alaska), the transition between kelp dominance and sea-urchin dominance has been rapid, frequent, widespread, and often long lasting. In southern California, where the diversity of predators, herbivores, and kelps is high, deforestation events have been rare or patchy in space and short in duration, and no single dominant sea-urchin predator exists (Steneck et al. 2002). Functional redundancy and response diversity may contribute to the resilience of kelp forests in California.

An important distinction should be made between real redundancy and apparent redundancy, which involves response diversity within functional groups. Functional redundancy refers to the number of species that perform the same function. Adding more species does not lead to increased system performance where there is real functional redundancy. Furthermore, if this set of functionally redundant species does not exhibit any response diversity, they do not contribute to the insurance value.

We argue that the biodiversity insurance metaphor needs to be revived with a focus on how to sustain ecosystem resilience and the services it generates in the context of multiple-equilibrium systems and human-dominated environments (Folke et al. 1996). Ecosystems with high response diversity increase the likelihood for renewal and reorganization into a desired state after disturbance (Chapin et al. 1997, Elmqvist et al. 2003).

Biodiversity in Ecosystem Renewal and Reorganization

Recovery after disturbance has often been measured as return time to the equilibrium state. Frequently, the sources of ecosystem recovery have been taken for granted and the phases of ecosystem development that prepare the system for succession and recovery largely neglected. Disturbance releases the climax state and is followed by renewal and reorganization. We refer to those phases as the back-loop of ecosystem development (Gunderson & Holling 2002). Functional roles in the back-loop and sources of resilience are critical for sustaining an ecosystem within a desired state in the face of change (Nyström & Folke 2001).

In coral reef systems, three functional groups of herbivores, dominated by fishes, play different and complementary roles in renewing and reorganizing reefs into a coral-dominated state after disturbance. These groups—grazers, scrapers, and bioeroders—prepare the reef for recovery. Bioeroding fishes remove dead corals and other protrusions, which exposes the hard reef matrix for new settlement of coralline algae and corals. Grazers remove seaweed, which reduces coral overgrowth and shading by macroalgae. Scrapers remove algae and sediment by close cropping, which facilitates settlement, growth, and survival of coralline algae and corals. Without bioeroders, recovery may be inhibited by extensive stands of dead staghorn and tabular coral that can remain intact for years before collapsing and taking with them attached coral recruits. Without grazers, algae can proliferate and limit coral settlement and survival of juvenile and adult colonies. Without scrapers, sediment-trapping algal turfs develop that smother coral spat and delay or prevent recovery. The extents to which reefs possess these functional groups are central to their capacity to renew and reorganize into a coral-dominated state in the face of disturbance (Bellwood et al. 2004).

The biological sources of renewal and reorganization for ecosystem resilience consist of functional groups of biological legacies and mobile link species and their support areas in the larger landscape or seascape. For example, large trees serve as biological legacies after fire and storms in forest ecosystems (Elmqvist et al. 2001, Franklin & MacMahon 2000), and seed banks and vegetative propagules play the same role in tundra ecosystems (Vavrek et al. 1999). Mobile link species connect habitats, sometimes widely separated in space and time (Lundberg & Moberg 2003). For example, vertebrates that eat fruit, such as flying foxes, play a key functional role in the regeneration of tropical forests hit by disturbances such as hurricanes and fire by bringing in seeds from sur- rounding ecosystems that result in renewal and reorganization (Cox et al. 1991, Elmqvist et al. 2001). The functional group of grazers on coral reefs connect a wide range of spatial scales from centimeters, such as amphipods and sea urchins, to thousands of kilometers, such as green turtles (Elmqvist et al. 2003). By operating at different spatial and tem- poral scales, competition among species within the guild of grazers is minimized, and the robustness over a wider range of environmental conditions is enhanced (Peterson et al. 1998). Such response diversity plays a significant role in the capacity of ecosystems to renew and reor- ganize into desired states after disturbance.

Metapopulation analyses have largely focused on dispersal, connectiv- ity, recovery, and life history of species, populations, and communities. Great potential lies in redirecting this knowledge into addressing the role of functional groups and response diversity in ecosystem resilience that considers the central role of human actors in this context.

A number of observations suggest that biodiversity at larger spatial scales (i.e., landscapes and regions) ensures that appropriate key species for ecosystem functioning are recruited to local systems after distur- bance or when environmental conditions change (Bengtsson et al. 2003, Nyström & Folke 2001, Peterson et al. 1998). The current emphasis on setting aside "hot spot" areas of diversity and protecting species richness in reserves will in itself not be a viable option in human-dominated environments (Folke et al. 1996). Present static reserves should be complemented with dynamic reserves, such as ecological fallows and dynamic successional reserves (Bengtsson et al. 2003), and serve as one important tool that contributes to sustaining the configuration of functional groups and response diversity required for renewing and reorganizing desired ecosystem states after disturbance. In this sense, biological diversity provides insurance, flexibility, and risk spreading across scales in dynamic landscapes and seascapes.

Hence, spatial and temporal relations of functional groups that renew and help reorganize ecosystem development after disturbance, and their response diversity, will influence the ability of ecosystems to remain within desired states.

Managing Resilience for Development

Archaeological research indicates that over time human societies have degraded the capacity of ecosystems to sustain societal development (Redman 1999, van der Leeuw 2000). Historical overfishing of coastal waters has created simplified, leaky, and weedy ecosystems that rapidly respond to external influences in an unpredictable fashion (Jackson et al. 2001). A shifting baseline, an incremental lowering of environmental standards over time (Pauly 1995), may occur, and each new human generation may adapt to the new conditions of less diverse and less productive ecosystems.

Our review clearly illustrates that regime shifts in ecosystems are, to a large extent, driven by human actions. A combination of top-down impacts, such as fishing down food webs and losing response diversity and functional groups of species, and bottom-up impacts, such as accumulation of nutrients, soil erosion, or redirection of water flows, as well as altered disturbance regimes, such as suppression of fire and increased frequency and intensity of storms, have shifted several ecosystems into less desired states with diminished capacities to generate ecosystem services.

Shifts from desired to less desired states may often follow gradual loss of ecosystem resilience. Resilience has multiple attributes, but four aspects are critical (Walker et al. 2004):

1. *Latitude* is the maximum amount the system can be changed before losing its ability to reorganize within the same state;

basically it is the width of the stability domain or the basin of attraction.

2. *Resistance* is the ease or difficulty of changing the system; deep basins of attraction indicate that greater disturbances are required to change the current state of the system away from the attractor.

3. *Precariousness* is how close the current trajectory of the system is to a threshold that, if breached, makes reorganization difficult or impossible.

4. *Cross-scale relations* (i.e., panarchy) is how the above three attributes are influenced by the states and dynamics of the (sub)systems at scales above and below the scale of interest.

Ecosystem management of resilience, biodiversity, and regime shifts needs to address those attributes. Such an initiative will require adaptability among the actors involved in ecosystem management (Berkes et al. 2003). Adaptability is the capacity of actors in a system to manage resilience in the face of uncertainty and surprise (Gunderson & Holling 2002). Humans are a part of, and not apart from, the trajectory and stability domain of the system and will, to a large extent, determine their own paths through management of the ecosystem. Human actors can (*a*) move thresholds away from or closer to the current state of the system by altering latitude, (*b*) move the current state of the system away from or closer to the threshold by altering precariousness, or (*c*) make the threshold more difficult or easier to reach by altering resistance. Actors can also manage cross-scale interactions to avoid or instigate loss of resilience at larger and more catastrophic scales (Holling et al. 1998).

Human actions have often altered slowly changing ecological variables, such as soils or biodiversity, with disastrous social consequences that did not appear until long after the ecosystems were first affected. A current major problem in this context is the large-scale salinization of land and rivers in Australia. About 5.7 million hectares are currently at risk for dryland salinity, and the amount of land at risk could rise to over 17 million hectares by 2050. Extensive land clearing during the past 200 years has removed native woody vegetation to make way for agricultural crops and pasture grasses that transpire much less water. Thus, more water is infiltrating the soils and causing groundwater tables to rise.

The rising water mobilizes salts and causes problems with salinity both in rivers and in the soils, which severely reduces the capacity for plant growth (Gordon et al. 2003). Increased vulnerability, as a consequence of loss of resilience, places a region on a trajectory of greater risk of the panoply of stresses and shocks that occur over time (Kasperson et al. 1995).

Most semiarid ecosystems have suffered from severe overexploitation by excessive grazing and agriculture that resulted in depletion of vegetation biomass and soil erosion. These changes are often difficult to reverse because of positive feedbacks that stabilize the new situation. According to one hypothesis, rainy periods associated with El Niño can be used in combination with grazer control to restore degraded ecosystems (Holmgren & Scheffer 2001). Removing grazers to regenerate vegetation during normal years will not be sufficient, because conditions are too dry. Also, wet years do not allow regeneration if grazers remain present. However, removing grazers during a wet year pulse may tip the balance and allow reorganization into a more desired state, and this pulse management may be sufficient for the state to remain intact, subject to grazing. Clearly, responding to El Niño as an opportunity for shifting an ecosystem back to the desired state demands a highly responsive social system, organized for rapid and flexible adaptation (Berkes et al. 2003).

At times, human societies or groups may find themselves trapped in an undesired basin of attraction that is becoming so wide and so deep that movement to a new basin, or sufficient reconfiguration of the existing basin, becomes extremely difficult. A major challenge in ecosystem management is to develop social and ecological capacity to transform such an undesired basin into a fundamentally new and more desirable configuration, a new stability landscape defined by different state variables or old state variables supplemented by new ones. We call this challenge transformability, that is, the capacity to create untried beginnings from which to evolve a fundamentally new way of living when existing ecological, economic, and social conditions make the existing system untenable (Walker et al. 2004). The new way of living requires social-ecological resilience to cope with future change and unpredictable events (Olsson et al. 2004).

Resilience-building management needs to be flexible and open to learning. It attends to slowly changing, fundamental variables such as experience, memory, and diversity in both social and ecological systems

(Folke et al. 2003). The crucial slow variables that determine the underlying dynamic properties of the system and that govern the supply of essential ecosystem services need to be identified and assessed. The processes and drivers that determine the dynamics of this set of crucial variables need to be identified and assessed. The role of biological diversity in ecosystem functioning and response to change should be explicitly accounted for in this context and acknowledged in resilience-building policies.

Conclusions

Ecosystems can be subject to sharp regime shifts. Such shifts may more easily occur if resilience has been reduced as a consequence of human actions. Human actions may cause loss of resilience through the following methods:

- removal of functional groups of species and their response diversity, such as the loss of whole trophic levels (top-down effects),
- impact on ecosystems via emissions of waste and pollutants (bottom-up effects) and climate change, and
- alteration of the magnitude, frequency, and duration of disturbance regimes to which the biota is adapted.

Loss of resilience through the combined and often synergistic effects of those pressures can make ecosystems more vulnerable to changes that previously could be absorbed. As a consequence, they may suddenly shift from desired to less desired states in their capacity to sustain ecosystem services to society. In some cases, these shifts may be irreversible (or too costly to reverse). Irreversibility is a reflection of changes in variables with long turnover times (e.g., biogeochemical, hydrological, or climatic) and loss of biological sources and interactions for renewal and reorganization into desired states.

In light of these changes and their implication for human well-being, the capacities for self-repair of ecosystems can no longer be taken for granted. Active adaptive management and governance of resilience will be required to help sustain or create desired states of ecosystems. A first step in this direction is to understand better the interactions between regime shifts, biological diversity, and ecosystem resilience.

Acknowledgments

This work is a product of the Resilience Alliance and has been supported by a grant from the JS McDonnell Foundation. In addition, the work of Carl Folke and Thomas Elmqvist is partly funded by grants from the Swedish Research Council for the Environment, Agricultural Sciences and Spatial Planning (Formas). The Beijer Institute has supported the collaboration with the review article.

Literature Cited

Algeo TJ, Scheckler SE. 1998. Terrestrial marine teleconnections in the Devonian: Links between the evolution of land plants, weathering processes, and marine anoxic events. *Philos. Trans. R. Soc. London Ser. B* 353:113–28.

Anderies JM, Janssen MA, Walker BH. 2002. Grazing management, resilience, and the dynamics of a fire-driven rangeland system. *Ecosystems* 5:23–44.

Barkai A, McQuaid C. 1988. Predator-prey role reversal in a marine benthic ecosystem. *Science* 242:62–64.

Beisner BE, Haydon DT, Cuddington K. 2003. Alternative stable states in ecology. *Front. Ecol. Environ.* 1:376–82.

Bellwood DR, Hughes TP, Folke C, Nyström M. 2004. Confronting the coral reef crisis. *Nature* 429:827–33.

Bengtsson J, Angelstam P, Elmqvist T, Emanuelsson U, Folke C, et al. 2003. Reserves, resilience, and dynamic landscapes. *Ambio* 32:389–96.

Berger R, Henriksson E, Kautsky L, Malm T. 2003. Effects of filamentous algae and deposited matter on the survival of *Fucus vesiculosus* L. germlings in the Baltic Sea. *Aquat. Ecol.* 37:1–11.

Berkes F, Colding J, Folke C, eds. 2003. *Navigating Social-Ecological Systems: Building Resilience for Complexity and Change.* Cambridge: Cambridge Univ. Press.

Bisigato AJ, Bertiller MB. 1997. Grazing effects on patchy dryland vegetation in Northern Patagonia. *J. Arid Environ.* 36:639–53.

Blindow I, Anderson G, Hargeby A, Johansson S. 1993. Long-term pattern of alternative stable states in two shallow eutrophic lakes. *Freshw. Biol.* 30:159–67.

Bonan GB, Pollard D, Thompson SL. 1992. Effects of boreal forest vegetation on global climate. *Nature* 359:716–18.

Brown JH, Valone TJ, Curtin CG. 1997. Reorganization of an arid ecosystem in response to recent climate change. *Proc. Natl. Acad. Sci. USA* 94:9729–33.

Caraco NF, Cole JJ, Likens GE. 1991. A cross-system study of phosphorus release from lake sediments. In *Comparative Analysis of Ecosystems*, ed. J Cole, G Lovett, S Findlay, pp. 241–58. New York: Springer-Verlag.

Carpenter SR. 2003. *Regime Shifts in Lake Ecosystems: Pattern and Variation*. Excellence in Ecology Series 15. Oldendorf/Luhe, Germany: Ecol. Inst.

Carpenter SR, Cole JJ, Hodgson JR, Kitchell JF, Pace ML, et al. 2001a. Trophic cascades, nutrients, and lake productivity: whole-lake experiments. *Ecol. Monogr.* 71:163-86.

Carpenter SR, Ludwig D, Brock WA. 1999. Management of eutrophication for lakes subject to potentially irreversible change. *Ecol. Appl.* 9:751-71.

Carpenter SR, Walker B, Anderies JM, Abel N. 2001b. From metaphor to measurement: Resilience of what to what? *Ecosystems* 4:765- 81.

Chapin FS, Walker BH, Hobbs RJ, Hooper DU, Lawton JH, et al. 1997. Biotic control over the functioning of ecosystems. *Science* 277:500-4.

Chapin FS, Zavaleta ES, Eviner VT, Naylor RL, Vitousek PM, et al. 2000. Consequences of changing biodiversity. *Nature* 405:234-42.

Cooke GD, Welch EB, Peterson SA, Newroth PR. 1993. *Restoration and Management of Lakes and Reservoirs*. Boca Raton, FL: Lewis.

Cox PA, Elmqvist T, Rainey EE, Pierson ED. 1991. Flying foxes as strong interactors in South Pacific Island ecosystems: a conservation hypothesis. *Conserv. Biol.* 5:448-54.

Danell K, Bergström R, Edenius L, Ericsson G. 2003. Ungulates as drivers of tree population dynamics at module and genet levels. *Forest Ecol. Manag.* 181:67-76.

Daskalov GM. 2002. Overfishing drives a trophic cascade in the Black Sea. *Mar. Ecol. Prog. Ser.* 225:53-63.

Davis SM. 1989. Sawgrass and cattail production in relation to nutrient supply in the Everglades. In *Fresh Water Wetlands & Wildlife, 9th Annual Symposium, Savannah River Ecology Laboratory*, 24-27 March 1986, ed. RR Sharitz, JW Gibbons, pp. 325-42. Charleston, SC: US Dep. Energy.

de Roos A, Persson L. 2002. Size-dependent life history traits promote catastrophic collapses of top predators. *Proc. Natl. Acad. Sci. USA* 99:12907-12.

Diaz S, Cabido M. 2001. Vive la difference: plant functional diversity matters to ecosystem processes. *Trends Ecol. Evol.* 16:646-55.

Done TJ. 1992. Phase shifts in coral reef communities and their ecological significance. *Hydrobiologia* 247:121-32.

Dublin HT, Sinclair ARE, McGlade J. 1990. Elephants and fire as causes of multiple stable states in the Serengeti-Mara woodlands. *J. Anim. Ecol.* 59:1147-64.

Duffy JE. 2002. Biodiversity and ecosystem function: the consumer connection. *Oikos* 99:201-19.

Eakin CM. 1996. Where have all the carbonates gone? A model comparison of calcium carbonate budgets before and after the 1982–1983 El Nino at Uva Island in the eastern Pacific. *Coral Reefs* 15:109-19.

Elmgren R. 1989. Man's impact on the ecosystems of the Baltic Sea: energy flows today and at the turn of the century. *Ambio* 18:326–32.

Elmgren R. 2001. Understanding human impact on the Baltic ecosystem: changing views in recent decades. *Ambio* 30:222–31.

Elmqvist T, Folke C, Nyström M, Peterson G, Bengtsson J, et al. 2003. Response diversity and ecosystem resilience. *Front. Ecol. Environ.* 1:488–94 .

Elmqvist T, Wall M, Berggren AL, Blix L, Fritioff S, Rinman U. 2001. Tropical forest reorganization after cyclone and fire disturbance in Samoa: remnant trees as biological legacies. *Conserv. Ecol.* 5:10. http://www.consecol.org/vol5/iss2/art10.

Eriksson BK, Johansson G. 2003. Sedimentation reduces recruitment success of *Fucus vesiculosus* (Phaeophyceae) in the Baltic Sea. *Eur. J. Phycol.* 38:217–22.

Flannery T. 1994. *The Future Eaters: An Ecological History of the Australasian Lands and People.* Sydney: Reed New Holland.

Foley JA, Coe MT, Scheffer M, Wang G. 2003. Regime shifts in the Sahara and Sahel: interactions between ecological and climatic systems in Northern Africa. *Ecosystems* 6:524–39.

Folke C. 2003. Freshwater and resilience: a shift in perspective. *Philos. Trans. R. Soc. London Ser. B* 358:2027–36.

Folke C, Carpenter SR, Elmqvist T, Gunderson L, Holling CS, Walker B. 2002. Resilience and sustainable development: building adaptive capacity in a world of transformations. *Ambio* 31:437–40.

Folke C, Colding J, Berkes F. 2003. Synthesis: building resilience and adaptive capacity in social-ecological systems. See Berkes et al. 2003, pp. 352–87.

Folke C, Holling CS, Perrings C. 1996. Biological diversity, ecosystems and the human scale. *Ecol. Appl.* 6:1018–24.

Franklin JF, MacMahon JA. 2000. Enhanced: messages from a mountain. *Science* 288:1183–90.

Frelich LE, Reich PB. 1999. Neighborhood effects, disturbance severity and community stability in forests. *Ecosystems* 2:151–66.

Frost TM, Carpenter SR, Ives AR, Kratz TK. 1995. Species compensation and complementarity in ecosystem function. In *Linking Species and Ecosystems*, ed. CG Jones, JH Lawton, pp. 224–39. New York: Chapman and Hall.

Glynn PW. 1988. El Nino warming, coral mortality and reef framework destruction by echinoid bioerosion in the eastern Pacific. *Galaxea* 7:129–60.

Gordon L, Dunlop M, Foran B. 2003. Land cover change and water vapour flows: learning from Australia. *Philos. Trans. R. Soc. London Ser. B* 358:1973–84.

Gunderson LH. 2000. Ecological resilience: in theory and application. *Annu. Rev. Ecol. Syst.* 31:425–39 .

Gunderson LH. 2001. Managing surprising ecosystems in southern Florida. *Ecol. Econ.* 37:371–78.

Gunderson LH, Holling CS, eds. 2002. *Panarchy: Understanding Transformations in Human and Natural Systems*. Washington, DC: Island Press.

Gunderson LH, Pritchard L, eds. 2002. *Resilience and the Behavior of Large-Scale Ecosystems*. Washington, DC: Island Press.

Higgins PAT, Mastrandrea MD, Schneider SH. 2002. Dynamics of climate and ecosystem coupling: abrupt changes and multiple equilibria. *Philos. Trans. R. Soc. London Ser. B* 357:647–55.

Hilborn R, Quinn TP, Schindler DE, Rogers DE. 2003. Biocomplexity and fisheries sustainability. *Proc. Natl. Acad. Sci.USA* 100:6564–68.

Holling CS. 1973. Resilience and stability of ecological systems. *Annu. Rev. Ecol. Syst.* 4:1–23.

Holling CS. 1978. The spruce-budworm/forest-management problem. In *Adaptive Environmental Assessment and Management*. International Series on Applied Systems Analysis, ed. CS Holling, 3:143–82. New York: John Wiley & Sons.

Holling CS. 1986. The resilience of terrestrial ecosystems: local surprise and global change. In *Sustainable Development of the Biosphere*, ed. WC Clark, RE Munn, pp. 292–317. Cambridge: Cambridge Univ. Press.

Holling CS. 1996. Engineering resilience versus ecological resilience. In *Engineering within Ecological Constraints*, ed. PC Schulze, pp. 31–44. Washington, DC: Natl. Acad. Press.

Holling CS, Berkes F, Folke C. 1998. Science, sustainability, and resource management. In *Linking Social and Ecological Systems: Management Practices and Social Mechanisms for Building Resilience*, ed. F Berkes, C Folke, pp. 342–62. Cambridge: Cambridge Univ. Press.

Holling CS, Meffe GK. 1996. Command and control and the pathology of natural resource management. *Conserv. Biol.* 10:328–37.

Holmgren M, Scheffer M. 2001. El Niño as a window of opportunity for the restoration of degraded arid ecosystems. *Ecosystems* 4:151–59.

Hughes TP. 1994. Catastrophes, phase shifts, and large-scale degradation of a Caribbean coral reef. *Science* 265:1547–51.

Hughes TP, Baird AH, Bellwood DR, Card M, Connolly SR, et al. 2003. Climate change, human impacts, and the resilience of coral reefs. *Science* 301:929–33.

Jackson JBC, Kirb MX, Berher WH, Bjorndal KA, Botsford LW, et al. 2001. Historical overfishing and the recent collapse of coastal ecosystems. *Science* 293:629–38.

Jackson LJ. 2003. Macrophyte-dominated and turbid states of shallow lakes: evidence from Alberta lakes. *Ecosystems* 6:213–23.

Jansson B-O, Jansson AM. 2002. The Baltic Sea: Reversibly unstable or irreversibly stable? See Gunderson & Pritchard 2002, pp. 71–109.

Jeppesen E, Sondergaard M, Sondergaard M, Christofferson K, eds. 1998. *The Structuring Role of Submerged Macrophytes in Lakes*. Berlin: Springer-Verlag.

Kautsky N, Kautsky H, Kautsky U, Waern M. 1986. Decreased depth penetration of *Fucus vesiculosus* (L.) since the 1940s indicates eutrophication of the Baltic Sea. *Mar. Ecol. Prog. Ser.* 28:1–8.

Kasperson JX, Kasperson RE, Turner BL. 1995. *Regions at Risk: Comparisons of Threatened Environments.* New York: United Nations University Press.

Kelly RD, Walker BH. 1976. The effects of different forms of land use on the ecology of a semi-arid region in south-eastern Rhodesia. *J. Ecol.* 64:553–76.

Knowlton N. 1992. Thresholds and multiple stable states in coral reef community dynamics. *Am. Zool.* 32:674–82.

Knowlton N. 2001. The future of coral reefs. *Proc. Natl. Acad. Sci. USA* 98:5419–25.

Konar B, Estes JA. 2003. The stability of boundary regions between kelp beds and deforested areas. *Ecology* 84:174–85.

Levin S. 1999. *Fragile Dominion: Complexity and the Commons.* Reading, MA: Perseus Books.

Luck GW, Daily GC, Ehrlich PR. 2003. Population diversity and ecosystem services. *Trends Ecol. Evol.* 18:331–36.

Ludwig D, Jones DD, Holling CS. 1978. Qualitative analysis of insect outbreak systems: Spruce-budworm and forest. *J. Anim. Ecol.* 47:315–32.

Ludwig D, Mangel M, Haddad B. 2001. Ecology, conservation, and public policy. *Annu. Rev. Ecol. Syst.* 32:481–517.

Ludwig D, Walker B, Holling CS. 1997. Sustainability, stability, and resilience. *Conserv. Ecol.* 1:7. http://www.consecol.org/vol1/iss1/art7.

Lundberg J, Moberg F. 2003. Mobile link organisms and ecosystem functioning: implications for ecosystem resilience and management. *Ecosystems* 6:87–98.

McClanahan TR. 2002. The near future of coral reefs. *Environ. Conserv.* 29:460–83.

McCook LJ. 1999. Macroalgae, nutrients and phase shifts on coral reefs: scientific issues and management consequences for the Great Barrier Reef. *Coral Reefs* 18:357–67.

Naeem S, Wright JP. 2003. Disentangling biodiversity effects on ecosystem functioning: deriving solutions to a seemingly insurmountable problem. *Ecol. Lett.* 6:567–79.

Newman S, Schuette J, Grace JB, Rutchey K, Fontaine T, et al. 1998. Factors influencing cattail abundance in the northern Everglades. *Aquat. Bot.* 60:265–80.

Norberg J, Swaney DP, Dushoff J, et al. 2001. Phenotypic diversity and ecosystem functioning in changing environments: a theoretical framework. *Proc. Natl. Acad. Sci. USA* 98:11376–81.

Nürnberg GK. 1995. Quantifying anoxia in lakes. *Limnol. Oceanogr.* 40:1100–11.

Nyström M, Folke C. 2001. Spatial resilience of coral reefs. *Ecosystems* 4:406–17.

Nyström M, Folke C, Moberg F. 2000. Coral reef disturbance and resilience in a human-dominated environment. *Trends Ecol. Evol.* 15:413–17.

O'Dowd DJ, Green PT, Lake PS. 2003. Invasional 'meltdown' on an oceanic island. *Ecol. Lett.* 6:812–17.

Olsson P, Folke C, Hahn T. 2004. Social-ecological transformation for ecosystem management: the development of adaptive co-management of a wetland landscape in southern Sweden. *Ecol. Soc.* 9(4):2. http:// www.ecologyandsociety.org/vol9/iss4/art2.

Pace ML, Cole JJ, Carpenter SR, Kitchell JF. 1999. Trophic cascades revealed in diverse ecosystems. *Trends Ecol. Evol.* 14:483–88.

Paine RT, Tegner MJ, Johnson EA. 1998. Compounded perturbations yield ecological surprises. *Ecosystems* 1:535–45.

Pauly D. 1995. Anecdotes and the shifting baseline syndrome of fisheries. *Trends Ecol. Evol.* 10:430.

Peterson GD, Allen CR, Holling CS. 1998. Ecological resilience, biodiversity, and scale. *Ecosystems* 1:6–18.

Peterson GD. 2002. Forest dynamics in the Southeastern United States: managing multiple stable states. See Gunderson & Pritchard 2002, pp. 227–46.

Post JR, Sullivan M, Cox S, Lester NP, Walters CJ, et al. 2002. Canada's recreational fisheries: the invisible collapse? *Fisheries* 27:6–17.

Redman CL. 1999. *Human Impact on Ancient Environments.* Tucson, AZ: Univ. Arizona Press.

Scheffer M. 1997. *The Ecology of Shallow Lakes.* London: Chapman and Hall.

Scheffer M, Carpenter SR. 2003. Catastrophic regime shifts in ecosystems: linking theory to observation. *Trends Ecol. Evol.* 18:648–56.

Scheffer M, Carpenter SR, Foley J, Folke C, Walker BH. 2001. Catastrophic shifts in ecosystems. *Nature* 413:591–96.

Scheffer M, Hosper SH, Meijer ML, Moss B, Jeppesen E. 1993. Alternative equilibria in shallow lakes. *Trends Ecol. Evol.* 8:275–79.

Scheffer M, Szabo S, Gragnani A, van Nes EH, Rinaldi S, et al. 2003. Floating plant dominance as a stable state. *Proc. Natl. Acad. Sci. USA* 100:4040–45.

Scholes RJ, Walker BH. 1993. *Nylsuley: The Study of an African Savanna.* Cambridge: Cambridge Univ. Press.

Smith VH. 1998. Cultural eutrophication of inland, estuarine and coastal waters. In *Successes, Limitations and Frontiers of Ecosystem Science*, eds. ML Pace, PM Groffman, pp. 7–49. New York: Springer-Verlag.

Springer AM, Estes JA, van Vliet GB, Williams TM, Doak DF, et al. 2003. Sequential megafaunal collapse in the North Pacific Ocean: an ongoing legacy of industrial whaling? *Proc. Natl. Acad. Sci. USA* 100:12223–28.

Srivastava DS, Jefferies RL. 1995. Mosaics of vegetation and soil salinity: a consequence of goose foraging in an arctic salt marsh. *Can. J. Bot.* 73:75–83.

Stachowicz JJ, Whitlach RB, Osman RW. 1999. Species diversity and invasion resistance in marine ecosystems. *Science* 286:1577–79.

Steele JH. 1998. Regime shifts in marine ecosystems. *Ecol. Appl.* 8:S33–S36.

Steele JH. 1996. Regime shifts in fisheries management. *Fish. Res.* 25:19–23.

Steffen W, Sanderson A, Jäger J, Tyson PD, Moore B III, et al. 2004. *Global Change and the Earth System: A Planet Under Pressure.* Heidelberg: Springer-Verlag. 336 pp.

Steneck RS, Graham MH, Bourque BJ, Corbett D, Erlandson JM, et al. 2002. Kelp forest ecosystems: biodiversity, stability, resilience and future. *Environ. Conserv.* 29:436–59.

Terborgh J, Lopez L, Nunez P, Rao M, Shahabuddin G, et al. 2001. Ecological meltdown in predator-free forest fragments. *Science* 294:1923–26.

Trenbath BR, Conway GR, Craig IA. 1989. Threats to sustainability in intensified agricultural systems: analysis and implications for management. In *Agroecology: Researching the Ecological Basis for Sustainable Agriculture,* ed. SR Gliessman, pp. 337–65. Berlin: Springer-Verlag.

Van Cappellen P, Ingall ED. 1994. Benthic phosphorus regeneration, net primary production, and ocean anoxia: a model of the coupled marine biogeochemical cycles of carbon and phosphorus. *Paleoceanography* 9:677–92.

van de Koppel J, Rietkerk M, Weissing FJ. 1997. Catastrophic vegetation shifts and soil degradation in terrestrial grazing systems. *Trends Ecol. Evol.* 12:352–56.

van der Leeuw S. 2000. Land degradation as a socionatural process. In *The Way the Wind Blows: Climate, History, and Human Action,* eds. RJ McIntosh, JA Tainter, SK McIntosh, pp. 357–83, New York: Columbia Univ. Press.

Vavrek MC, Fetcher N, McGraw JB, Shaver GR, Chapin FS, Bovard B. 1999. Recovery of productivity and species diversity in Tussock tundra following disturbance. *Arct. Antarct. Alp. Res.* 31:254–58.

Vitousek PM, Walker LR, Whiteaker LD, Muellerdombois D, Matson PA. 1987. Biological invasion by *Myrica-Faya* alters ecosystem development in Hawaii. *Science* 238:802–4.

Walker BH, Holling CS, Carpenter SR, Kinzig AS. 2004. Resilience, adaptability and transformability. *Ecol. Soc.* In press.

Walker BH, Kinzig A, Langridge J. 1999. Plant attribute diversity, resilience, and ecosystem function: the nature and significance of dominant and minor species. *Ecosystems* 2:95–113.

Walker BH, Meyers JA. 2004. Thresholds in ecological and social-ecological systems: a developing database. *Ecol. Soc.* 9(2):3.

Walters CJ, Kitchell JF. 2001. Cultivation/depensation effects on juvenile survival and recruitment: implications for the theory of fishing. *Can. J. Fish. Aquat. Sci.* 58:1–12.

Wang GL, Eltahir EAB. 2000. Role of vegetation dynamics in enhancing the low-frequency variability of the Sahel rainfall. *Water Resour. Res.* 36:1013–21.

Wilson JB, Agnew ADQ. 1992. Positive feedback switches in plant communities *Adv. Ecol. Res.* 23:263–336.

Worm B, Lotze H, Boström C, Engkvist R, Labanauskas V, Sommer U. 1999. Marine diversity shift linked to interactions among grazers, nutrients and propagule banks. *Mar. Ecol. Prog. Ser.* 185:309–14.

Worm B, Lotze H, Hillebrand H, Sommer U. 2002. Consumer versus resource control of species diversity and ecosystem functioning. *Nature* 417:848–51.

Zaitsev Y, Mamaev V. 1997. *Marine Biological Diversity in the Black Sea: A Study of Change and Decline.* NewYork: United Nations Publications.

Zimov SA, Chuprynin VI, Oreshko AP, Chapin FS, Reynolds JF, Chapin MC. 1995. Steppe-tundra transition: a herbivore-driven biome shift at the end of the Pleistocene. *Am. Nat.* 146:765–94.

Biological Diversity, Ecosystems, and the Human Scale

CARL FOLKE, C. S. HOLLING, AND CHARLES PERRINGS

CARL FOLKE, *The Beijer International Institute of Ecological Economics,
The Royal Swedish Academy of Sciences, P.O. Box 50005, S-10405
Stockholm, Sweden, and Department of Systems Ecology, Stockholm
University, Stockholm, Sweden.*

C. S. HOLLING, *Department of Zoology, University of Florida, Gainesville,
Florida 32611, USA.*

CHARLES PERRINGS, *Department of Environmental Economics and
Environmental Management, University of York, Heslington, York YO1
5DD, United Kingdom.*

Ecological Applications, Vol. 6, No. 4 (Nov., 1996), pp. 1018–1024.

Reproduced by permission of The Ecological Society of America.

Abstract

THIS PAPER CONSIDERS the significance of biological diversity in relation to large-scale processes in complex and dynamic ecological–economic
systems. It focuses on functional diversity, and its relation to production

and maintenance of ecological services that underpin human societies. Within functional groups of organisms two important categories of species are identified: keystone process species and those essential for ecosystem resilience. The latter group represents "natural insurance capital." In addition to basic research on the interplay among biological diversity, functional performance, and resilience in complex self-organizing systems, we suggest that a functional approach has two main implications for a strategy for biodiversity conservation: (1) Biodiversity conservation to assure the resilience of ecosystems is required for all systems, no matter how heavily impacted they are. It should not be limited to protected areas. (2) The social, cultural, and economic driving forces in society that cause biodiversity loss need to be addressed directly. Specifically, (a) differences between the value of biological diversity to the private individual and its fundamental value to society as a whole need to be removed; (b) social and economic policies that encourage biodiversity loss should be reformed, especially where there is a risk of irreversible damage to ecosystems and diversity; and (c) institutions that are adaptive and work in synergy with ecosystem processes and functions are critical and should be created at all levels.

Key words: biodiversity; biological diversity, conservation of; critical ecosystem processes; disturbance, capacity to buffer; ecological services, maintenance of; ecosystem function and resilience; functional diversity; market externalities; multiple equilibria; nature reserves; transboundary effects.

Introduction

It is generally accepted that the significance of biodiversity includes much more than the mix of species. Wilson (1992:393) defines it to include ". . . the variety of ecosystems, which comprise both the communities of organisms within particular habitats and the physical conditions under which they live." This is also the position taken by the Convention on Biological Diversity of the United Nations Conference on Environment and Development in Rio de Janeiro 1992, signed by more than 150 nations (UNEP 1992).

Despite this, much research and policy on biodiversity conservation assumes that what matters is the number of genetically distinct organisms. There has been a tendency for biologists to estimate extinction rates, stress the importance of the abundance of taxa, and discuss the potential for preservation of genetic information (Olney et al. 1993, Prendergast et al. 1993, Smith et al. 1993). This approach has led to a conservation strategy dominated by the establishment of protected reserves in the megadiversity regions of the world.

We strongly support the ethical argument for conserving the uniqueness and diversity of life. However, we question whether the focus on genotypic diversity and megadiversity hot-spots is the most appropriate way to conceptualize, analyze, or respond to biodiversity loss. Nor is it clear to us that it actually satisfies the ethical goal. We are suggesting an approach that complements the conventional approach, but which has a different focus. We will discuss biological diversity in large-scale ecological processes, and the intensifying human driving forces behind its loss. Specifically, we will analyze biodiversity loss in terms of its impact on the ability of interdependent ecological–economic systems to maintain functionality under a range of environmental conditions. Human demographic, social, cultural, and economic trends are not seen as external to ecosystems, but as parts of the biogeochemical and hydrological flows of the ecosphere. That is, we take an ecological–economics approach.

Whether we like it or not the growing human impact on the planet is a fact. "Keeping humans out of nature" through a protected-area strategy may buy time, but it does not address the factors in society driving the loss of biodiversity. This paper stresses the importance of creating incentives for people and economies to act more in harmony than in conflict with essential processes that control the dynamics and structure of ecosystems, and of which biological diversity is a crucial part. Incentives need to be created to conserve biodiversity not just in protected areas but everywhere.

The approach requires that the functional relationship between the diversity of organisms and the set of ecological services on which humanity depends is addressed (Ehrlich and Mooney 1983, Perrings et al. 1992, 1995a). In addition to the fact that they house the genetic library, organisms help to sustain a flow of ecological services that are prerequisites for economic activities. These include photosynthesis, provision of food and other renewable resources, soil generation and preservation, pollination of crops, recycling of nutrients, filtering of pollutants and waste assimilation, flood control, climate moderation, operation of the hydrological cycle, and maintenance of the gaseous composition of the atmosphere. These functions sustain and protect human activities, and so human well being (Folke 1991, de Groot 1992, Ehrlich and Ehrlich 1992). In economic terms, ecosystems are fundamental "factors of production"—factors that are becoming increasingly scarce as a consequence of the rapid human population growth, and human behavior towards the natural capital base (Barbier et al. 1994, Jansson et al. 1994).

Increasing globalization of human activities and large-scale movements of people mean that humankind is in an era of novel co-evolution of ecological and socioeconomic systems at regional and even planetary scales (Holling 1994). Ensuring the capacity of ecosystems to continue to generate ecological services on which the well being of human societies depends is a major challenge. Developing a feasible and useful strategy for biodiversity conservation is a central component of this challenge.

Functional Diversity in Large-Scale Ecosystems

The critical aspect of the diversity of organisms and their environments in this context is functional diversity (Schulze and Mooney 1993). We know that for any ecosystem function to be sustained, a minimum composition of organisms is required to develop the relations between primary producers, consumers, and decomposers that mediate the flow of energy, the cycling of elements, and spatial and temporal patterns of vegetation. We also know that this composition varies with the environmental conditions in which the system operates. What is currently missing is detailed knowledge of the critical thresholds of diversity associated with different environmental conditions at different temporal and spatial scales. It is not clear, for example, how specific sets of genes, genotypes, species, populations, and communities influence ecosystem functions over the existing range of environmental conditions, or what the critical levels of diversity in communities and ecosystems and the factors that control them are (Solbrig 1991, Schulze and Mooney 1993). Nor is it clear how things would change with a change in environmental conditions. To a large extent this is a consequence of the fact that organisms and their environments are connected by a complex web of interrelations and feedbacks that are non-linear, and contain lags and discontinuities, thresholds, and limits (Kay 1991, Costanza et al. 1993).

Recent small-scale, multi-species experiments indicate that species deletion under given environmental conditions may lead to loss of function (Naeem et al. 1994, Tilman and Downing 1994). Such experiments have been undertaken in 1–4 m² plots or enclosures and contain species that live their life at those scales. It is not self-evident that results from small assemblages of species on small scales are valid for large-scale ecosystems.

Empirical findings from large-scale ecosystem studies of lakes, forests, marine and savanna ecosystems indicate that there are differences in the link between species and functional diversity and the generation of eco-system services. Studies of natural, managed, and disturbed or impacted ecosystems have shown that individual species population dynamics are more sensitive to stress than are ecosystem processes (Schindler 1990, Vitousek 1990). This implies that an ecosystem under stress may be expected to keep more of its functional performance than its species composition (Holling et al. 1995). This functional robustness is based on evidence that a relatively few processes, having distinct frequencies in space and time, structure ecosystems and set the rhythm of ecosys-tem dynamics (Holling 1992), a pattern that seems to be particularly true for terrestrial ecosystems (Holling et al. 1995). A limited number of organisms and groups of organisms seem to drive or control the criti-cal processes necessary for ecosystem functioning, while the remaining organisms exist in the niches formed by these keystone process species. Such organisms modify, maintain, and create habitats. Jones et al. (1994) refer to them as "ecosystem engineers." (We prefer the term "keystone process species," to avoid confusion with the field in applied ecology called "ecological engineering" [e.g., Mitsch and Jörgensen 1989].) Nor is the set of such species necessarily constant over time, since it depends on environmental conditions (Lawton and Brown 1993, Holling et al. 1995).

As ecosystems are complex self-organizing systems, they are char-acterized by multiple locally stable equilibria or persistent states, each of which may correspond to a distinct set of environmental conditions, and may be controlled by a distinct set of keystone process species (Schneider and Kay 1994, Perrings et al. 1995a). But keystone process species alone will not guarantee the continuation of the ecosystem in question. It is in this context that the concept of ecosystem resilience becomes crucial in biodiversity conservation. "Resilience," as the term is conventionally used in ecology, refers to resistance to disturbance and speed of return to a stable equilibrium state (Pimm 1984)—what might be termed "efficiency of function." Following the work of Holling (1973), we use resilience to capture the existence of function in systems that, at any given moment, will be away from any one of a number of locally stable equilibrium states. Resilience in this sense is a measure of the perturbation that can be absorbed before an ecosystem in the attractor domain of one equilibrium state is dislodged into that of another equilibrium state. It is the capacity of the system to buffer disturbance. As an example of multiple equilibrial states, forest land cleared for agriculture in Amazonia, once abandoned, tends to develop into grassland or savanna and may not revert to the original forest. Different sets of interacting physical and biological processes and organisms control ecosystem functioning in each of the equilibrium states. Such phenomena have been documented in savanna ecosystems, and are also observed in boreal forests, fisheries, and agriculture (Regier and Baskerville 1986, Westoby et al. 1989, Trenbath et al. 1990, Gunderson et al. 1995).

The significance of functional diversity involving the controlling species in such cases may be illustrated by the semi-arid grasslands of eastern and southern Africa. There are (at least) two equilibrium states in these ecosystems, one dominated by grasses, and one by shrubs. The grassland state is controlled by two groups of grasses, each of which has different functions. One group contains species that are tolerant to grazing and drought, with the capacity to hold soil and water because of deep roots. The second group is more productive in terms of plant biomass than the first group, which makes it attractive to grazers, but is less drought tolerant. The species of the second group have a competitive advantage over the first group during periods when grazing is less intensive, and when rainfall is higher. The reverse occurs during transient but intense grazing pulses of migrating herbivores like zebras or antelopes. In this way, a diversity of both types of grass species is maintained in a manner that assures that the ecological functions underpinning ecosystem productivity are preserved, over a range of climatic conditions.

When fixed management rules are applied, such as the stocking of ranched cattle at a moderate but "sustained" level, the functional diversity of the system is gradually lost. This is due to a shift from the natural intense pulses of grazing to more modest but persistent grazing pressure. Continuous moderate grazing shifts the competitive advantage to the productive group of grass species, at the expense of the drought-resistant grasses. As a consequence, it slowly reduces diversity to one type of function, with the consequence that the system loses its capacity to function under as wide a range of climatic conditions. That is, the resilience contracts. An episode of drought that previously could be absorbed can flip the ecosystem into a state that is dominated and controlled by woody shrubs of low value for cattle ranching (Perrings and Walker 1995).

This sort of behavior is also observed in other managed systems (Ludwig et al. 1993, Gunderson et al. 1995). Many ecosystems evolve through management to become more spatially uniform, less functionally diverse, and more sensitive to disturbances that otherwise could have been absorbed. They have lost resilience.

Biological Diversity as Insurance

The problem to be addressed in developing a strategy for biodiversity conservation is that current institutions in society, including markets, do not respond to environmental feedbacks (Berkes and Folke 1994). That is, many of the most important environmental effects of human behavior are not recognized in the set of market prices. They are external to the market. The implication of this is that individual users of biological resources will not take the true cost of their actions into account. The problem is frequently exacerbated by governmental policies that, by subsidizing users of biological resources, deepen the wedge between the private and social cost of their behavior. Indeed, the failure of markets and the inability of government policies to correct the failure of markets can be seen as the prime driving force behind the loss of biodiversity.

The problem is also exacerbated by the fact that ecosystems themselves often "fail to signal" the long-term consequences of loss of resilience, continuing to function in the short term even as resilience declines. That is, large-scale ecosystems continue to function even when the composition and number of organisms is reduced. It seems like ecosystems frequently signal loss of resilience only at the point at which external shocks, previously absorbed by a diversity of organisms with overlapping influences, now flip those systems into some other regime of behavior, as in the grassland case above (Holling et al. 1996).

An important ecological objective of a biodiversity conservation strategy to avoid such flips is the protection of those organisms that establish and maintain the niches formed by keystone process species. These species provide a buffer to rare and extreme events. The vulnerability of key structuring processes is a function of the number of organisms that can take over and run such processes when the system is perturbed (Holling et al. 1995). Such organisms are essential for ecosystem resilience. Their loss implies a reduction in ecosystem plasticity and capacity for self-organization and evolution, which, in turn, threatens the capacity of the system to produce valuable ecological services that are needed for human existence. These organisms can therefore be viewed as "natural insurance capital" for securing the generation of ecological services, both at present and in the future (Barbier et al. 1994).

The insurance function of biodiversity certainly includes its role in the genetic evolution of microbial, plant, animal, and human life. But it also includes the capacity of the ecosystems in which those reservoirs are contained to function under current and future environmental conditions. When loss of biodiversity reduces ecosystem resilience, it threatens the functions of that system, and hence the foundation for economic activity and human welfare (Perrings et al. 1995a). In most cases, however, the reduction in the "insurance value" of biodiversity is not signalled in the incentive structures of human society, including the price mechanism. Nor is it within the reach of the policies of any one government.

The problem may be illustrated by the link between migratory insectivorous bird populations and changes in insect outbreaks in boreal regions of Canada. A set of insectivorous birds is one of the controlling factors of the forest renewal patterns produced by budworm population cycles. Their existence contributes to the resilience of the boreal forest. Simulations based on long-term studies of budworm–forest systems dynamics indicate that the total bird population would have to be reduced by ≈75% before the system would flip to a different pattern of behavior (Holling 1988). A large proportion of the bird species spend the winter in Central America and parts of South America. Radar images of flights of migratory birds across the Gulf of Mexico over a roughly 20-yr period have revealed that the frequency of trans-Gulf flights has declined by almost 50% (Gauthreaux 1992), approaching the range of uncertainty in the simulation estimate above. Hence, in addition to regional forest fragmentation and its negative effects on nesting success of migratory birds (Robinson et al. 1995), Canadian boreal forests and the economic activities dependent upon their functioning seem to be threatened by increasing land-use pressures in neo-tropical countries and along the migration paths of insectivorous birds (Holling 1994).

Similarly, the widespread cutting of mangrove ecosystems in Southeast Asia and South America for shrimp farming causes the loss of resilience of these coastal ecosystems to provide spawning and nursery grounds for fish and shellfish. In this case the degradation takes place in the coastal area of one country, but can cause reduced or lost yields of adult fish harvested in feeding grounds that belong to other countries. In neither case is it possible for the government of the country that is most affected to address the problem directly (Barbier et al. 1994).

Biodiversity loss through ecosystem modification is a new form of international environmental impact, a new type of "transboundary pollution." Many of the ecosystem modifications and the environmental effects that they cause, which today are regarded as of only local or national concern, are in fact of regional and ultimately global concern. The web of connections linking one ecosystem with the next is intensifying across all scales in both space and time. Local human influences on air, land, and oceans slowly accumulate to trigger sudden abrupt changes when thresholds are reached, directly affecting the generation of ecological services and the vitality of human societies elsewhere. Local environmental problems may have their cause half a world away, as illustrated by the above examples. Everyone is now in everyone else's backyard.

Implications for Biodiversity Conservation

What implication does this have for a strategy of biodiversity conservation? In addition to basic research to increase our understanding of the interplay among biological diversity, functional performance, and resilience in complex and nested self-organizing systems, we suggest that there are two main implications.

First, preserving biodiversity through nature reserves and other protected areas is an important short-term step, which we endorse, but it is not sufficient to solve the problem of biodiversity loss. Nature reserves are embedded in their larger environments. Most reserves cannot alone deal with ecological attributes that cover larger scales, such as those discussed in this article. For example, wading birds in the Everglades National Park depend on areas more than 10 times the size of the park.

Conservation efforts should be planned at the scale of the regional landscape. Small reserves will lose their distinctive species if they are surrounded by a hostile landscape (Askins 1995). A hostile landscape is, generally, the result of increasing and intensifying human activities. Human influences at scales from the local ecosystem to the planet as a whole are becoming so extensive and pervasive that conditions that justified locating a protected area in one place originally may disappear over decades and move elsewhere. A biodiversity-conservation strategy that only focuses on preserving as much of the genetic library as possible in a few small areas may not be effective even in its own terms.

Second, we have argued for a strategy that aims at conserving the capacity of ecosystems to continue to deliver life-support and other ecological services to humanity under a wide range of environmental conditions. The role of biological diversity in the functioning of ecosystem performance is not limited to protected areas. Hence the conservation of biodiversity should be addressed everywhere. From this perspective, the heavily debated concept of "sustainability" might be interpreted as the maintenance of a level of biological diversity that will guarantee the resilience of ecosystems that sustain human societies. The goal of a conservation strategy should be to protect not all biodiversity in some areas, but biodiversity thresholds in all areas. Conserving essential self-organizing processes and the resilient patterns that they produce is the foundation for preservation of the biodiversity heritage (Holling et al. 1996). This approach has several implications.

1) To do this, it is necessary to identify the major social and economic forces that are currently driving the loss of functional diversity and to create incentives to redirect those forces. This includes both the proximate and the underlying forces (Perrings et al. 1992). The proximate forces refer to the direct reduction of biodiversity because of over-exploitation of species, land-use changes, and landscape fragmentation. The underlying forces include inappropriate government policies, the structure of property rights, pressure of human population growth and poverty, patterns of consumption and production, and the values of society. It is the underlying forces that need to be addressed in a strategy for biodiversity conservation (Barbier et al. 1994, Perrings et al. 1995b).

2) Following (1), it is necessary to create economic incentives that reduce the differences between the value of biological diversity to the private individual and its value to society as a whole. That is, it is necessary to internalize the external costs of biodiversity loss (in the language of economists). Internalizing only a few of the external costs of biodiversity loss is often enough to motivate the conservation of biodiversity.

3) A precondition for this is the development of effective institutions for biodiversity conservation. Institutions are the humanly devised constraints that shape human interaction. They structure incentives in human exchange, whether political, social, or economic, and shape the way societies evolve through time (North 1990). Institutions provide the framework for human actions, but to be effective they have to be adaptive (Gunderson et al. 1995). Being adaptive means, among other things, being able to respond to environmental feedbacks before those effects challenge the resilience of the entire resource base and the economic activities that depend on it. That is, it is necessary to frame the level of economic activity in a way that minimizes the risk of irreversible damage to the systems on which human activity depends (Perrings and Opschoor 1994). While many recent institutional innovations tend to do just the opposite, there are success stories from both traditional and contemporary societies from which we can learn (Feeny et al. 1990, Ostrom 1990, Berkes et al. 1995).

4) Rather than focusing attention on areas with the highest count of genetically distinct organisms, we should be stimulating the development of institutions, policies, and patterns of human consumption and production that work in synergy with ecosystem functions and processes. We should be reversing the trend towards large-scale and intensive monoculture. These simplified ecosystems are characterized by very low levels of diversity and even lower levels of resilience. They need to be redirected to mimic functional ecosystems (Jordan et al. 1987, Mitsch and Jörgensen 1989, Soulé and Piper 1992). In this way biodiversity may be conserved and even enhanced, while ecosystem resilience and the insurance value of biological resources may be increased. Nature reserves in both terrestrial

and aquatic environments become important in this context, as a source of immigrant organisms for redevelopment and restoration of degraded areas.

Conclusion

Biodiversity loss includes not just the extinction of species, but any change in the mix of species and ecosystems that results from human activity and compromises the structuring processes upon which human well being and survival depend (Perrings et al. 1992, 1995a, b). This approach to biodiversity conservation supports a strategy that reorients biodiversity research and policy away from genetic information and the preservation of species for tourism and recreation in nature reserves, and towards conservation for the protection of ecosystem function and resilience. This complementary view of biodiversity implies that the benefits derived from its conservation are much wider and more fundamental than previously conceived both in science and in policy. It illuminates the fact that people depend on biological diversity for their well being and survival. The emphasis in research and conservation work should not be as closely tied, as it currently is, to protected areas in the mega-diversity zones. Biodiversity loss is a matter of consequence in any ecosystem in which resilience is threatened by the deletion of populations, irrespective of the species head-count in such systems. Moreover, the most powerful instrument for biodiversity conservation is not the park fence in isolation, but policies and reforms that make conservation a matter of private as well as social interest.

Acknowledgements

This article is one of the products of the research program on the Ecology and Economics of Biodiversity Loss, of the Beijer International Institute of Ecological Economics, The Royal Swedish Academy of Sciences. Perrings has served as the Director of the program. The article has been discussed in the Swedish Scientific Committee on Biological Diversity, and at courses in Systems Ecology and Ecological Economics. We are grateful to constructive criticisms we have received during those discussions, and to the reviewers for their input. Folke's work was in part financed by the Swedish Council for Forestry and Agricultural Research (SJFR).

Literature Cited

Askins, R. A. 1995. Hostile landscapes and the decline of migratory songbirds. Science 267:1956–1957.

Barbier, E. B., J. Burgess, and C. Folke. 1994. Paradise lost? The ecological economics of biodiversity. Earthscan, London, England.

Berkes, F, and C. Folke. 1994. Investing in cultural capital for a sustainable use of natural capital. Pages 128–149 in A. M. Jansson, M. Hammer, C. Folke, and R. Costanza, editors. Investing in natural capital: the ecological economics approach to sustainability. Island Press, Washington, D.C., USA.

Berkes, F, C. Folke, and M. Gadgil. 1995. Traditional ecological knowledge, biodiversity, resilience and sustainability. Pages 281–299 in C. A. Perrings, K.-G. Maler, C. Folke, C. S. Holling, and B.-O. Jansson, editors. Biodiversity conservation: problems and policies. Kluwer Academic Publishers, Dordrecht, The Netherlands.

Costanza, R., L. Waigner, C. Folke, and K.-G. Maler. 1993. Modeling complex ecological economic systems: toward an evolutionary, dynamic understanding of people and nature. BioScience 43:545–555.

de Groot, R. S. 1992. Functions of nature. Wolters Noordhoff BV, Groningen, The Netherlands.

Ehrlich, P. R., and A. E. Ehrlich. 1992. The value of biodiversity. Ambio 21:219–226.

Ehrlich, P. R., and H. A. Mooney. 1983. Extinction, substitution and ecosystem services. BioScience 33:248–254.

Feeny, D., F. Berkes, B. J. McCay, and J. M. Acheson. 1990. The tragedy of the commons: twenty-two years later. Human Ecology 18: 1–19.

Folke, C. 1991. Socioeconomic dependence on the life-supporting environment. Pages 77–94 in C. Folke and T. Kåberger, editors. Linking the natural environment and the economy: essays from the Eco-Eco group. Kluwer Academic Publishers, Dordrecht, The Netherlands.

Gauthreaux, S. A. 1992. The use of weather radar to monitor long-term patterns of trans-Gulf migration in spring. Pages 96–100 in J. M. Hagen III and D. W. Johnston, editors. Ecology and conservation of neotropical migrant landbirds. Smithsonian Institution Press, Washington, D.C., USA.

Gunderson, L., C. S. Holling, and S. Light, editors. 1995. Barriers and bridges to the renewal of ecosystems and institutions. Columbia University Press, New York, New York, USA.

Holling, C. S., 1973. Resilience and stability of ecological systems. Annual Review of Ecology and Systematics 4:1–23.

———, 1988. Temperate forest insect outbreaks, tropical deforestation and migratory birds. Memoirs of the Entomological Society of Canada 146:21–32.

———, 1992. Cross-scale morphology, geometry, and dynamics of ecosystems. Ecological Monographs 62:447–502.

———, 1994. An ecologists view of the Malthusian conflict. Pages 79–103 in K. Lindahl-Kiessling and H. Landberg, editors. Population, economic development, and the environment. Oxford University Press, Oxford, UK.

Holling, C. S., G. Peterson, P. Marples, J. Sendzimir, K. Redford, L. Gunderson, and D. Lampert. 1996. Self-organization in ecosystems. Pages 346–384 in B. H. Walker and W. L. Steffen, editors. Global change and terrestrial ecosystem. Cambridge University Press, Cambridge, England.

Holling, C. S., D. W. Schindler, B. H. Walker, and J. Roughgarden. 1995. Biodiversity in the functioning of ecosystems: an ecological primer and synthesis. Pages 44–83 in C. A. Perrings, K.-G. Mäler, C. Folke, C. S. Holling, and B.-O. Jansson, editors. Biodiversity loss: ecological and economic issues. Cambridge University Press, Cambridge, England.

Jansson, A. M., M. Hammer, C. Folke, and R. Costanza, editors. 1994. Investing in natural capital: the ecological economics approach to sustainability. Island Press, Washington, D.C., USA.

Jones, C. G., J. H. Lawton, and M. Shachak. 1994. Organisms as ecosystem engineers. Oikos 69:373–386.

Jordan, W. R., M. E. Gilpin, and J. D. Aber. 1987. Restoration ecology: a synthetic approach to ecological research. Island Press, Washington, D.C., USA.

Kay, J. J. 1991. A nonequilibrium thermodynamic framework for discussing ecosystem integrity. Environmental Management 15:483–495.

Lawton, J. H., and V. K. Brown. 1993. Redundancy in ecosystems. Pages 255–270 in E.-D. Schulze and H. A. Mooney, editors. Biodiversity and ecosystem function. Springer, New York, New York, USA.

Ludwig, D., R. Hilborn, and C. Walters. 1993. Uncertainty, resource exploitation, and conservation: lessons from history. Science 260:17, 36.

Mitsch, W. J., and S. E. Jörgensen, editors. 1989. Ecological engineering: an introduction to ecotechnology. John Wiley & Sons, New York, New York, USA.

Naeem, S., L. J. Thompson, S. P. Lawler, J. H. Lawton, and R. M. Woodfin. 1994. Declining biodiversity can alter the performance of ecosystems. Nature 368:734–737.

North, D. C. 1990. Institutions, institutional change and economic performance. Cambridge University Press, Cambridge, England.

Olney, P. J. S., G. M. Mace, and A. T. C. Feistner, editors. 1993. Creative conservation. Chapman & Hall, London, England.

Ostrom, E. 1990. Governing the Commons: the evolution of institutions for collective action. Cambridge University Press, Cambridge, England.

Perrings, C. A., C. Folke, and K.-G. Mäler. 1992. The ecology and economics of biodiversity loss: the research agenda. Ambio 21:201–211.

Perrings, C. A., and H. Opschoor. 1994. The loss of biological diversity: some policy implications. Environmental and Resource Economics 4:1–11.

Perrings, C. A., K.-G. Mäler, C. Folke, C. S. Holling, and B.-O. Jansson, editors. 1995a. Biodiversity loss: ecological and economic issues. Cambridge University Press, Cambridge, England.

Perrings, C. A., K.-G. Mäler, C. Folke, C. S. Holling, and B.-O. Jansson, editors. 1995b. Biodiversity conservation: problems and policies. Kluwer Academic Publishers, Dordrecht, The Netherlands.

Perrings, C., and B. H. Walker. 1995. Biodiversity loss and the economics of discontinuous change in semi-arid rangelands. Pages 190–210 in C. A. Perrings, K.-G. Mäler, C. Folke, C. S. Holling, and B.-O. Jansson, editors. Biodiversity loss: ecological and economic issues. Cambridge University Press, Cambridge, England.

Pimm, S. L. 1984. The complexity and stability of ecosystems. Nature 307:321–326.

Prendergast, J. R., R. M. Quinn, J. H. Lawton, B. C. Eversham, and D. W. Gibbons. 1993. Rare species, the coincidence of diversity hotspots and conservation strategies. Nature 365:335–337.

Regier, H. A., and G. L. Baskerville. 1986. Sustainable redevelopment of regional ecosystems degraded by exploitive development. Pages 75–101 in W. C. Clark and R. E. Munn, editors. Sustainable development of the biosphere. Cambridge University Press, Cambridge, England.

Robinson, S. K., F. R. Thompson III, T. M. Donovan, D. R. Whitehead, and J. Faaborg. 1995. Regional forest fragmentation and the nesting success of migratory birds. Science 267:1987–1990.

Schindler, D. W. 1990. Experimental perturbations of whole lakes as tests of hypotheses concerning ecosystem structure and function. Oikos 57:25–41.

Schneider, E., and J. J. Kay. 1994. Complexity and thermodynamics: towards a new ecology. Futures 24:626–647.

Schulze, E.-D., and H. A. Mooney, editors. 1993. Biodiversity and ecosystem function. Springer, New York, New York, USA.

Smith, F, D. M., R. M. May, P. Pellew. T. H. Johnson, and K. S. Walter. 1993. Estimating extinction rates. Nature 364:494–496.

Solbrig, O. T. 1991. The origin and function of biodiversity. Environment 33:16–38.

Soulé, J. D., and J. K. Piper. 1992. Farming in nature's image: an ecological approach to agriculture. Island Press, Washington D.C., USA.

Tilman, D., and J. A. Downing. 1994. Biodiversity and stability in grasslands. Nature 367:363–365.

Trenbath, B., G. Conway, and I. Craig. 1990. Threats to sustainability in intensified agricultural systems: analysis and implications for management. Pages 337–370 in S. R. Gliessman, editor. Agroecology. Springer, New York, New York, USA.

Vitousek, P. M. 1990. Biological invasions and ecosystem processes: towards an integration of population biology and ecosystem studies. Oikos 57:7–13.

Westoby, M., B. H. Walker, and I. Noy-Meir. 1989. Opportunistic management for rangelands not at equilibrium. Journal of Rangeland Management 42:266–274.

Wilson, E. O. 1992. The diversity of life. Belknap, Cambridge, Massachusetts, USA.

UNEP [United Nations Environment Programme]. 1992. Convention on biological diversity. United Nations Environment Programme, Nairobi, Kenya.

ARTICLE 6

Ecological Resilience, Biodiversity, and Scale

GARRY PETERSON, CRAIG R. ALLEN, AND C. S. HOLLING

C. S. HOLLING AND GARRY PETERSON, Department of Zoology, Box 118525, University of Florida, Gainesville, FL 32611.

CRAIG R. ALLEN, Department of Wildlife Ecology and Conservation, 117 Newins-Zeigler Hall, University of Florida, Gainesville, FL 32611.

With kind permission from Springer Science+Business Media: Ecosystems, Ecological Resilience, Biodiversity, and Scale (1998) 1:6–18, Garry Peterson, Craig R. Allen, and C. S. Holling.

Abstract

WE DESCRIBE EXISTING MODELS of the relationship between species diversity and ecological function, and propose a conceptual model that relates species richness, ecological resilience, and scale. We suggest that species interact with scale-dependent sets of ecological structures and processes that determine functional opportunities. We propose that ecological resilience is generated by diverse, but overlapping, function within a scale and by apparently redundant species that operate at different scales, thereby reinforcing function across scales. The distribution of functional diversity with and across scales enables regeneration and renewal to occur following ecological disruption over a wide range of scales.

Introduction

One of the central questions in ecology is how biological diversity relates to ecological function. This question has become increasingly relevant as anthropogenic transformation of the earth has intensified. The distribution and abundance of species have been radically transformed as massive land-use changes have eliminated endemic species (Turner and others 1993), and the expansion of global transportation networks has spread other species (McNeely and others 1995). This biotic reorganization is co-occurring with a variety of other global changes, including climate change, alteration of nutrient cycles, and chemical contamination of the biosphere. Maintaining the ecological services that support humanity, and other life, during this extensive and rapid ecological reorganization requires understanding how ecological interactions among species produce resilient ecosystems.

Species perform diverse ecological functions. A species may regulate biogeochemical cycles (Vitousek 1990; Zimov and others 1995), alter disturbance regimes (Dublin and others 1990; D'Antonio and Vitousek 1992), or modify the physical environment (Jones and others 1994; Naiman and others 1994). Other species regulate ecological processes indirectly, through trophic interactions such as predation or parasitism (Kitchell and Carpenter 1993; Prins and Van der Jeud 1993), or functional interactions such as pollination (Fleming and Sosa 1994) and seed dispersal (Brown and Heske 1990). The variety of functions that a species can perform is limited, and consequently ecologists frequently have proposed that an increase in species richness also increases functional diversity, producing an increase in ecological stability (Tilman and others 1996).

The idea that species richness produces ecological stability was originally proposed by Darwin (1859), reiterated by MacArthur (1955), and modeled by May (1973). Recently, Tilman and colleagues (Tilman 1996; Tilman and others 1996) experimentally demonstrated that in small systems, over ecologically brief periods, increased species richness increases the efficiency and stability of some ecosystem functions, but decreases population stability. Despite the demonstrated link between species richness and ecological stability over small scales, the nature of this connection remains uncertain.

Models of Ecological Organization

Many competing models attempt to describe how an increase in species richness increases stability. Following previous authors, we divide these models into four classes: "species richness–diversity" (MacArthur 1955), "idiosyncratic" (Lawton 1994), "rivet" (Ehrlich and Ehrlich 1981), and "drivers and passengers" (Walker 1992). These models all explicitly or implicitly assume that a species has ecological function, and that the function of a species can be represented as occupying an area of multidimensional ecological function space (Grinnell 1917; Hutchinson 1957; Sugihara 1980). For illustrative purposes, we compress multidimensional functional space into one dimension in which breadth represents the variety of a species' ecological function (Clark 1954). For example, a species such as a beaver, that strongly influences tree populations, hydrology, and nutrient cycles, has a broad function, whereas a fig wasp that pollinates a single species of fig would have a narrow function. We represent the intensity of a species' ecological function by height. For example, a "keystone species" (Paine 1969; Power and others 1996) has a stronger influence than a "passenger" species (Walker 1992).

We emphasize the differences between these models before discussing their similarities. We then present our model of "cross-scale resilience," which incorporates scale into an expanded model of the relationship between diversity and ecological function.

Species Diversity

Darwin (1859) proposed that an area is more ecologically stable if it is occupied by a large number of species than if it is occupied by a small number. This idea was formalized by MacArthur (1955), who proposed that the addition of species to an ecosystem increases the number of ecological functions present, and that this increase stabilizes an ecosystem (Figure 1).

Although many experimental studies have demonstrated that increasing the number of species increases the stability of ecosystem function (Schindler 1990; Naeem and others 1994; Frost and others 1995; Holling and others 1995; Ewel and Bigelow 1996; Tilman 1996), apparently no

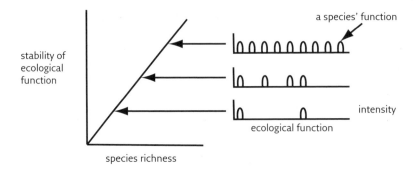

FIGURE 1: *A representation of the Darwin/MacArthur model: increasing species richness increases the stability of ecological function. This model, and the other models we discuss, implicitly represents species ecological function as occupying a portion of a multidimensional ecological function space that is analogous to niche space (MacArthur 1955). As species accumulate, they fill this space. The width and height dimensions of the inset diagrams represent the breadth and intensity of a species' ecological function. This model assumes that function space is relatively empty and therefore species can be continually added to a community without saturating it. It also assumes that the strength and breadth of ecological functions do not vary among species.*

investigations of the relationship between species richness and stability have indicated that additional species continue to increase stability at a constant rate, indicating that the species-diversity model is excessively simplistic. Consequently, we focus our attention upon models that propose more complex relationships between species richness and ecological stability.

Idiosyncratic

A competing model of the relationship between species and ecological function proposes that strong ecological interactions among species result in an ecosystem that is extremely variable, and contingent on the particular nature of interspecific interactions (Lawton 1994). This model proposes that the degree of stability in a community depends idiosyncratically upon which species are present (Figure 2). For example, fire ants have had great impacts on ecosystems of the southeastern United States (Porter and Savignano 1990; Allen and others 1995), but have a much different role in the Pantanal of Brazil and Paraguay (Orr and oth-

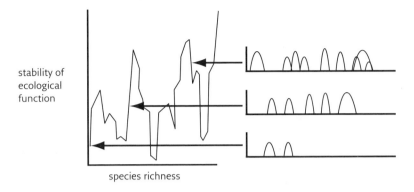

FIGURE 2: *A representation of the idiosyncratic model (Lawton 1994). In this model, ecological function varies idiosyncratically as species richness increases. This model argues that the contribution of each species to ecological function is strongly influenced by interactions among species. Therefore, the effects of the introduction or removal of species to an ecosystem can be either insignificant or major, depending upon the nature of the species introduced or removed and the nature of the species with which it interacts.*

ers 1995). Such situations suggest that ecosystem function is contingent on the ecological history of a region and the evolutionary history of interacting species. However, ecosystems are not only products of historical contingency, ecosystem ecology has demonstrated that many ecosystems are similarly organized.

Many ecosystem studies have revealed that despite dissimilar species compositions, ecosystems can have striking ecological similarities. For example, lake studies have demonstrated that similar ecological function can be maintained over a wide mix of species and population densities (Schindler 1990; Frost and others 1995). Mediterranean climate ecosystems provide a good example of functional convergence. The world's five Mediterranean climate regions, despite geographic and evolutionary isolation that has produced radically different floras and faunas, are extremely similar in ecological structure and function (Di Castri and Mooney 1973; Kalin Arroyo and others 1995). This convergence suggests that species are organized into functional groups, and that these groups are determined by regional ecological processes. Both the "rivet" (Ehrlich and Ehrlich 1981) and "drivers and passengers" (Walker 1992) models

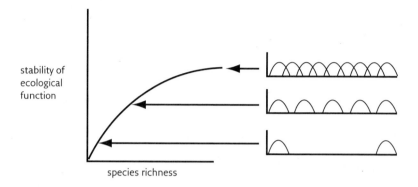

FIGURE 3: *The "rivet" model of ecological function (Ehrlich and Ehrlich 1981) presumes that ecological function space is relatively small. Therefore, as species are added to an ecosystem, their functions begin to overlap or complement one another. This overlap allows ecological function to persist despite the loss of a limited number of species, since species with similar functions can compensate for the elimination or decline of other species. However, the increase of stability gained by adding new species decreases as species richness increases and functional space becomes increasingly crowded.*

of functional diversity assume that some sort of functional redundancy exists, but they differ in the importance they assign to functional groups.

Rivets

Empirical evidence suggests that the effect of species removal from or addition to an ecosystem varies. Ehrlich and Ehrlich's (1981) rivet hypothesis, which is similar to Frost and colleagues' (1995) model of compensating complementarity, likens the ecological function of species to the rivets that attach a wing to a plane. Several rivets can be lost before the wing falls off. This model proposes that the ecological functions of different species overlap, so that even if a species is removed, ecological function may persist because of the compensation of other species with similar functions (Figure 3).

In the rivet model, an ecological function will not disappear until all the species performing that function are removed from an ecosystem. Overlap of ecological function enables an ecosystem to persist. Compensation masks ecosystem degradation, because while a degraded system may function similarly to an intact system, the loss of redundancy decreases the system's ability to withstand disturbance or further species removal.

Drivers and Passengers

Walker's "drivers and passengers" hypothesis accepts the notion of species complementarity and extends it by proposing that ecological function resides in "driver" species or in functional groups of such species (Walker 1992, 1995). It is similar to Holling's (1992) "extended keystone hypothesis." Walker defines a driver as a species that has a strong ecological function. Such species significantly structure the ecosystems in which they and passenger species exist. Passenger species are those that have minor ecological impact. Driver species can take many forms. They may be "ecological engineers" (Jones and others 1994), such as beavers (Naiman and others 1994), or gopher tortoises (Diemer 1986), which physically structure their environments. Or drivers may be "keystone species" (Paine 1969), such as sea otters (Estes and Duggins 1995) or asynchronously fruiting trees (Terborgh 1986), that have strong interactions with other species (Power and others 1996). Walker (1995) proposes that since most ecological function resides in the strong influence of driver species, it is their presence or absence that determines the stability of an ecosystem's ecological function (Figure 4).

Model Synthesis

Whereas the "rivet" hypothesis assumes that ecological function is evenly partitioned among species, Walker's model assumes there are large differences between drivers that have strong ecological function and passengers that have weak ecological function (Figure 4). Both hypotheses recognize that different types of ecological functionality are required to produce ecological stability, and that as additional species are added to an ecosystem the increasing redundancy of function decreases the rate at which ecological stability increases. The existence of some type of ecological redundancy is supported by experiments conducted in Minnesota grasslands, tropical rainforests, artificial mesocosms, and lakes (Schindler 1990; Naeem and others 1994; Ewel and Bigelow 1996; Tilman and others 1996).

Tilman, for example, demonstrated that more diverse plots (4 x 4 m) have greater plant cover and more efficiently utilize nitrogen (Tilman

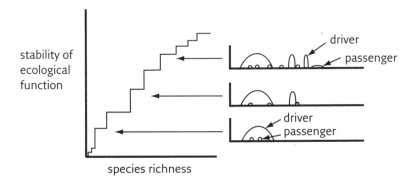

FIGURE 4: *Walker's "drivers and passengers" model of redundant ecological function (1992, 1995) proposes that ecological function is unevenly distributed among species. Drivers have a large ecological impact, while passengers have a minimal impact. The addition of drivers increases the stability of the system, while passengers have little or no effect.*

1996). Tilman and colleagues demonstrated that ecological function was more stable in diverse communities despite, or perhaps because of, large fluctuations in populations of species (Tilman and others 1996). These results echo those of Frank and McNaughton (1991), who demonstrated that more diverse natural grass communities recovered faster than less diverse communities following drought.

In a series of experiments, Ewel and coworkers constructed a set of tropical ecosystems with different levels of species richness and compared their functioning to adjacent rainforest. They demonstrated that relatively few species, if drawn from different functional groups, can duplicate the ecological flows of a diverse rainforest (Ewel and others 1991). Herbivory per leaf area was lower and less variable in species-rich plots (Brown and Ewel 1987). They also demonstrated that a variety of ecosystem variables, such as soil organic matter, increase rapidly as one adds different functional types to a plot (Ewel and Bigelow 1996), and that simple agroecosystems function quite similarly to much more species-rich rainforests, at least in areas of about 1/3 ha (80 x 40 m) for 5 years (Ewel and others 1991).

Naeem and coworkers (1994) assembled replicate artificial ecosystems at a number of levels of species richness. They demonstrated that carbon dioxide consumption, vegetative cover, and productivity increased with species richness. These increases were greater between 9 and 15 species than between 15 and 31 species, providing support for the hypothesis that an increase in species richness increases ecological redundancy. Water and nutrient retention did not vary with species richness.

Frost and coworkers (1995) demonstrated that ecological function is preserved if population declines of zooplankton species are compensated for by population increases in other species with similar ecological functions. Their results suggest that lakes with fewer species in a functional group would exhibit decreased ability to compensate for population declines in other species. Similarly, Schindler (1990) observed that the largest changes in ecological processes and food-web organization occurred when species that were the only remaining member of a functional group were eliminated.

These studies demonstrate that the stability of many, but not all, ecological processes increases with species richness. They also suggest that the ecological stability is generated more by a diversity of functional groups than by species richness. These results suggest a possible synthesis of the various models relating stability to species richness.

The model that best describes an ecosystem appears to depend upon the variety of functional roles that are occupied in that system, and the evenness of the distribution of ecological function among species. An ecosystem consisting of species that each perform different ecological functions will be less redundant than an ecosystem consisting of the same number of species that each perform a wide variety of ecological functions. Similarly, if there is little difference between the ecological impact of different species, there is little point in differentiating driver and passenger species; they can all be considered rivets. We propose that these models of how species richness influences the stability of ecological function can be collapsed into a simple model that can produce specific versions of these models by varying the degree of functional overlap and the degree of variation in ecological function among species (Figure 5).

The experimental results just discussed suggest ecosystems possess considerable functional redundancy. Indeed, it is difficult to envision how ecosystems without redundancy could continue to persist in the face of disturbance. We assume that since no species are identical, redundancy does not reside in groups of species, but rather it emerges from the interactions of species. Therefore, it is not possible to substitute species for

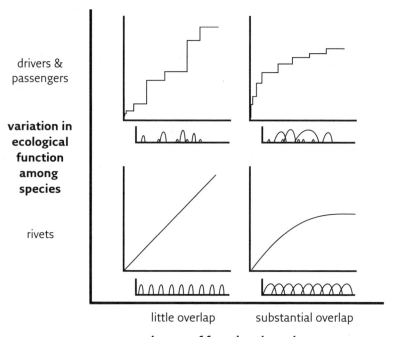

FIGURE 5: *The relationship between stability and species richness varies with the degree of overlap that exists among the ecological function of different species and the amount of variation in the ecological impact of species ecological function. Overlap in ecological function leads to ecological redundancy. If the ecological impact of different species is similar they are "rivets," whereas if some species have relatively large ecological impact they are "drivers" and others are "passengers."*

one another; rather, there are many possible combinations and organizations of species that can produce similar ecological functions. Redundancy quickly emerged in the experimental ecosystems, but these experiments were all conducted over relatively small areas and short time periods. Ewel and his coworkers (1991) conducted the longest and largest experimental manipulations of diversity, but even 5 years and a 1/3 ha are small in comparison to the spatial and temporal dynamics of an ecosystem, or even the life span and home range of a medium-sized mammal.

Understanding of stability and ecological function developed at small scales cannot be easily extended to larger scales, since the type and effect of ecological structures and processes vary with scale. At different scales, different sets of mutually reinforcing ecological processes leave their imprint on spatial, temporal and morphological patterns. Change may cause an ecosystem, at a particular scale, to reorganize suddenly around a set of alternative mutually reinforcing processes. For example, Hughes (1994) described an epidemic that caused a 99% decline in the population of an algae-eating fish in a Jamaican near-shore coral community. The loss of these herbivores caused the community to shift from being dominated by corals to being dominated by fleshy macroalgae. Similar reorganizations are demonstrated in paleo-ecological (Carpenter and Leavitt 1991), historical (Prins and Jeud 1993), and long-term ecological research (Hughes 1994).

Resilience

Assessing the stability of ecosystems that can reorganize requires more than a single metric. One common measure, what we term engineering resilience (Holling 1996), is the rate at which a system returns to a single steady or cyclic state following a perturbation. Engineering resilience assumes that behavior of a system remains within the stable domain that contains this steady state. When a system can reorganize (that is, shift from one stability domain to another), a more relevant measure of ecosystem dynamics is ecological resilience (Holling 1973). Ecological resilience is a measure of the amount of change or disruption that is required to transform a system from being maintained by one set of mutually reinforcing processes and structures to a different set of processes and structures. Note that this use of resilience is different from its use by others [for example, Pimm (1984)], who define resilience as what we term engineering resilience (Holling 1996).

The difference between ecological and engineering resilience can be illustrated by modeling an ecological "state" as the position of a ball on a landscape. Gravity pulls the ball downward, and therefore pits in the surface of the landscape are stable states. The deeper a pit, the more stable it is, because increasingly strong disturbances are required to move an ecological state away from the bottom of the pit. The steepness of the sides of a stability pit corresponds to the strength of negative feedback processes maintaining an ecosystem near its stable state, and consequently engineering resilience increases with the slope of the sides of a pit (Figure 6).

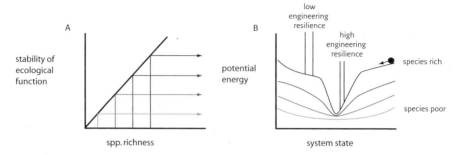

FIGURE 6: *The relationship between stability and species richness can be repre-sented by a set of stability landscapes. The dynamics of a system are expressed by a landscape, and its "state" is represented by a ball that is pulled into pits. Different landscape topographies may exist at different levels of species richness. In this model, the stability of a state increases with the depth of a pit. Zones of the stability surface that have low slopes have less engineering resilience than do areas that have steep slopes.*

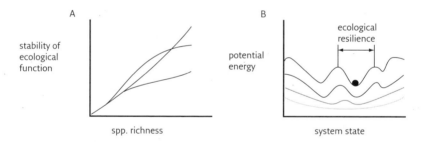

FIGURE 7: *A system may be locally stable in a number of different states. Disturbance that moves the system across the landscape and slow systemic changes that alter the shape of the landscape both drive the movement of a system between states. The stability of a state is a local measure. It is deter-mined by the slope of the landscape at its present position. The resilience of a state is a large-scale measure, as it corresponds to the width of the pit the system is currently within.*

Ecological resilience assumes that an ecosystem can exist in alterna-tive self-organized or "stable" states. It measures the change required to move the ecosystem from being organized around one set of mutually reinforcing structures and processes to another. Using the landscape met-aphor, whereas engineering resilience is a local measure of slope of the stability landscape, ecological resilience is a measure of regional topography. The ecological resilience of a state corresponds to the width of its stability pit. This corresponds to the degree to which the system would have to be altered before it begins to reorganize around another set of processes (Figure 7).

Ecological and engineering resilience reflect different properties. Ecological resilience concentrates on the ability of a set of mutually reinforcing structures and processes to persist. It allows ecologists or managers to focus upon transitions between definable states, defined by sets of organizing processes and structures, and the likelihood of such occurrence. Engineering resilience, on the other hand, concentrates on conditions near a steady state where transient measurements of rate of return are made following small disturbances. Engineering resilience focuses upon small portions of a system's stability landscape, whereas ecological resilience focuses upon its contours. Engineering resilience does not help assess either the response of a system to large perturbations or when gradual changes in a system's stability landscape may cause the system to move from one stability domain to another. For these reasons we concentrate on ecological resilience.

Scale

Ecosystems are resilient when ecological interactions reinforce one another and dampen disruptions. Such situations may arise due to compensation when a species with an ecological function similar to another species increases in abundance as the other declines (Holling 1996) or as one species reduces the impact of a disruption on other species. However, different species operate at different temporal and spatial scales, as is clearly demonstrated by the scaling relationships that relate body size to ecological function (Peters 1983).

We define a scale as a range of spatial and temporal frequencies. This range of frequencies is defined by resolution below which faster and smaller frequencies are noise, and the extent above which slower and larger frequencies are background. Species that operate at the same scale interact strongly with one another, but the organization and context of these interactions are determined by the cross-scale organization of an ecosystem. Consequently, understanding interactions among species requires understanding how species interact within and across scales.

Many disturbance processes provide an ecological connection across scales. Contagious disturbance processes such as fire, disease, and insect outbreaks have the ability to propagate themselves across a landscape, which allows small-scale changes to drive larger-scale changes. For example, the lightning ignition of a single tree can produce a fire that spreads across thousands of square kilometers. Such disturbances are not external to ecological organization, but rather form integral parts of ecological organization (Holling 1986). Disturbance dynamics affect and are affected by species and their ecological functions (D'Antonio and Vitousek 1992). Consequently, the processes regulating contagious disturbances are as much determinants of ecological resilience as are more local interactions among species.

Current models of the relationship between species richness and stability implicitly model species and their ecological functions at the same scale; however, ecological systems are not scale invariant. A growing body of empirical evidence, theory, and models suggests that ecological structure and dynamics are primarily regulated by a small set of plant, animal, and abiotic processes (Carpenter and Leavitt 1991; Levin 1992; Holling and others 1995). Processes operate at characteristic periodicities and spatial scales (Holling 1992). Small and fast scales are dominated by biophysical processes that control plant physiology and morphology. At the larger and slower scale of patch dynamics, interspecific plant competition for nutrients, light, and water influences local species composition and regeneration. At a still larger scale of stands in a forest, mesoscale processes of fire, storm, insect outbreak, and large mammal herbivory determine structure and successional dynamics from tens of meters to kilometers, and from years to decades. At the largest landscape scales, climate, geomorphological, and biogeographical processes alter ecological structure and dynamics across hundreds of kilometers and over millennia (Figure 8). These processes produce patterns and are in turn reinforced by those patterns; that is, they are self-organized (Kauffman 1993).

Ecological processes produce a scale-specific template of ecological structures that are available to species (Morse and others 1985; Krummel and others 1987; O'Neill and others 1991). Ecological structure and patterns vary across landscapes and across scales. Many species may inhabit a given area, but if they live at different scales they will experience that

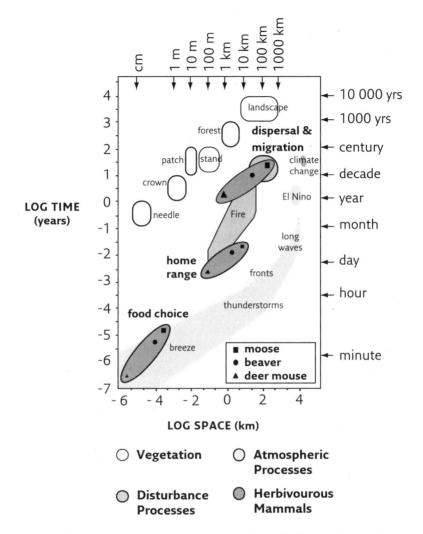

FIGURE 8: *Time and space scales of the boreal forest (Holling 1986) and their relationship to some of the processes that structure the forest. These processes include insect outbreaks, fire, atmospheric processes, and the rapid carbon dioxide increase in modern times (Clark 1985). Contagious mesoscale disturbance processes provide a linkage between macroscale atmospheric processes and microscale landscape processes. Scales at which deer mouse, beaver, and moose choose food items, occupy a home range, and disperse to locate suitable home ranges vary with their body size (Holling 1992; Macdonald 1985; Nowak and Paradiso 1983).*

area quite differently. For example, a wetland may be inhabited by both a mouse and a moose, but these species perceive and experience the wetland differently. A mouse may spend its entire life within a patch of land smaller than a hectare, while a moose may move among wetlands over more than a thousand hectares (Figure 8). This scale separation reduces the strength of interactions between mice and moose relative to interactions among animals that operate at similar scales (Allen and Hoekstra 1992). In the next section, we propose a conceptual model that relates species richness, ecological resilience, and scale.

Species, Scale, and Ecological Function

Species can be divided into functional groups based upon their ecological roles (Clark 1954; Körner 1996). Species can be also be divided into groups based upon the specific scales that they exploit. The ecological scales at which species operate often strongly correspond with average species body mass, making this measure a useful proxy variable for determining the scales of an animal's perception and influence (Holling 1992). We propose that the resilience of ecological processes, and therefore of the ecosystems they maintain, depends upon the distribution of functional groups within and across scales.

We hypothesize that if species in a functional group operate at different scales, they provide mutual reinforcement that contributes to the resilience of a function, while at the same time minimizing competition among species within the functional group (Figure 9). This cross-scale resilience complements a within-scale resilience produced by overlap of ecological function among species of different functional groups that operate at the same scales. Competition among members of a multitaxa functional group may be minimized if group members that use similar resources exploit different ecological scales. Ecological resilience does not derive from redundancy in the traditional engineering sense; rather, it derives from overlapping function within scales and reinforcement of function across scales.

We illustrate these two features of resilience by summarizing the effects of two functional groups on ecosystem dynamics and diversity in two different systems. The first example summarizes the results of field and modeling investigations of the role of avian predators in the dynamics of spruce/fir forests of eastern North America. The second summarizes

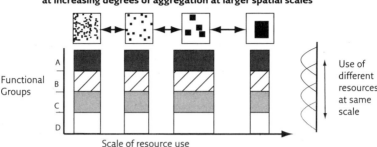

FIGURE 9: *Our hypothetized relationship between the scale of species interactions and their membership in a functional group. Different species use resources at different spatial and temporal scales. Members of a functional group use similar resources, but species that operate at larger scales require those resources to be more aggregated in space than do species that operate at smaller scales. Within scales, the presence of different functional groups provides robust ecological functioning, whereas replication of function across scales reinforces ecological function. The combination of a diversity of ecological functional specific scales and the replication of function across a diversity of scales produces resilient ecological function.*

field and modeling studies of the role of mammalian seed dispersers in the tropical forests of East Africa.

Avian Predation of Insect Defoliators

The combination of within-scale and across-scale resilience enables an ecological function such as predation of keystone defoliators to be maintained despite sudden variations in resource availability or environmental conditions. It is well known that if a particular insect becomes more common, species that would not normally exploit it may switch to using it (Murdoch 1969). This occurs as the increasing relative abundance of a resource makes its utilization less costly. We argue that as resources become increasingly aggregated they become available to larger animals that are unable to exploit dispersed resources efficiently. This mechanism introduces strong negative feedback regulation of resource abundance over a wide range of resource densities.

A well-studied example of such a situation is found in the forests of New Brunswick, Canada, where outbreaks of a defoliating insect, spruce budworm (*Choristoneura fumiferana*) periodically kill large areas of mature boreal fir forest. The initiation of these outbreaks is controlled by the interactions between the slowly changing volume of a growing forest susceptible to budworm, the more quickly changing densities and feeding responses of budworm's avian predators, and rapidly changing weather conditions (Morris 1963; Clark and Holling 1979).

Avian predation on budworm regulates the timing of budworm outbreaks by having its largest influence when budworm densities are low and forests stands are young. At least 31 species of birds prey upon budworm (Holling 1988). These bird species can be divided into five distinct body-mass classes or body-mass lumps, separated by gaps in their body-mass distributions (Holling 1992). The existence of budworm predators in these different body-size classes makes the influence of predation robust over a broad range of budworm densities. This robustness emerges not because the predators exhibit redundant functional forms of predation, but rather because the scales at which predators are effective overlap, spreading their impact over a wide range of densities and spatial aggregations of budworms.

The predatory effectiveness of a bird is largely determined by its body size. The amount of food that a bird can consume—its functional response (Holling 1959)—is a function of its body size, and a bird's search rate is greatly influenced by the scale at which it searches. Kinglets (*Regulus* sp.), chickadees (*Parus* sp.) and warblers (Emberizidae), small birds with an average body mass of about 10 g, concentrate on recognizing prey at the scale of needles or tufts of needles. Medium-sized birds focus their foraging upon branches, while larger birds such as evening grosbeaks (*Coccothraustes vespertinus*, 45 g) react to standlevel concentrations of food such as irruptions of seeds during good mast years or stand-level budworm outbreaks. The movement of birds over a landscape also is scaled to its body size. Larger birds forage over wider areas than do smaller birds. Consequently, both the body mass of birds attracted to budworm and the distance from which they are attracted will increase as the size of local aggregations of budworm increase. A diversity of foraging strategies within and across scales thus provides a strong and highly resilient predation on budworm populations (Holling 1988), particularly at low densities of budworm within stands of young trees (<30 years old).

Members of functional groups maintain and therefore determine the resilience of ecosystems by spreading their influence over a range of scales. When a functional group consists of species that operate at different scales, that group provides cross-scale functional reinforcement that greatly increases the resilience of its function. This interpretation of the partitioning of ecological function suggests that what is often defined as redundancy, is not. The apparent redundancy of similar function replicated at different scales adds resilience to an ecosystem: because disturbances are limited to specific scales, functions that operate at other scales are able to persist. The production of resilience by cross-scale functional diversity can be illustrated in a model of seed dispersal.

Mammalian Seed Dispersal in an African Tropical Forest

In Uganda's Kibale National Park, seed dispersers vary in size from small mice that range over areas of less than a hectare, to chimpanzees that range over tens of square kilometers. In a simple model of seed dispersal, when the area disturbed annually and the total amount of dispersal are held constant, the population growth rate of mammal-dispersed trees is determined by the distance over which its seeds are dispersed and the size of disturbance. A diverse set of dispersers, functioning at different scales, enables the tree population to persist despite disturbance. If, however, large, long-distance seed dispersers are absent, the tree population declines, especially when large disturbances occur (Figure 10). Mammal-dispersed trees are more aggregated when dispersal is only by small mammals that move the seeds small distances. When disturbance sizes are large, this limited dispersal is unable to maintain populations of mammal-dispersed trees (G. Peterson and C. A. Chapman, unpublished data).

Due to cross-scale functional reinforcement, and the nonlinear fashion in which ecosystem behavior can suddenly flip from one set of mutually reinforcing structures and processes to another, the gradual loss of species in a functional group may initially have little apparent effect, but their loss would nevertheless reduce ecological resilience. This decrease in resilience would be recognized only at specific spatial and temporal scales, and even then may be compensated for within or across scales. However, the ecosystem would become increasingly vulnerable to perturbations that previously could have been absorbed without changes in function or structure.

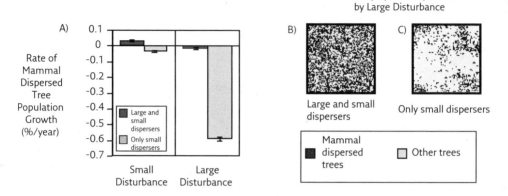

FIGURE 10: *Results from a simple model of forest dynamics and seed dispersal by mammalian frugivores in Kibale National Park, Uganda. (A) Forest distur- bance size interacts with the disperser community to determine the success of mammal-dispersed trees. When both large and small seed dispersers are present, the mammal-dispersed trees are resilient to both small and large disturbance events. When large dispersers are absent, mammal-dispersed trees slowly decline after small disturbances, but rapidly decline after large disturbances. Large differ- ences in landscape pattern can be seen after 200 years, when the forest is subjected to large disturbances, between (B) a forest containing both large and small seed dispersers and (C) a forest with only small seed dispersers. The model demonstrates that seed dispersal at a diversity of scales is more resilient to distur- bance than is seed dispersal over small scales. The model assumes lottery colonization of disturbed sites (Hubbell 1979) by either mammal-dispersed or other tree species (Chapman and Chapman 1996). Total mammal seed dispersals assumed to be constant. Dispersal range was estimated for large mammals (1010 m for* Cercocebus albigena, *and 1930 m for* Pan trogolyptes) *and small mammals (355 m for* Cercopithecus mitus, *245 m for* Cercopithecus ascanius, *and 30 m for various* Rodentia) *(C.A. Chapman, unpublished data). The distur- bance rate was held constant at 1.5%/year in the model, with only the spatial scale of disturbance varying between the small (0.04 ha) and large (10.24 ha) disturbance regimes.*

An indirect consequence of species loss is that it limits the potential number of ways a system can reorganize. Especially troubling is the possibility that the loss of large species, such as moose (Pastor and others 1993) or elephants (Dublin and others 1990), that generate mesoscale ecological structure may also eliminate forms of ecological organization. This may have occurred during the Pleistocene extinctions of megaherbivores (Owen-Smith 1989; Flannery 1994; Zimov and others 1995). These losses appear to be particularly difficult to reverse even with large-scale ecological engineering projects (Flannery 1994).

Potential Tests of Cross-Scale Resilience

Our model expands theory relating biodiversity to ecological resilience by incorporating scale. The scaling relationships we propose can be tested through the analysis of empirical data, simulation, and field experimentation.

The proposition that ecological function is distributed across scales can be tested by analyzing the distribution of ecological function of an ecosystems species, and determining whether species belonging to the same guild or functional group are dispersed across scales as we predict. The proposition that competition within a scale drives the dispersion of guilds across scales can be tested by determining whether species are more evenly morphologically dispersed within a scale than across scales.

Our model of cross-scale resilience can be tested by creating simulations that use various assemblages of species, divided by function and scale, to assess the resilience of a system to a fluctuating environment. We advocate two approaches, one focusing on the role of scale in function, and the other focusing on the plausibility of our model of ecological organization. The first approach is the one followed in the model of Kibale Forest that was just described. An ecological function that is performed by a number of speciesat different scales can be modeled, and then this model may be perturbed by disrupting function and species composition to analyze ecological resilience. Our idea that ecological resilience derives from cross-scale functional redundancy resulting from strong within-scale interactions can be tested by simulating an evolving community of organisms that compete for a set of resources. Allowing the resource preference and scale of the organisms to evolve allows one to evaluate our hypothesis that competitive interactions could lead to the distribution of similar function across scales and functional diversity within scales.

Finally, field experiments can be designed to test the response of species to resource availability at different scales. We hypothesize that limited, nonaggregated resources will be used by species that live at small scales (for example, small birds such as warblers), whereas if resources are aggregated they will be used by larger species. We predict that resource utilization by animals is determined by the density of resources at their foraging scale. Since density is a scale-dependent measure, as resources are increasingly aggregated we expect that they will be used by larger animals.

These tests will provide partial evaluation of our model. To test our theory more fully, and better understand ecological resilience in general, requires long-term and extensive experiments that manipulate species composition and ecological structure at different scales.

Conclusions

We argue that ecosystems are usefully considered not as fixed objects in space, but as interacting, self-organized sets of processes and structures that vary across scales. Our approach integrates existing models of the relationship between species and ecological function, and extends these models to incorporate scale. Ecological organization at a specific scale is determined by interactions between species and processes operating within that scale. Competitive interactions are strongest among species that have similar functions and operate at similar scales. These interactions encourage functional diversity within a scale, and the distribution of ecological function across scales, enhancing cross-scale resilience. We suggest that it is possible to identify critical scales of landscape change that may be altered by species extinctions or introductions, or alternatively to identify which species may be affected by changes in landscape structure. Ultimately, we argue that understanding interactions between the scaling of species and scaling of ecological processes should be a central goal of ecology.

Our model of cross-scale resilience has several consequences for ecological policy. The history of resource exploitation and development reveals that ecological crisis and surprises often emerge from unexpected cross-scale interactions (Holling 1986; Regier and Baskerville 1986; Gunderson and others 1995). Management of natural resources often produces high short-term yields and, either purposefully or unintentionally, creates ecosystems that are less variable and diverse over space and time. Management channels ecological productivity into a reduced number of ecological functions and eliminates ecological functions at many scales. This simplification reduces cross-scale resilience, leaving systems increasingly vulnerable to biophysical, economic, or social events that otherwise could have been absorbed—disease, weather anomalies, or market fluctuations. In Jamaica, for example, off-shore fishing reduced the diversity of herbivorous fish species, leading to the replacement of coral reefs by macroalgae (Hughes 1994). Similarly, in New Brunswick, forestry eliminated landscape and age-class diversity, leading to a long period of chronic spruce budworm infestation (Regier and Baskerville 1986). In both of these cases, management reduced the resilience of these ecosystems, leaving the existing people and biota vulnerable to abrupt ecological reorganization. To avoid repeating the ecological management disasters of the past, it is necessary that ecologists understand how the scale-dependent organization of ecosystems and functional reinforcement across scales combine to produce ecological resilience.

We propose that ecological resilience is generated by diverse, but overlapping, function within a scale and by apparently redundant species that operate at different scales. The distribution of functional diversity within and across scales allows regeneration and renewal to occur following ecological disruption over a wide range of scales. The consequences of species loss may not be immediately visible, but species loss decreases ecological resilience to disturbance or disruption. It produces ecosystems that are more vulnerable to ecological collapse and reduces the variety of possible alternative ecological organizations. Ecological resilience must be understood if humanity is to anticipate and cope with the ecological crises and surprises that accelerating global change will bring.

Acknowledgments
We appreciate the support of a NASA/EOS Interdisciplinary Scientific Investigations of the Earth Observing Systems grant (NAG 2524), a NASA Terrestrial Ecology grant (NAG 3698), and a NASA Earth System Science Fellowship to G.P. Our manuscript was improved by comments from S. Bigelow, C. Chapman, K. Sieving, F. Putz, T. Allen, T. Frost, and S. Carpenter.

References

Allen CR, Lutz RS, Demarais S. 1995. Red imported fire ant impacts on Northern Bobwhite populations. Ecol Appl 5:632–8.

Allen TFH, Hoekstra TW. 1992. Toward a unified ecology. New York: Columbia University.

Brown BJ. Ewel JJ. 1987. Herbivory in complex and simple tropical successional ecosystems. Ecology 68:108–16.

Brown JH, Heske EJ. 1990. Control of a desert-grassland by a keystone rodent guild. Science 250:1705–7.

Carpenter SR, Leavitt PR. 1991. Temporal variation in paleolimnological record arising from a trophic cascade. Ecology 72:277–85.

Chapman CA, Chapman LJ. 1996. Frugivory and the fate of dispersed and non-dispersed seeds of 6 African tree species. J Trop Ecol 12:491–504.

Clark GL. 1954. Elements of ecology. New York: John Wiley.

Clark WC. 1985. Scales of climate impacts. Climatic Change 7:5–27.

Clark WC, Holling CS. 1979. Process models, equilibrium structures, and population dynamics: on the formulation and testing of realistic theory in ecology. Popul Ecol 25:29–52.

D'Antonio CM, Vitousek PM. 1992. Biological invasions by exotic grasses, the grassifire cycle and global change. Annu Rev Ecol Syst 23:63–87.

Darwin C. 1859. On the origin of species by means of natural selection or the preservation of favoured races in the struggle for life [reprinted 1964]. Cambridge (MA): Harvard University.

Di Castri F, Mooney HA. 1973. Mediterranean type ecosystems: origins and structure. New York: Springer-Verlag.

Diemer JE. 1986. The ecology and management of the gopher tortoise in the southeastern United States. Herpetologica 42:125–33.

Dublin HT, Sinclair ARE, McGlade J. 1990. Elephants and fire as causes of multiple stable states in the Serengeti-Mara woodlands. J Anim Ecol 59:1147–64.

Ehrlich PR, Ehrlich AH. 1981. Extinction: the causes and consequences of the disappearance of species. New York: Random House.

Estes JA, Duggins DO. 1995. Sea otters and kelp forests in Alaska: generality and variation in a community ecological paradigm. Ecol Monogr 65:75–100.

Ewel JJ, Bigelow SW. 1996. Plant life-forms and tropical ecosys- tem functioning. In: Orians GH, Dirzo R, Cushman JH, editors. Biodiversity and ecosystem processes in tropical forests. Heidelberg: Springer-Verlag, p. 101–26.

Ewel JJ, Mazzarino MJ, Berrish CW. 1991. Tropical soil fertility changes under monocultures and successional communities of different structure. Ecol Appl 1:289–302.

Flannery T. 1994. The future eaters: an ecological history of the Australasian lands and people. New York: George Braziller.

Fleming TH, Sosa VJ. 1994. Effects of nectarivorous and frugivorous mammal on reproductive success of plants. J Mammal 75:845–51.

Frank DA. McNaughton SJ. 1991. Stability increases with diversity in plant communities: empirical evidence from the 1988 Yellowstone drought. Oikos 62:360–62.

Frost TM, Carpenter SR, Ives AR, Kratz TK. 1995. Species compensation and complementarity in ecosystem function. In: Jones CG, Lawton JH, editors. Linking species and ecosystems. New York: Chapman and Hall. p. 224–39.

Grinnell J. 1917. The niche relations of the California thrashers. Auk 34:427–33.

Gunderson L, Holling C, Light S. 1995. Barriers and bridges to the renewal of ecosystems and institutions. New York: Columbia University.

Holling CS. 1959. The components of predation as revealed by a study of small mammal predation of the European pine sawfly. Can Entomol 91:293–320.

Holling CS. 1973. Resilience and stability of ecological systems. Annu Rev Ecol Syst 4:1–23.

Holling CS. 1986. The resilience of ecosystems: local surprise and global change. In: Clark WC, Munn RE, editors. Sustainable development of the biosphere. Cambridge (UK): Cambridge University. p. 292–317.

Holling CS. 1988. Temperate forest insect outbreaks, tropical deforestation and migratory birds. Mem Entomol Soc Can 146:21–32.

Holling CS. 1992. Cross-scale morphology, geometry and dynamics of ecosystems. Ecol Monogr 62:447–502.

Holling CS. 1996. Engineering resilience versus ecological resilience. In: Schulze P, editor. Engineering within ecological constramts. Washington (DC): National Academy, p. 31–44.

Holling CS, Schindler DW, Walker BW, Roughgarden J. 1995. Biodiversity in the functioning of ecosystems: an ecological synthesis. In: Perrings C, Mäler K-G, Folke C, Holling CS, Jansson B-O, editors. Biodiversity loss: economic and ecological issues. New York: Cambridge University. p. 44–83.

Hubbell SP. 1979. Tree dispersion, abundance, and diversity in a tropical dry forest. Science 203:1299–309.

Hughes TP. 1994. Catastrophes, phase shifts, and large-scale degradation of a Caribbean coral reef. Science 265:1547–51.

Hutchinson GE. 1957. Concluding remarks. Cold Spring Harbor Symp Quant Biol 22:415–27.

Jones CG, Lawton JH, Shachak M. 1994. Organisms as ecosystem engineers. Oikos 69:373–86.

Kalin Arroyo MT, Zedler PH, Fox MD. 1995. Ecology and biogeography of Mediterranean ecosystems in Chile, California, and Australia. New York: Springer-Verlag.

Kauffman SA. 1993. Origins of order: self-organization and selection in evolution. Oxford: Oxford University.

Kitchell JF, Carpenter SR. 1993. Synthesis and new directions. In: Carpenter SR, Kitchell JF, editors. The trophic cascade in lakes. New York: Cambridge University. p. 332–50.

Körner C. 1996. Scaling from species to vegetation: the usefulness of functional groups. In: Schulze E-D, Mooney HA, editors. Biodiversity and ecosystem function. New York: Springer-Verlag. p. 117–40.

Krummel JR, Gardner RH, Sugihara G, O'Neill RV, Coleman PR. 1987. Landscape patterns in a disturbed environment. Oikos 48:321–24.

Lawton JH. 1994. What do species do in ecosystems? Oikos 71 :367–74.

Levin SA. 1992. The problem of pattern and scale in ecology. Ecology 73:1943–67.

MacArthur RH. 1955. Fluctuations of animal populations and a measure of community stability. Ecology 36:533–6.

Macdonald D. 1985. The encyclopedia of mammals. New York: Facts On File.

May RM. 1973. Stability and complexity in model ecosystems. Princeton (NJ): Princeton University.

McNeely JA, Gadgil M, Leveque C, Padoch C, Redford K. 1995. Human influences on biodiversity. In: Heywood V, editor. Global biodiversity assessment. Cambridge (UK): Cambridge University. p. 711–822.

Morris RF. 1963. The dynamics of epidemic spruce budworm populations. Mem Entomol Soc Can 31:1–322.

Morse DR. Lawton JH, Dodson MM. 1985. Fractal dimension of vegetation and the distribution of arthropod body lengths. Nature 314:731–3.

Murdoch WW. 1969. Switching in general predators: experiments on predator specificity and stability of prey populations. Ecol Monogr 39:335–54.

Naeem S, Thompson LJ, Lawler SP, Lawton JH, Woodfin RM. 1994. Declining biodiversity can alter the performance of ecosystems. Nature 368:734–7.

Naiman RJ, Pinay G, Johnston CA, Pastor J. 1994. Beaver influences on the long-term biogeochemical characteristics of boreal forest drainage networks. Ecology 75:905–21.

Nowak RM, Paradiso JL. 1983. Walker's mammals of the world. Baltimore: John Hopkins University.

O'Neill R, Tumer SJ, Cullinam VI, Coffin DP, Cook T, Conley W, Brunt J, Thomas JM, Conley MR, Gosz J. 1991. Multiple landscape scales: an intersite comparison. Landscape Ecol 5:137–4.

Orr, MR, Seike SH, Benson WW, Gilbert LE. 1995. Flies suppress fire ants. Nature 373:292–3.

Owen-Smith N. 1989. Megafaunal extinctions: the conservation message from 11,000 years B.C. Conserv Biol 3:405–11.

Paine RT. 1969. A note on trophic complexity and community stability. Am Nat 103:91–3.

Pastor J, Dewey B, Naiman RJ, McInnes PF, Cohen Y. 1993. Moose browsing and soil fertility in the boreal forests of Isle Royale National Park. Ecology 74:467–80.

Peters RH. 1983. The ecological implications of body size. Cambridge (UK): Cambridge University.

Pimm SL. 1984. The complexity and stability of ecosystems. Nature 307:321–6.

Porter SD, Savignano DA. 1990. Invasion of polygyne fire ants decimates native ants and disrupts arthropod community. Ecology 71:2095–116.

Power ME, Tilman D, Estes JA, Menge BA, Bond WJ, Mills LS, Daily G, Castilla JC, Lubchenco J, Paine R. 1996. Challenges in the quest for keystones. BioScience 46:609–20.

Prins HHT, Van der Jeud HP. 1993. Herbivore population crashes and woodland structure in East Africa. J Ecol 81:305–14.

Regier HA, Baskerville GL. 1986. Sustainable redevelopment of regional ecosystems degraded by exploitive development. In: Munn WC, Munn RE, editors. Sustainable development of the biosphere. Cambridge (MA): Cambridge University. p. 75–101.

Schindler DW. 1990. Experimental perturbations of whole lakes as tests of hypotheses concerning ecosystem structure and function. Oikos 57:25–41.

Sugihara G. 1980. Minimal community structure: an explanation for species abundance patterns. Am Nat 116:770–87.

Terborgh J. 1986. Keystone plant resources in the tropical forest. In: Soulé ME, editor. Conservation biology: the science of scarcity and diversity. Sunderland (UK): Sinauer. p. 330–44.

Tilman D. 1996. Biodiversity: population versus ecosystem stability. Ecology 77:350–63.

Tilman D, Wedin D, Knops J. 1996. Productivity and sustainability influenced by biodiversity in grasslands ecosystems. Nature 379:718–20.

Turner BL, Clark WC, Kates RW, Richards JF, Mathews JT, Meyer WB. 1993. The earth as transformed by human action. New York: Cambridge University.

Vitousek PM. 1990. Biological invasions and ecosystem processes: towards an integration of population biology and ecosystem studies. Oikos 57:7–13.

Walker B. 1992. Biological diversity and ecological redundancy. Conserv Biol 6:18–23.

Walker B. 1995. Conserving biological diversity through ecosystem resilience. Conserv Biol 9:747–52.

Zimov SA, Chuprynin VI, Oreshko AP, Chapin IFS, Reynolds JF, Chapin MC. 1995. Steppe-tundra transition: a herbivore-driven biome shift at the end of the Pleistocene. Am Nat 146:765–94.

Ecological Examples

Commentary on
Part Two Articles

CRAIG R. ALLEN, LANCE H. GUNDERSON, AND C. S. HOLLING

EVEN THOUGH THE CONCEPT was introduced in 1973, the ecological research community took almost two decades to seriously grapple with the idea that alternative ecological states could exist. Part of that delay is because ecology is, as is much of science, conservative. Novel ideas and paradigm shifts may take decades to gain acceptance. And ecosystem science lacks sufficiently controlled replicates as well as tight laws governing system behaviors. Thus, we are left with comparative studies, time series analyses, and a search for pattern. A reductionist emphasis and adherence to a strict hypothesis-testing framework with an emphasis on reducing Type I error make the detection of large-scale patterns and trends in ecosystems problematic at best (Holling and Allen 2002).

The ecological examples presented in this section all date from the mid-1990s. Hughes (1994) presents empirical data on phase shifts (regime shifts) in coral reef communities of Jamaica. This work demonstrates how a disturbance type (hurricanes) that coral reefs had adapted to over millennia caused a regime shift because ecological resilience had been eroded following overfishing. Estes and Duggin's (1995) article is also included in this section because it too demonstrates a regime shift in a marine system—the kelp-dominated ecosystem in the northeastern Pacific Ocean. In the second example, Estes and Duggin demonstrate that the overharvest of a key species—the sea otter—led to the phase shift. The final article in this section, by Allen et al. (1999), also demonstrates how phase shifts in terrestrial systems subject to land use transformations may manifest in animal communities. They demonstrate how transformations may differentially effect species near "scale breaks" following the erosion of resilience, and how the replacement of declining species by invasive species occurs nonrandomly. It took more than two decades for resilience thinking to gain a degree of acceptance. Now there is a wide array of contemporary examples to call upon; in addition to those reprinted here, many are listed in the bibliography at the end of this book.

One of the more compelling examples of regime shifts is found in tropical coral reefs. These important communities are complex and reliant on a symbiotic relationship between coral polyps and algae. Coral reefs have withstood numerous challenges in ecological time that have built their resilience. They are adapted to hurricanes, changes in ocean currents and temperatures, and shifts in species abundances. But over the past fifty years, coral reefs have been increasingly threatened by human activities, including overfishing, deforestation leading to increased nutrient and sediment inputs, dredging and mining, and more.

Hughes (1994) documents the phase shift that occurred rapidly in Jamaican reefs once their resilience was exceeded. This documentation was only possible because of intensive monitoring and research over decades. Such long-term data are critical, as the phenomenon of shifting baselines demonstrates (Pauly 1995), and can otherwise make multiple stable states difficult to detect. Phase shifts in coral reefs as first reported by Hughes in 1994 have subsequently been documented by many other coral reefs researchers (Hughes et al. 2003, Jackson et al. 2001, McClanahan et al. 2002). Although slight deviations from the pattern occur, similar pathologies can be found throughout the tropical areas of the planet.

Many coral reefs occur in countries where people rely on reefs for critical protein resources and revenue. In Jamaica, overfishing led to a reduction in fish biomass by the early 1970s. Continued fishing also reduced populations of large predators and large, herbivorous fish. As a result, herbivorous fish populations were dominated by small, nonreproductive individuals. As is typical in ecosystems with compromised resilience, the impact of overfishing did not become evident for some time and coral cover and diversity remained high. Despite the loss of herbivorous fishes, macroalgae remained in check due to grazing by sea urchins. Hence, the functional group of grazers remained present, but functional reinforcement (Peterson et al. 1998) was lost, reducing the resilience of the reef ecosystem.

The reduction of resilience was not manifest until a disease in 1983 caused a dramatic decline in the sea urchins, the lone remaining member of the grazing functional group. Following a 99 percent decline in sea urchin density, macroalgae—without other significant herbivores present— rapidly increased in abundance and coverage. This greatly decreased the recruitment of corals due to the loss of suitable substrate. Additionally, many of the macroalgae form extensive mats, which overgrow and kill existing remnant corals. Prior to 1983, two hurricanes damaged corals. Following the loss of sea urchins and the occurrence of the hurricanes, the shift from diverse coral communities to macroalgae-dominated communities was extremely rapid. Coral cover was reduced from 52 percent to 3 percent while fleshy macroalgae increased from 4 percent to 92 percent cover. The resilience-eroding events had occurred over decades, but the catastrophic shift was sudden and not anticipated.

The loss of corals has made Jamaica and other island nations more vulnerable to the effects of future hurricanes. The loss of herbivorous species has also enormously reduced the economic potential of the reefs. Terrestrial land cover changes adjacent to reefs have ensured continued influxes of nutrients and sediments, thus creating a continued strong positive feedback for the macroalgae-dominated state. The resilience of this altered state is strong, and management to force a regime shift back to a more desired coral-dominated state will be difficult to implement (Hughes et al. 2003, Jackson et al. 2001).

Near-shore kelp ecosystems on the Pacific coast of North America can exist in two "stable" states: kelp dominated or sea urchin dominated. Estes and Duggins (1995) first described these alternative states and the differences between them in the Gulf of Alaska. Which of the two states is present is largely driven by the interactions among three key species: sea otters, invertebrate herbivores, and macroalgae. Otters eat urchins, urchins in turn eat macroalgae (kelp), and the abundance of each determines the system state.

As was the case in the Caribbean reef study by Hughes (1994), Estes and Duggins's data set is unusual because of its temporal and spatial extent. Most field studies occur in one-square-meter plots over one or two years and at such small scales that cross-scale dynamics are unlikely to be detected. Discontinuities, thresholds, and other nonlinear behavior make it purely speculative, at best, to extrapolate such small studies to ecosystem behavior or state. It is studies such as those by Estes and Duggins (1995) and Hughes (1994) that provide the understanding of complex dynamics that is impossible to discern over short time periods and small extents. The generation time of the species involved is not long even by ecological measures but greatly exceeds the size and duration of most ecological studies.

Sea otters actively select for large sea urchins, so population distributions of sea otter size (and thus age) affect the otter-urchin-kelp relationship. When urchin recruitment is fairly constant, small urchins are abundant. Where urchin recruitment is episodic, urchins of larger size classes are more abundant. These populations are much more vulnerable to otter predation, resulting in trophic cascades and regime shifts that are much more pronounced. Estes and Duggins were able to sample a broad array of sites and to sample all three key species simultaneously where all three were present. The long duration of the study meant that long-term monitoring was present both before and after otter colonization of new sites. Such time series analyses were used in conjunction with comparisons of sites with and without otters. Subsequent work by Steneck et al. (2003) indicate similar system flips in the north Atlantic over many centuries of human intervention.

The state of the system is driven by the presence or absence of otters. Otter presence and abundance over the past two hundred years have largely been controlled by human harvests. Regime shifts resulting from otter colonization are fast, especially on sites with episodic urchin recruitment. Estes and Duggins effectively ended much of the controversy existing among community ecologists (Foster 1990, Foster and Schiel 1988) regarding the role of sea otters and cross-trophic interactions in driving and structuring Pacific near-shore marine communities. Because the transition between states is usually rapid, intermediate states are very rare and transitory or are absent. Alternative stable states have been documented for other herbivore systems such as rangelands (Noy-Mier 1975, Walker et al. 1981).

The existence of alternative stable states, a requisite for resilience theory, has now been thoroughly documented, but for most systems the catalyst for abrupt system change is unknown, as are the structure and function of possible alternate states. Because there is usually a very large uncertainty regarding the configuration of alternative states, it is usually prudent to manage for resilience in systems and to avoid crossing critical thresholds. The presence of hysteresis in many systems further reinforces this approach as a reasonable management goal.

Ecosystems are structured by the actions of a limited number of key processes, each of which acts on a specific spatial and temporal scale (Holling 1992). The limited scales of structure and pattern available in a given system provide a limited number of scales of opportunity for species inhabiting a system. Therefore, there should be groupings of species that correspond to the scales of opportunity available in a system. Positive interactions within ranges of scale—feedbacks between plants and animals and abiotic processes—create strong self-organization within scales. This self-organization within limited ranges of scales is manifest in vertebrates in the discontinuous distribution of body sizes. Contention that animal body mass distributions are discontinuous was initially met with skepticism (Manly 1996, Seimann and Brown 1999). More recently, however, most ecologists accept that animal body mass distributions are discontinuous, but many mechanisms have been forwarded to explain these discontinuities (Allen et al. 2006).

Allen et al. (1999) investigated the phenomena of invasions and extinctions in relation to discontinuities in animal body mass distributions. First, they compiled data on body masses of mammals, birds, and herpetofauna to determine whether those body mass distributions were distributed continuously or discontinuously. Then they utilized data independent of body size, species status as nonnative invasive or declining or extinct native, to investigate whether those two biological extremes were randomly or nonrandomly distributed in terms of the body mass distributions. All three taxa had body mass distributions with multiple discontinuities (or "gaps"). For invasive and declining species, hypotheses of random distribution, morphological overdispersion, repellency, and scale-specific success or failure failed to account for their distribution in terms of the overall body mass distributions. Both declining species and invasive species tended to have body masses that were close to discontinuities.

Further analysis published later (Forys and Allen 2002) documented that, despite substantial turnover in species due to invasions and probable future extinctions, the body mass distributions of mammals, birds, and herpetofauna of the Everglades ecosystem remained little changed. These analyses provided independent tests confirming the discontinuities in body mass distributions and, like Havlicek and Carpenter (2001), confirmed that the discontinuous patterns were highly conservative to species turnover. Such conservatism, we believe, is due to the strong positive feedbacks present within ranges of scale. It is the strength of this scale-specific self-organization that, in part with the distribution of function, generates resilience in ecological systems. Resilience, then, is best viewed as an emergent property of complex systems.

Discontinuity analysis provides a method to determine the key scales of structure, process, and pattern present in a given system. Discontinuity analysis provides insight into the underlying panarchy (Gunderson and Holling 2002). It is likely that in addition to the distribution of structure within and across scales, the number of scales present (the levels of hierarchy present) in a system provides clues to the overall system resilience (Allen et al. 2005). The independent tests confirm the link between measures of a species' population status and discontinuities, and, with the theoretical model presented by Peterson et al. (1998), were the beginning of a body of work confirming discontinuities in body mass distributions and of other variables in complex systems (Allen and Holling 2008). The association between species' body mass distributions and populations has been recently affirmed (Allen 2006, Skillen and Maurer 2008, Wardwell and Allen 2009), and the link between scale-specific distributions in resource distributions and body mass discontinuities has been confirmed with models (Szabo and Meszena 2006). Other independent attributes associated with the heightened variability in resource distributions have been documented, including phenomena as diverse as nomadism (Allen and Saunders 2002, 2006) and population variability (Wardwell and Allen 2009). The picture emerges of ecosystems as complex systems with scale-specific reinforcement, discontinuities characterized by heightened variability separating scales of structure and process, and emergent properties such as resilience.

Literature Cited

Allen, C. R. 2006. Predictors of introduction success in the South Florida avifauna. *Biological Invasions* 8:491–500.

Allen, C. R., and C. S. Holling, eds. 2008. *Discontinuities in ecosystems and other complex systems*. New York: Columbia University Press.

Allen, C. R., and D. A. Saunders. 2002. Variability between scales: Predictors of nomadism in birds of an Australian Mediterranean-climate ecosystem. *Ecosystems* 5:348–59.

———. 2006. Multi-model inference and the understanding of complexity, discontinuity and nomadism. *Ecosystems* 9:694–99.

Allen, C. R., E. A. Forys, and C. S. Holling. 1999. Body mass patterns predict invasions and extinctions in transforming landscapes. *Ecosystems* 2:114–21.

Allen, C. R., L. Gunderson, and A. R. Johnson. 2005. The use of discontinuities and functional groups to assess relative resilience in complex systems. *Ecosystems* 8:958–66.

Allen, C. R., A. Garmestani, T. Havlicek, P. Marquet, G. D. Peterson, C. Restrepo, C. Stow, and B. Weeks. 2006. Patterns in body mass distributions: Sifting among alternative competing hypotheses. *Ecology Letters* 9:630–43.

Estes, J. A., and D. O. Duggins. 1995. Sea otters and kelp forests in Alaska: Generality and variation in a community ecological paradigm. *Ecological Monographs* 65 (1): 75–100.

Forys, E. A., and C. R. Allen. 2002. Functional group change within and across scales following invasions and extinctions in the Everglades ecosystem. *Ecosystems* 5:339–47.

Foster, M. S. 1990. Organization of macroalgal assemblages in the northeast Pacific: The assumption of homogeneity and the illusion of generality. *Hydrobiologia* 192:21–33.

Foster, M. S., and D. R. Schiel. 1988. Kelp communities and sea otters: Keystone species or just another brick in the wall? In *The community ecology of sea otters*, ed. G. R. VanBlaricom and J. A. Estes, 92–108. Berlin: Springer-Verlag.

Gunderson, L. H., and C. S. Holling. 2002. *Panarchy: Understanding transformations in human and natural systems*. Washington, DC: Island Press.

Havlicek, T., and S. R. Carpenter. 2001. Pelagic size distributions in lakes: Are they discontinuous? *Limnology and Oceanography* 46:1021–33.

Holling, C. S. 1992. Cross-scale morphology, geometry, and dynamics of ecosystems. *Ecological Monographs* 62:447–502.

Holling, C. S., and C. R. Allen. 2002. Adaptive inference for distinguishing credible from incredible patterns in nature. *Ecosystems* 5:319–28.

Hughes, T. P. 1994. Catastrophes, phase shifts, and large-scale degradation of a Caribbean coral reef. *Science* 265:1547–51.

Hughes, T. P., A. H. Baird, D. R. Bellwood, M. Card, and S. R. Connolly. 2003. Climate change, human impacts, and the resilience of coral reefs. *Science* 301:929–33.

Jackson, J. B. C., M. X. Kirb, W. H. Berher, K. A. Bjorndal, L. W. Botsford, et al. 2001. Historical overfishing and the recent collapse of coastal ecosystems. *Science* 293:629–38.

Manly, B. F. J. 1996. Are there clumps in body-size distributions? *Ecology* 77:81–86.

McClanahan, T. R., N. V. C. Polunin, and T. Done. 2002. Ecological states and the resilience of coral reefs. *Conservation Ecology* 6 (2): 18. http://www.consecol.org/vol16/iss2/art18.

Noy-Mier, I. 1975. Stability in grazing systems: An application of predator-prey graphs. *Journal of Ecology* 63:459–81.

Pauly, D. 1995. Anecdotes and the shifting baseline syndrome of fisheries. *Trends in Ecology and Evolution* 10:430.

Peterson, G., C. R. Allen, and C. S. Holling. 1998. Ecological resilience, biodiversity and scale. *Ecosystems* 1:6–18.

Siemann, E., and J. H. Brown. 1999. Gaps in mammalian body size distributions reexamined. *Ecology* 80:2788–92.

Skillen, J. J., and B. A. Maurer. 2008. The ecological significance of discontinuities in body mass distributions. In *Discontinuities in ecosystems and other complex systems*, ed. C. R. Allen and C. S. Holling, 193–218. New York: Columbia University Press.

Steneck, R. S., J. Vavrinec, and A. V. Leland. 2003. Accelerating trophic-level dysfunction in kelp forest ecosystems of the western North Atlantic. *Ecosystems* 7:323–32.

Szabo, P., and G. Meszena. 2006. Spatial ecological hierarchies: Coexistence on heterogeneous landscapes via scale niche diversification. *Ecosystems* 9:1009–16.

Walker, B. H., D. Ludwig, C. S. Holling, and R. M. Peterman. 1981. Stability of semi-arid savanna grazing systems. *Journal of Ecology* 69:473–98.

Wardwell, D., and C. R. Allen. 2009. Variability in population abundance is associated with thresholds between scaling regimes. *Ecology and Society* (in press).

Catastrophes, Phase Shifts, and Large-Scale Degradation of a Caribbean Coral Reef

TERENCE P. HUGHES

TERENCE P. HUGHES, *Department of Marine Biology, James Cook University, Townsville, QLD 4811, Australia.*

From: Hughes, T. P. 1994. Catastrophes, phase shifts, and large-scale degradation of a Caribbean coral reef. *Science* 265:1547–51. Reprinted with Permission from AAAS.

Abstract

MANY CORAL REEFS have been degraded over the past two to three decades through a combination of human and natural disturbances. In Jamaica, the effects of overfishing, hurricane damage, and disease have combined to destroy most corals, whose abundance has declined from more than 50 percent in the late 1970s to less than 5 percent today. A dramatic phase shift has occurred, producing a system dominated by fleshy macroalgae (more than 90 percent cover). Immediate implementation of management procedures is necessary to avoid further catastrophic damage.

Coral reefs are renowned for their spectacular diversity and have significant aesthetic and commercial value, particularly in relation to fisheries and tourism. However, many reefs around the world are increasingly threatened, principally from overfishing and from human activities causing excess inputs of sediment and nutrients such as pollution, deforestation, reef mining, and dredging (1). There is a pressing need to monitor coral reefs to assess the spatial and temporal scale of any damage that may be occurring and to conduct research to understand the mechanisms involved.

Here I describe dramatic shifts in reef community structure that have largely destroyed coral reefs around Jamaica. The results presented here summarize the most comprehensive reef monitoring program yet conducted in the Caribbean, in which annual censusing has been carried out for 17 years at multiple sites and depths along 300 km of coastline. In addition, Jamaican reefs are among the best studied in the world, with a wealth of information available on marine ecology and reef status since the 1950s (2). These long-term observations provide a basis for evaluating the role of rare events such as hurricanes and for quantifying gradual trends in coral cover and diversity over a decadal time scale.

Jamaica (18°N, 77°W) is the third largest island in the Caribbean and lies at the center of coral diversity in the Atlantic Ocean (2). Over 60 species of reef-building corals occur there, four of which are spatial dominants: branching elkhorn and staghorn corals, *Acropora palmata* and *Acropora cervicornis*, which form two distinctive zones on the shallow fore-reef; massive or platelike *Montastrea annularis*, the most important framework coral; and encrusting or foliose *Agaricia agaricites* (3). Reefs fringe most of the north Jamaican coast along a narrow (<1 to 2 km) belt and occur sporadically on the south coast on a much broader (>20 km) shelf. Sea-grass beds and mangrove are often closely associated with reefal areas and provide significant nurseries for commercially important reef fisheries (4). Similar ecosystems, with minor variations in community composition, occur throughout the Caribbean (2).

Jamaica's population growth trajectory is typical of most Third World countries (Fig. 1). The population was less than half a million before 1870, then doubled by 1925 and again by 1975, rising to 2.5 million today. Exponential growth continues, with a further 20% increase expected in the next 15 years (5). Environmental changes on land are conspicuous, with virtually all of the native vegetation having been cleared for agriculture and urban development. Major transformations are also occurring on Jamaica's coral reefs.

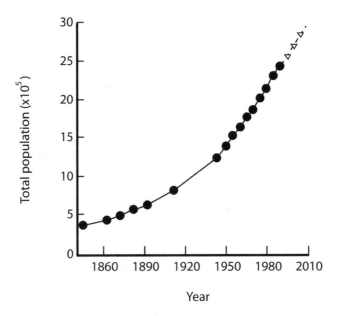

FIGURE 1: *Population growth of Jamaica, based on numerous sources (5).*

Overfishing (1960s to Present)

Chronic overfishing is an ever increasing threat to coral reefs worldwide as coastal populations continue to grow (for example, Fig. 1) and exploit natural resources (6). Extensive studies in Jamaica by Munro (7) showed that by the late 1960s fish biomass had already been reduced in preceding decades by up to 80% on the extensive (but narrow) fringing reefs of the north coast, mainly a result of intensive artisanal fish-trapping (Fig. 2). By 1973, the number of fishing canoes deploying traps on the north coast was approximately 1800 (or 3.5 canoes per square kilometer of coastal shelf), which was two to three times above sustainable levels (7). The taxonomic composition of fish has changed markedly over the past 30 to 40 years. Large predatory species, such as sharks, lutjanids (snappers), carangids (jacks), ballistids (triggerfish), and serranids (groupers) have virtually, disappeared, while turtles and manatees are also extremely rare. The remaining fish, including herbivores such as scarids (parrotfish) and acanthurids (surgeonfish), are small, so that fully half of the species caught in traps recruit to the fishery below the minimum reproductive size. Indeed, because adult stocks on the northern coast of Jamaica have

FIGURE 2: *(A) Healthy reefs are characterized by a high degree of habitat heterogeneity, which provides habitat for fish and invertebrates. (B) A Z-shaped fish trap commonly used throughout the Caribbean (7). (C) Removal of fish is likely to have promoted population growth of the echinoid* Diadema antillanum *which became the dominant macroherbivore on overfished reefs throughout the Caribbean (13). (D) After the mass mortality of* Diadema *from disease in 1983, spectacular algal blooms ensued on overfished reefs. In Jamaica, abundance of macroalgae has increased steadily for the past decade (see Figure. 3B). (E and F) Macroalgal overgrowth and preemption of space for larval recruitment has caused a dramatic decline in abundance of corals. Here, a massive coral has been partially smothered by* Lobophora *(E), killing tissue overlying the white coral skeleton as revealed by peeling away the algae (F).*

been sharply reduced for several decades, populations today may rely heavily on larval recruitment from elsewhere in the Caribbean (7). This sequence of changes was repeated more recently along the southern coast of Jamaica. There, the broader coastal shelf has become increasingly accessible to a modernizing fishing fleet, with the number of motorized canoes almost doubling from the 1970s to the mid-1980s (8). Despite this increased fishing effort, the catch from the south coast remained the same over this 15-year period (that is, the catch per unit effort declined by half). The species composition of the fishery has also changed markedly, indicative of severe overfishing nationwide (6–8).

The ecological effects of the drastic reduction in fish stocks on Jamaica's coral reefs as a whole were not immediately obvious. Throughout the 1950s to the 1970s the reefs appeared to be healthy; coral cover and benthic diversity were high (3) (Figs. 2 and 3). There were relatively few macroalgae throughout this period despite the paucity of large herbivorous fish as a result mainly of grazing by huge numbers of the echinoid *Diadema antillmum* (9, 10). The major predators of adult *Diadema* are fish [for example, ballistids, sparids (porgies), and batrachoidids (toadfish) (11)] that are now rare in Jamaica. Other fish (such as scarids and acanthurids) compete strongly with *Diadema* for algal resources, as evidenced by competitor removal experiments (12). Therefore, the unusually high abundance of *D. antillmum* on overfished reefs such as Jamaica's was almost certainly a result of the over-exploitation of reef fisheries. Hay (13) investigated this hypothesis on a geographic scale and found that densities of echinoids were much greater on overfished than on pristine reefs throughout the Caribbean. A mass mortality of *Diadema* in 1983 had far-reaching consequences, in part because of the prior reduction (for several decades) of stocks of herbivorous and predatory fish.

Hurricane Damage (1980)

Hurricanes, typhoons, or cyclones are predictable, recurrent events and an integral part of the natural dynamics of a coral reef (14). The regeneration of a healthy reef system is facilitated by rapid colonization of larval recruits, but in Jamaica this crucial recovery mechanism has been hindered by human influences (that is, by overfishing, which contributed to a prolonged macroalgal bloom causing recruitment failure in corals).

Extensive damage was inflicted on Jamaican coral reefs by Hurricane Allen, a category 5 hurricane that struck in 1980, following a period of

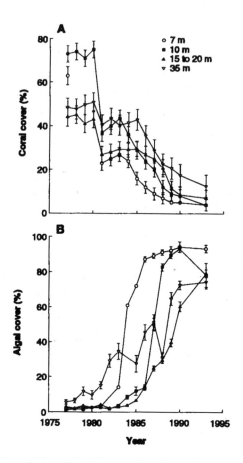

FIGURE 3: *Degradation of Jamaican coral reefs over the past two decades. Small-scale changes in (A) coral cover and in (B) macroalgal cover over time at four depths near Discovery Bay (32).*

almost four decades without a major storm (15). Damage was the greatest at shallow sites (Fig. 3A). The hurricane smashed shallow-water branching species, most notably the elkhorn and staghorn corals (*Acropora palmata* and *A. cervicornis*). In addition, beds of the soft coral *Zoanthus*, which occupied large areas of the inner reef flat, were damaged by *A. palmata* rubble pushed shoreward by storm waves. Corals with more robust morphologies or living in deeper water (>10 to 15 m, Fig. 3A) were much less susceptible to physical destruction, so the hurricane increased the rela-

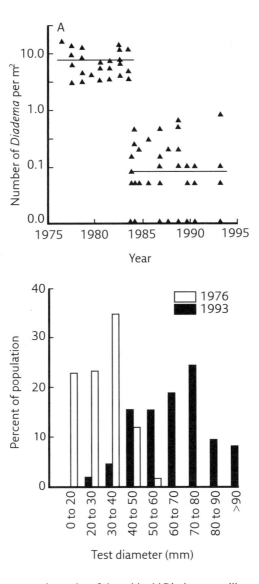

FIGURE 4: *Long-term dynamics of the echinoid* Diadema antillarum *on Jamaican reefs. (A) Abundances over time based on estimates at 14 sites along >100 km of coastline over nearly two decades. Note the 99% drop in 1983 (from a mean of 9 to 0.09 per square meter), with no recovery after 10 years. (B) Population structure of* Diadema *(33) before and after the 1983 die-off.*

tive abundances of species with encrusting or massive-shaped colonies (15, 16). Immediately following Hurricane Allen, there was a short-lived algal bloom (primarily composed of the ephemeral Rhodophyte *Liagora*) probably caused by a pulse of nutrient release from terrestrial runoff and suspended reef sediments and from a temporary depression of herbivory by *Diadema* and other herbivores. Within a few months, however, the algae disappeared and substantial coral recruitment began (16). Recruitment by *Acropora* was minimal and broken fragments survived poorly (17), but other corals, notably brooding agaricids and *Porites*, settled in large numbers onto free space generated by the hurricane (18). For the next 3 years up to 1983, cover increased slowly as the reef began to recover (Fig. 3A). However, recovery from Hurricane Allen was short-lived and was soon reversed by biological events that were less selective and ultimately more destructive and widespread than even this powerful hurricane.

Disease and Algal Blooms (1983 to Present)

The echinoid species *Diadema antillarum* suffered mass mortality from a species-specific pathogen throughout its entire geographic range from 1982 to 1984 (18). In Jamaica, densities of *Diadema* were reduced by 99% from pre–die-off estimates of close to 10 per square meter on shallow fore-reefs, and there has been no significant recovery in the subsequent 10 years (Fig. 4A). Before 1983, *Diadema* were small (19–21), presumably because of food limitation caused by the prevailing high densities of this species (20). Following the die-off, the mean and maximum size of individuals increased greatly, whereas individuals in smaller size classes became uncommon, indicative of low rates of recruitment (Fig. 4B). Individuals today are large, well fed, and have well-developed gonads. However, densities may be too low for effective spawning success because fertilization in *Diadema* is strongly density-dependent (21).

Without *Diadema*, and with the continued depression of herbivorous fish from trapping, the entire reef system of Jamaica has undergone a spectacular and protracted benthic algal bloom that began in 1983 and continues today at all depths (up to 40 m or deeper) (Fig. 3B). Before the echinoid die-off, cover of fleshy macroalgae was typically less than 5% except intertidally, within damsel fish territories, or in very deep water (>25 m) where *Diadema* were scarce (9, 10, 13) (Fig. 3B). In the initial

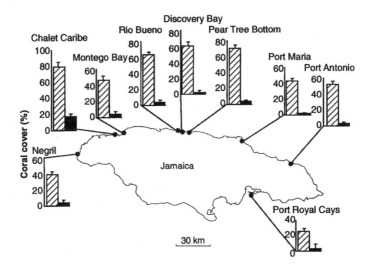

FIGURE 5: *Large-scale changes in community structure at fore-reef sites along >300 km of the Jamaican coastline, surveyed in the late 1970s (1977, hatched bars) and the early 1990s (1993, solid bars) (34).*

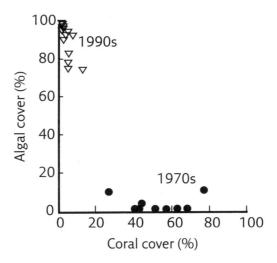

FIGURE 6: *Large-scale community phase shifts on Jamaican reefs, from coral- to algal-dominated systems (34).*

stages of the bloom, algae were small and ephemeral, but within 2 to 3 years weedy species were replaced by longer lived, late successional taxa (notably *Sargassum, Lobophma, Dictyota,* and *Halimeda*) that formed extensive mats up to 10 to 15 cm deep (10, 22). As a result of this preemption of space, larval recruitment by all species of corals has failed for the past decade (16). Most adult colonies that survived Hurricane Allen have been killed by algal overgrowth, especially low-lying species with encrusting or platelike morphologies (16). Additional mortality occurred following bleaching events in 1987, 1989, and 1990 (23). The most abundant coral on the fore-reef today is mound-shaped *Montastrea annularis,* but even this robust, dominant species has declined to 0 to 2% cover at a depth of 10 m in 1993 (24). This decline in a long-lived coral such as *Montastrea* is particularly significant because it is resistant to hurricanes and is the chief frame-builder of Jamaican reefs. Its slow recruitment and growth rate (25) ensure that the decline of the past 10 years will not be reversed for many decades.

The scale of damage to Jamaican reefs is enormous. Censuses at sites 5 to 30 km apart along >300 km of coastline in 1977 to 1980 and again in 1990 to 1993 show a decline in coral cover from a mean of 52 to 3% and an increase in cover by fleshy macroalgae from 4 to 92% (Fig. 5). Indeed, the classic zonation patterns of Jamaican reefs, described by Goreau and colleagues just two to three decades ago (3), no longer exist. A striking phase shift has occurred from a coral-dominated to an algal-dominated system (Fig. 6).

Implications and Prospects for the Future

This spectacular sequence of events highlights the dynamic and complex nature of coral reefs; points to the fundamental importance of fish, herbivory, and recovery of the reefs from physical disturbance to their functioning; and provides a clear demonstration of how quickly (one to two decades) a seemingly healthy coral reef can be severely damaged on a spatial scale similar to the size of most tropical island-nations (hundreds of kilometers). Although it was not widely recognized at the time, Jamaica's reefs were already extensively damaged by the late 1970s (from direct and indirect effects of overfishing) to the extent that the synergistic effects of two subsequent hurricanes and the *Diadema* die-off were sufficient to cause a radical phase shift to algae (Fig. 6). Paradoxically, the changes have occurred although reef systems have demonstrable robustness on a geological time scale. For example, coral reefs have continued to flourish despite major fluctuations in sea level occurring on a time scale of 10^3 to 10^5 years (26). However, the ability of coral reefs to cope with such disturbances in the past is no guarantee of continued resilience in the face of unprecedented and much more rapid anthropogenic stresses. It is highly probable that global reef growth is currently being outpaced by reef degradation (1), with unknown consequences for the future.

A great deal has been learned about the functioning of coral reefs from the litany of disasters described here, and the opportunity should be seized to implement scientifically based management procedures that would facilitate processes of recovery. Clearly, the Jamaican reef system needs more herbivory to allow coral recruitment to resume (27). Herbivorous fish (mostly juvenile scarids) responded immediately to the *Diadema* die-off by changing their spatial distribution and increasing their grazing rates in shallow water (28). However, this behavioral response is unlikely to be reflected later in increased fish abundance because of continued overfishing. Clearly, current stocks of herbivorous fish are not capable of reducing algal abundance in the absence of *Diadema* (Fig. 3B). Similarly, other echinoids have not increased in abundance to compensate for the loss of *Diadema* (10, 29). Recovery of *Diadema* has not yet taken place and is likely to be slow if densities have fallen below some threshold level required for successful spawning (21). Even a full recovery of *Diadema* would leave the reef reliant once more on a single dominant herbivore and vulnerable to a recurrence of disease. Future hurricanes will reinforce rather than reverse the phase shift, as illustrated by the more recent impact of Hurricane Gilbert in 1988. Also a category 5 hurricane, it swept much of the algal covering off the reef and caused further damage to corals. However, the algae recovered fully within a few weeks of Hurricane Gilbert (Fig. 3B), mainly from regenerating filaments and holdfasts, long before successful recruitment of corals could resume. Thus, further hurricanes are likely to act in a ratchet fashion, further depressing coral abundances and favoring the phase shift to algae (Fig. 6).

There is an urgent need, therefore, to control overfishing, a call that had first been made by Munro 20 years ago (7), before more recent events demonstrated the key role of fish and echinoid herbivores in the overall functioning of Jamaica's coral reefs. On the basis of our knowledge of the demography and life histories of fish (7, 8, 30) and corals (25, 31), it will take far longer to rebuild stocks than the two to three decades it has taken to destroy them. Severe, long-term damage has already occurred, and the trajectories of coral and algal abundance (Figs. 3 and 6) predict a gloomy future unless action is taken immediately.

References and Notes

1. H. A. Lessios, P. W. Glynn, D. R. Robertson, *Science* 222, 715 (1983); C. S. Rogers, *Proc. 5th Int. Coral Reef Symp.* 6, 491 (1985); B. E. Brown, *Mar. Polut. Bull.* 18, 9 (1987); B. Salvat, Ed., *Human Impacts on Coral Reefs: Facts and Recommendations* (Antenne Museum Ecole Pratique des Hautes Etudes, French Polynesia, 1987); C. F. d'Elia, R. W. Buddemeier, S. V. Smith, Eds., *Workshop on Coral Bleaching, Coral Reef Ecosystems and Global Change: Report of Proceedings* (Maryland Sea Grant College, College Park, 1991); T. J. Done, *Hydrobiologia* 247, 121 (1992); *Global Aspects of Coral Reefs: Health, Hazards and History* (University of Miami, Miami, FL, 1993).

2. S. M. Wells, Ed., *Coral Reefs of the World*, vol. 1 of *United Nations Environment Program Regional Seas Directories and Bibliographies* (International Union for the Conservation of Nature, Cambridge, 1988). There are over 500 refereed publications since the 1950s based on coral reef research conducted at the Discovery Bay and Port Royal Marine Laboratories, which are on the north and south Jamaican coasts, respectively.

3. T. F. Goreau, *Ecology* 40, 67 (1959); J. Lang, *Am. Sci.* 62, 272 (1973); W.D. Liddell and S. L. Ohlhorst, *Bull. Mar. Sci.* 40, 311 (1987).

4. J. D. Parrish , *Mar Ecol. Prog. Ser.* 58, 143 (1989).

5. D. Hall, *Free Jamaica, 1838–1865: An Economic History* (Yale Univ. Press, New Haven, CT, 1959); B. T. Walsh, *Economic Development and Population Control: A Fifty-Year Projection for Jamaica* (Praeger, New York, 1970); D. Watts, *The West Indies: Patterns of Development, Culture and Environmental Change Since 1492* (Cambridge Univ. Press, Cambridge, 1987); B. T. Walsh, *The Sex and Age Distribution of the World Populations* (United Nations, Department of Economic and Social Development, New York, 1993).

6. G. R. Russ, in *The Ecology of Coral Reef Fishes*, P. F. Sale, Ed. (Academic Press, New York, 1991), chap. 20.

7. J. L. Munro, *ICLARM Stud. Rev.* 7, 1 (1983); *Jam. J.* 3, 16 (1969).

8. J. A. Koslow, F. Hanley, R. Wicklund, *Mar. Ecol. Prog. Ser.* 43, 201 (1988).

9. J. C. Ogden, R. A. Brown, N. Salesky, *Science* 182, 715 (1973); P. W. Sammarco, *J. Exp. Mar. Biol. Ecol.* 45, 245 (1980); R. C. Carpenter, *J. Mar. Res.* 39, 49 (1981); P. W. Sammarco, *J. Exp. Mar. Biol. Ecol.* 65, 83 (1982).

10. T. P. Hughes, D. C. Reed, M. J. Boyle, *J. Exp. Mar. Biol. Ecol.* 113, 39 (1987).

11. J. E. Randall, *Caribb. J. Sci.* 4, 421 (1964); *Stud. Trop. Oceanogr.* 5, 665 (1967); D. R. Robertson, *Copeia* 1987, 637 (1987).

12. A. H. Williams, *Ecology* 62, 1107 (1981); M. E. Hay and P. R. Taylor, *Oecologia* 65, 591 (1985).

13. M. E. Hay, *Ecology* 65, 446 (1984).

14. J. H. Connell, *Science* 199, 1302 (1978); T. P. Hughes, Ed., "Disturbance: Effects on Coral Reef Dynamics," (Special Issue of *Coral Reefs* 12 (no. 3 and 4), 115 (1993).

15. J. D. Woodley et al., *Science* 214, 749 (1981); J. Porter et al., *Nature* 294, 249 (1981).

16. T. P. Hughes, *Ecology* 70, 275 (1989).

17. N. Knowlton, J. C. Lang, M. C. Rooney, P. Clifford, *Nature* 294, 251 (1981); N. Knowlton, J. C. Lang, B. D. Keller, *Smithson. Contrib. Mar. Sci.* 31, 1 (1990).

18. R. P. M. Bak, M. J. E. Carpay, E. D. De Ruyter Van Steveninck, *Mar. Ecol. Prog. Ser.* 17, 105 (1984); H. A. Lessios, D. R. Robertson, J. D. Cubit, *Science* 226, 335 (1984); T. P. Hughes, B. D. Keller, J. B. C. Jackson, M. J. Boyle, *Bull. Mar. Sci.* 36, 377 (1985): W. Hunte, I . Cote, T. Tomascik, *Coral Reefs* 4, 135 (1986); H. Lessios, *Annu. Rev. Ecol. Syst.* 19, 371 (1988); R. C. Carpenter, *Mar. Biol.* 104, 67 (1990).

19. A. H. Williams, *J. Exp. Mar. Biol. Ecol.* 75, 233 (1984).

20. D. R. Levitan, *Ecology* 70, 1419 (1989).

21. R. H. Karlson and D. R. Levitan, *Oecologia* 82, 44 (1990); D. R. Levitan, *Biol. Bull.* 181, 261 (1991).

22. W. D. Liddell and S. L. Ohlhorst, *J. Exp. Mar. Biol. Ecol.* 95, 271 (1986); E. D. de Ruyter Van Steveninck and R. P. M. Bak, *Mar. Ecol. Prog. Ser.* 34, 87 (1986); E. D. de Ruyter Van Steveninck and A. M. Breeman, ibid. 36, 81 (1987); D. R. Levitan, *J. Exp. Mar. Ecol.* 119, 167 (1988).

23. R. D. Gates, *Coral Reefs* 8, 193 (1990); T .J. Goreau and A. H. Macfarlane, ibid., p. 211; J. D. Woodley, unpublished data.

24. Based on estimates of coral cover in 1993 at Rio Bueno, Discovery Bay, Pear Tree Bottom, and Ocho Rios (spanning 40 km of the north Jamaican coast). Twenty replicate 10-m line-intercept transects were run at a depth of 10 m at each site.

25. P. Dustan, *Mar. Biol.* 33, 101 (1975); R. P. M. Bak and M. S. Engel, ibid. 54, 341 (1979); R. P. M. Bak and B. E. Luckhurst, *Oecologia* 47, 145 (1980); K. W. Rylaarsdam, *Mar. Ecol. Prog. Ser.* 13, 249 (1983); C. S. Rogers, H. C. Fitz III, M. Gilnack, J. Beets, J. Hardin, *Coral Reefs* 3, 69 (1984); T. P. Hughes and J. B. C. Jackson, *Ecol. Monogr.* 55, 141 (1985); T. P. Hughes, *6th Int. Coral Reef Symp.* 2, 721 (1988).

26. K. J. Mesolella, *Science* 156, 638 (1967); N. D. Newell, *Sci. Am.* 226, 54 (June 1971); R. W. Buddemeier and D. Hopley, *Proc. 5th Int. Coral Reef Symp.* 1, 253 (1988); J. B. C. Jackson, *Am. Zool.* 32, 719 (1992).

27. There is no evidence that the nationwide algal bloom in Jamaica was caused by increased nutrients, because it occurred throughout the Caribbean immediately following the *Diadema* die-off (16, 20), usually far from sources of pollution. Some groundwater input does occur into the shallow margins of the back-reef at Discovery Bay, which enhances nitrates and reduces salinity close to the shore [C. F. D'Elia, K. L. Webb, J. W. Porter, *Bull. Mar. Sci.* 31, 903 (1981)]. These conditions

produce localized areas around submarine springs, typically 2 to 3 m in diameter, which contain characteristic brackish-water algal assemblages (dominated by *Chaetomorpha*, *Enteromorpha*, and *Ulva*) that are quite unlike those occurring on the reef further offshore. None of the sites in Figs. 3 to 6 are located close to urban areas or point sources of pollution, with the exception of the Port Royal cays on the south coast near Kingston.

28. R. C. Carpenter, *Proc. Natl. Acad. Sci. U.S.A.* 85, 511 (1988); *Mar. Biol. 704*, 79 (1990); D. Morrison, *Ecology* 69, 1367 (1988).

29. Densities of *Echinometra viridis*, *Eucidaris tribuloides*, *Lytochinus williamsi*, and *Trypneustes ventricosus* in 1973 were reported for two Jamaican patch reefs by P.W. Sammarco [*J.Exp.Mar. Biol. Ecol.* 61, 31 (1982)]. The combined total then was 27.5 and 54.0 per square meter, respectively. By 1986, the combined total had fallen two- to threefold (10). In 1993, mean densities (number per square meter ± SE) on these same reefs were 14.0 ± 1.5 and 14.4 ± 1.2.

30. P. F. Sale, Ed., *The Ecology of Coral Reef Fishes* (Academic Press, New York, 1991).

31. J. H. Connell, in *Biology and Geology of Coral Reefs*, O.A. Jones and R. Endean, Eds. (Academic Press, New York, 1973), chap. 7; J. B. C. Jackson, *BioScience* 41, 475 (1991); T. P. Hughes, D. J. Ayre, J. H. Connell, *Trends Ecol. Evol.* 7, 292 (1992).

32. Coral and algal abundance (percent cover) shown here were measured from annual photographs of 10 to 20 permanent 1-m² plots at each depth (7, 10, and 15 to 20 m at Rio Bueno; 35 m at Pinnacle 1). All corals (approximately 38,000 records over 17 years) were traced and digitized to obtain relative abundances, while algal cover was estimated by superimposing a grid of dots on each image (100 per square meter) and counting those covering algae. The small-scale trends reported here for permanent plots mirror almost exactly the results from a larger scale program that was based on replicate 10-m line-intercept transects. For example, in 1993 mean coral cover (± SE) estimated from 20 random transects at each of the 7-, 10-, 15- to 20-, and 35-m stations in Fig. 3A was 5.0 ± 0.8, 5.4 ± 1.2, 5.6 ± 0.9, and 12.8 ± 2.4, respectively. Reef degradation at an even larger scale is shown in Fig. 5.

33. Data for 1976 are from (19), based on a random collection of 97 *Diadema antillarum* from the East Back Reef at Discovery Bay, Jamaica. Data for 1993 are based on 207 individuals from the same site.

34. Coral and macroalgal cover in Figs. 5 and 6 is based on 5 to 10 10-m line-intercept transects run at 10 m from 1976 to 1980 (mostly in 1977 and 1978) on fore-reefs at Negril, Chalet Caribe, Rio Bueno, Discovery Bay (two locations), Pear Tree Bottom, Port Maria, Port Antonio (on the north coast), and Port Royal (on the south coast). These measurements were repeated in 1990 to 1993 with 20 transects, with the addition of five more north coast sites.

35. I thank J. H. Connell, F. Jeal, and J. B. C. Jackson for providing encouragement over 20 years; J. D. Woodley and the staff of Discovery Bay Marine Laboratory for excellent logistic support; M. J. Boyle, G. Bruno, M. Carr, L. Dinsdale, F. Jeal, M. Gleason, S. Pennings, D. Reed, L. Sides, L. Smith, J. Tanner, C. Tyler, and many others for field and lab assistance; and D. Bellwood, H. Choat, T. Done, B. Willis, and the Coral Group at James Cook University (JCU), whose comments improved an early draft

of the manuscript. Supported by the National Science Foundation; the National Geographic Society, the Whitehall Foundation, the Australian Research Council, and JCU. This is contribution no. 133 of the Coral Group at JCU.

Sea Otters and
Kelp Forests in Alaska

Generality and Variation in a
Community Ecological Paradigm

JAMES A. ESTES AND DAVID O. DUGGINS

JAMES A. ESTES, *National Biological Survey, Institute of Marine Sciences. University of California, Santa Cruz, California 95064 USA.*

DAVID O. DUGGINS, *Friday Harbor Laboratories, University of Washington, Friday Harbor, Washington 98250 USA.*

Ecological Monographs, Vol. 65, No. 1. (1995), pp. 75–100.
Reproduced by permission of the Ecological Society of America.

Abstract

MULTISCALE PATTERNS of spatial and temporal variation in density and population structure were used to evaluate the generality of a three-trophic-level cascade among sea otters (*Enhydra lutris*), invertebrate herbivores, and macroalgae in Alaska. The paradigm holds that where sea otters occur herbivores are rare and plants are abundant, whereas when sea otters are absent herbivores are relatively common and plants are rare. Spatial patterns were based on 20 randomly placed quadrats at 153 randomly selected sites distributed among five locations with and four locations without sea otters. Both sea urchin and kelp abundance differed significantly among locations with vs. without sea otters in the Aleutian Islands and southeast Alaska. There was little (Aleutian Islands) or no (southeast Alaska) overlap between sites with and without sea otters, in plots of kelp density against urchin biomass. Despite intersite variation in the abundance of kelps and herbivores, these analyses demonstrate that sea otter predation has a predictable and broadly generalizable influence on the structure of Alaskan kelp forests. The percent cover of algal turf and suspension feeder assemblages also differed significantly (although less dramatically) between locations with and without sea otters.

Temporal variation in community structure was assessed over periods of from 3 to 15 yr at sites in the Aleutian Islands and southeast Alaska where sea otters were 1) continuously present, 2) continuously absent, or 3) becoming reestablished because of natural range expansion. Kelp and sea urchin abundance remained largely unchanged at most sites where sea otters were continuously present or absent, the one exception being at Torch Bay (southeast Alaska), where kelp abundance varied significantly through time and urchin abundance varied significantly among sites because of episodic and patchy disturbances. In contrast, kelp and sea urchin abundances changed significantly, and in the expected directions, at sites that were being recolonized by sea otters. Sea urchin biomass declined by 50% in the Aleutian Islands and by nearly 100% in southeast Alaska following the spread of sea otters into previously unoccupied habitats. In response to these different rates and magnitudes of urchin reduction by sea otter predation, increases in kelp abundance were abrupt and highly significant in southeast Alaska but much smaller and slower over similar time periods in the Aleutian Islands.

The different kelp colonization rates between southeast Alaska and the Aleutian Islands appear to be caused by large-scale differences in echinoid recruitment coupled with size-selective predation by sea otters for larger urchins. The length of urchin jaws (correlated with test diameter, $r^2 = 0.968$) in sea otter scats indicates that sea urchins <15–20 mm test diameter are rarely eaten by foraging sea otters. Sea urchin populations in the Aleutian Islands included high densities of small individuals

(<20 mm test diameter) at all sites and during all years sampled, whereas in southeast Alaska similarly sized urchins were absent from most populations during most years. Small (<30–35 mm test diameter) tetracycline-marked urchins in the Aleutian Islands grew at a maximum rate of ≈10 mm/yr; thus the population must have significant recruitment annually, or at least every several years. In contrast, echinoid recruitment in southeast Alaska was more episodic, with many years to perhaps decades separating significant events. Our findings help explain regional differences in recovery rates of kelp forests following recolonization by sea otters.

Key words: Aleutian Island; bottom-up forces; community structure; growth; herbivory; random sampling; recruitment; scale; size-selective predation; southeast Alaska; top-down forces.

Introduction

Ecological field studies have provided definitive evidence for the population- and community-level influences of many biotic and abiotic processes. Rarely, however, has variation in these results been assessed over large areas or long periods of time. Moreover, the experimental units (i.e., field plots) to which treatments are applied in manipulative, natural, or measurative experiments (sensu Hurlbert 1984) often are purposely selected in field ecological studies so as to reduce natural variation. The tendency to design field experiments around small areas, short time periods, and homogeneous experimental units has the desirable effects of increasing both logistical feasibility and the power of hypothesis testing. However, the cost of these benefits is an inability to rigorously generalize results for larger areas or longer intervals. The problem is illustrated in the largely unresolved and often contested status of broad conceptual issues, such as the importance of competition (Roughgarden 1983), top-down vs. bottom-up forces in biologically structured food webs (Hunter and Price 1992, Matson and Hunter 1992, Menge 1992), and forces acting on larval vs. adult life history stages (Underwood and Denley 1984, Roughgarden et al. 1988), as well as species- or guild-specific paradigms such as sea star predation in mussel beds (Foster 1990, Paine 1992), fish and crustacean predation on tropical gastropods (Palmer 1979, Ortega 1986), and plant benefits from herbivory (McNaughton 1983, Belsky 1986), among many. Thus, one of the most daunting challenges and contentious issues in ecology concerns increasing the extent to which patterns and processes are general.

The generality of sea otter predation as a top-down force in structuring North Pacific kelp forests is a case in point. Several short term studies, conducted over relatively small spatial scales, describe a three-level trophic interaction among sea otters (*Enhydra lutris*), sea urchins (*Strongylocentrotus* spp.), and macroalgae (Estes and Palmisano 1974, Estes et al. 1978, Duggins 1980, Breen et al. 1982, Laur et al. 1988; see Table 1). The evidence is mainly comparative, contrasting nearby sites with and without sea otters that exist because of the fragmentation of a once continuously distributed population across the North Pacific rim. Another approach used in one published account (Laur et al. 1988), which controls for spatial variation not attributable to sea otter predation, has been to measure community structure at specific sites before, during, and after the reestablishment of sea otters. These spatial and temporal comparisons provide the evidence for a well-known ecological paradigm which holds that sea otter predation limits herbivorous invertebrate populations, unlimited herbivore populations limit kelp and other macroalgal populations, and thus, areas inhabited by sea otters support kelp forests whereas areas lacking sea otters are deforested by sea urchin grazing (VanBlaricom and Estes 1988). This paradigm was criticized by Foster and Schiel (1988) and Foster (1990). These authors contended that kelp forest communities in California (and perhaps in the larger North Pacific Ocean region) are organized by numerous biotic and abiotic processes, and that the role of sea otter predation has been overemphasized and overgeneralized. Strong (1992: 749) also questioned the paradigm stating "there is a distinct possibility that this is a donor-controlled system, without effective carnivore suppression of the herbivore."

Published studies, while individually providing unequivocal small-scale evidence for the sea otter paradigm, must be used cautiously to evaluate its generality because in each case study sites were purposefully selected and the number and geographical range of sites is small. Conversely, the large number of studies (Table 1), their extensive collective geographical range, and the near-unanimity of their conclusions argues for the paradigm.

TABLE 1: *Summary of prior field study results on sea otter–herbivore–kelp interactions in the North Pacific Ocean.*

Literature source	Study area	Study period	Conclusions*
Benech 1977	Central California	1975–1976	A
Bowlby et al. 1988	Washington	1987	A
Breen et al. 1982	Vancouver Island	1979	B
Cowen et al. 1982	Central California	1977–1979	D
Duggins 1980	Southeast Alaska	1975–1979	B
Duggins et al. 1989	Aleutian Islands	1985–1987	B
Ebeling et al. 1985	Central California	1979–1983	C
Ebert 1968b	Central California	1967	A
Estes et al. 1978	Aleutian Islands	1970–1972	B
Estes and Palmisano 1974	Aleutian Islands	1970–1972	B
Kvitek et al. 1989	Washington	1987	A
Kvitek et al. 1992	Kodiak Island	1986–1988	A
Kvitek and Oliver 1992	Southeast Alaska	1988–1990	A
Laur et al. 1988	Central California	1976–1981	B
Lowry and Pearse 1973	Central California	1973	A
McLean 1962	Central California	1959–1961	B
Oshurkov et al. l988	Commander Islands	1972–1986	B
Ostfeld 1982	Central California	1977–1979	A
Pearse and Hines 1979	Central California	1972–1977	C
Simenstad et al. 1978	Aleutian Islands	1976–1977	B
Watanabe and Harrold 1991	Central California	1984–1988	C
Watson 1993	Vancouver Island	1986–1991	B
Wild and Ames 1974	Central California	1965–1972	A

* Key to conclusions: A = sea otter predation greatly reduced sea urchin populations (but no consideration of effect on algae); B = sea otter predation on urchins greatly increased algal abundance (or biomass); C = changes in urchin populations due to reasons other than otter predation, caused changes in algal abundance (or biomass); D = sea urchin effect on algae overshadowed by effect of physical disturbance.

The role of otters in California is particularly controversial (Foster and Schiel 1988, Foster 1990). While the critics agree that otters have important effects on kelp forest community structure, they also argue that these effects are overshadowed in most places by other, more important factors, such as physical disturbance. The controversy concerns the role of herbivores in limiting algal assemblages rather than the role of sea otters in limiting herbivore populations. There is ample evidence from California studies that otters rapidly and predictably decimate grazer (particularly urchin) populations (McLean 1962, Ebert 1968a, b, Lowry and Pearse 1973, Wild and Ames 1974, Gotshall et al. 1976, Benech 1977, Laurent and Benech 1977, Pearse and Hines 1979, Ostfeld 1982, Laur et al. 1988).

One purpose of our paper is to provide a more rigorous and detailed account of generality in the sea otter paradigm for kelp forests in Alaska. As with previously published studies, our approach was to compare the abundance and population structure of kelp forest plants (kelps and other macroalgae) and their principal herbivores (sea urchins) across space or through time where the abundance and status of sea otter populations varied naturally. However, our data were taken from randomly selected plots at numerous randomly selected sites in two widely separate regions, thus permitting the analysis of variation over small (<100 m), intermediate (<100 km), and large (>1000 km) spatial scales. As such, our study is the first truly rigorous census of areas differing in sea otter influence.

A second purpose of our paper is to examine temporal variation in the structure of kelp-dominated and urchin-dominated communities. By following a given site over relatively long periods (≤15 yr), three issues were addressed. (1) Time-course data, gathered during periods of sea otter recolonization, provided another rigorous test of the otter–kelp paradigm. (2) Rates of transition between urchin- and kelp-dominated communities following sea otter recolonization were compared among regions. Regional differences were discovered and we examined their causes and consequences. (3) Long-term observation of both the urchin- and otter (kelp)-dominated communities provides an opportunity to evaluate the "persistence stability" (sensu Orians 1975: 141) of these purported equilibria (Simenstad et al. 1978).

We propose that the effects of sea otter predation on sea urchins and kelps can be generalized to large areas. We further propose that kelp- and urchin-dominated assemblages exist as stable equilibria, define the "domains of attraction" (Holling 1973: 4) for each, and describe patterns of temporal variation within and between these domains. Our findings show that although kelp forests and urchin-dominated deforested habitats define two locally stable domains of attraction in Alaskan rocky-reef habitats, the trajectories of change between these domains following the recolonization of sea otters differ markedly among geographic regions. Two ecological processes—size-selective predation by sea otters on sea urchins, and large-scale variation in sea urchin recruitment—appear responsible for these different trajectories. We propose a model based on 1) variation in the intensity and predictability of sea urchin recruitment and 2) the dynamics of plant production and herbivory, to explain differences between the Aleutian Islands and southeast Alaska in recovery rates of kelp forests following the reestablishment of sea otters.

History and Status of Sea Otter Populations in Study Regions

Sea otters, once distributed across the Pacific Rim from northern Japan to central Baja California and numbering in the hundreds of thousands of individuals, were hunted to near-extinction during the 18th and 19th centuries (Kenyon 1969). By the early 1900s about a dozen colonies containing ≤1000 sea otters remained (Lensink 1962). Remnant colonies survived in the Rat and Andreanof island groups of the western and central Aleutian archipelago (Fig. 1). Following their protection in the early 1900s, these colonies increased at rates of 17–20% yr (Chapman 1981, Estes 1990) until resources (food or space) apparently became limiting. The population at Amchitka Island (Rat Island group) contained an estimated 5500–8500 sea otters during the time our study was conducted (Estes 1990). The population at Adak Island (Andreanof Island group), although more poorly known, is roughly comparable in history, size, and density to the Amchitka population (J. A. Estes, unpublished data). Both populations have existed at or near equilibrium density for the past 20–40 yr (Kenyon 1969).

Sea otters were exterminated from the Near Islands, westernmost of the Aleutian archipelago, although records from the fur trade indicate they once were abundant in the area (Lensink 1962). Broad, deep passes separating the Near Islands from the Rat Islands to the east and the Commander Islands to the west apparently prevented recolonization until the mid-1960s, at which time R. D. Jones (1965) counted 13 sea

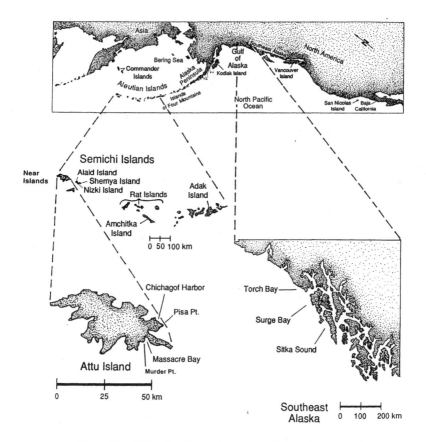

FIGURE 1: *Map of North Pacific Ocean showing study locations, sample sites, and place names referred to in the text.*

otters at Attu Island. The population at Attu increased rapidly thereafter, and when last surveyed in 1991 it had spread around the entire perimeter of the island and contained an estimated 3000–4000 individuals (J. Estes and A. DeGange, unpublished data). The spreading population first appeared at our study sites on the west side of the west coast of Massacre Bay in the mid-1980s (Estes 1990). Two sites at Chichagof Harbor were occupied by sea otters before benthic sampling was begun in 1976. Although there have been occasional sea otter sightings east of Attu in the Semichi Islands (Alaid, Nizke, and Shemya islands), these animals apparently were transients and the Semichi Islands remained uninhabited by otters through the time of our studies there (1987).

In 1968–1971, the Alaska Department of Fish and Game relocated 411 otters (Jameson et al. 1982) from Prince William Sound to southeast Alaska (Fig. 1). The relocated populations had expanded into Surge Bay (on the west coast of Yakobi Island) by the early to mid-1970s and several hundred individuals occurred there in 1978 when we first surveyed the benthic habitat (Duggins 1980). Otters spread into Torch Bay in 1985 (Vequist 1987).

Methods

Spatial Patterns of Community Structure

To assess spatial variation in the structure and composition of rocky-reef communities with and without sea otters, we conducted extensive surveys in the Aleutian Archipelago and southeast Alaska. Henceforth we refer to the Aleutian Islands and southeast Alaska as "regions," islands or place name locations within regions as "locations," study areas within locations as "sites," and sample points within sites as "quadrats."

To assess the generality of the otter–kelp paradigm in Alaska, we sampled 153 randomly selected sites from seven locations at which the history and status of sea otter populations were known: four locations with otters (Amchitka and Adak islands, Torch and Surge bays), two without otters (the Semichi Islands, including Alaid, Nizki, and Shemya islands, and Sitka Sound), and one in transition (Attu Island; Fig. 1, Tables 2 and 3). The Aleutian Islands were sampled in July 1987, Torch and Surge bays in May 1988, and Sitka Sound in July 1988. At each location we first delineated a 10–20 km stretch of coast as the study area, determined by the maximal distances we could travel safely from the bases of operation in a small boat. A grid pattern was then superimposed over a navigational chart of the study area and the grid intersections with shore marked and numbered as potential sites. The sites we sampled were selected at random from among these with no prior knowledge of the benthic flora or fauna. At each location our sample of sites included a broad range of conditions, from the exposed outer coast to protected bays. The substratum was typically consolidated rock or large, stable boulders at all of our sample sites in the Aleutian Islands and southeast Alaska. Cryptic habitats capable of harboring small sea urchins were rare or absent.

Sampling at each site was conducted in the following way. A scuba diver placed 20 0.25-m² quadrats on the sea floor along the 6–7 m (shallow) and 12–13 m (deep) depth contours. Distances between quadrat placements were determined by a prearranged, random number of kicks. We counted individuals of each kelp and sea urchin species and estimated percent cover of fleshy red algae and suspension feeding invertebrates in the quadrats. Although the latter two groups contain numerous species (see Lebednick and Palmisano 1977 and O.Clair 1977 for species lists of marine algae and invertebrates at Amchitka Island) we did not identify these in the field samples because of 1) the time required to do so; 2) difficulties with in situ species identifications; and 3) varying levels of skill and training among the numerous observers.

Percent cover was visually estimated as 1 of 6 categories (i.e., 1 = 0–5%, 2 = 6–25%, 3 = 26–50%, 4 = 51–75%, 5 = 76–95%, 6 = 96–100%) and the site means calculated from the midpoints. Visual estimates can be used quickly and effectively underwater and have been shown to provide measures that are both unbiased and consistent among observers in littoral assemblages (Dethier 1984, Dethier et al. 1993; but see Foster et al. 1991). Although we did not rigorously analyze for inter-observer variation, we doubt that it contributed significantly to areal or regional patterns because our sampling procedures purposefully confounded possible inter-observer biases nearly equally among areas and we (J. Estes and D. Duggins) together gathered much of the data from the Aleutian Islands and most of the data from southeast Alaska.

We identified a single species of sea urchin (*S. polyacanthus*, the green urchin) in our samples from the Aleutian Islands and 3 species (*S. franciscanus, S. droebachiensis*, and *S. purpuratus*; the red, green, and purple urchins, respectively) in southeast Alaska. Densities and size distributions of sea urchin populations were estimated by collecting and measuring animals from randomly placed 0.25-m² quadrats at each of the 153 sites described above. Sampling was terminated at each site after obtaining 200 animals or sampling 20 quadrats. These plots were searched carefully, and individuals ≤2–3 mm test diameter, when present, were easily seen and collected. Sea urchin biomass per 0.25 m² was estimated for each site from that site's urchin density, size distribution, and a regression function of live mass vs. test diameter.

TABLE 2: *Summary statistics from habitat surveys at islands with and without sea otters in the western and central Aleutian Islands.**

Parameter	Location						Statistics		
	Alaid Island[1]	Nizki Island[1]	Shemya Island[1]	Attu Island[2]	Amchitka Island[3]	Adak Island[3]	F	df	P
Understory kelp density (inds./0.25 m²)	0.23[C]	0.79[C]	0.27[C]	1.46[BC]	3.87[AB]	6.56[A]	10.07	5,128	<0.0001
Surface canopy kelp density (inds./0.25 m²)	0.35	0.14	0.37	0.65	0.81	0.85	1.90	5,128	0.098
Turf cover (%)	16.3[C]	12.8[C]	18.1[C]	26.3[C]	98.0[A]	50.4[B]	39.74	5,128	<0.0001
Suspension feeder cover (%)	31.6[CD]	16.0[D]	21.7[CD]	64.8[A]	50.3[BC]	88.2[AB]	9.40	5,128	<0.0001
Sea urchin density (inds./0.25 m²)	28.5	26.3	29.6	39.8	24.6	25.4	0.98	5,124	0.4315
Sea urchin biomass (g/0.25 m²)	336.4[B]	471.4[A]	434.7[AB]	208.6[C]	38.8[D]	38.4[D]	28.00	5,124	<0.0001
Sea urchin max. size (test diameter, mm)	66.4[B]	71.4[A]	77.3[A]	40.5[C]	26.9[D]	26.7[D]	295.87	5,125	<0.0001

* Similarities among locations tested with 1-way ANOVA using locations as treatments and site means within locations as replicates. When H_o was rejected, selected treatment comparisons were done with Duncan's multiple range test ($\alpha = 0.05$). Means with the same letter could not be shown to differ significantly. Analysis of percent cover data done on arcsine transformed values.

[1] Sea otters absent.
[2] Sea otters recently reestablished; population < equilibrium.
[3] Sea otters long established; population at or near equilibrium.

TABLE 3: *Summary statistics from habitat surveys at locations with and without sea otters in southeastern Alaska. Explanations as for Table 2.*

Parameter	Location			Statistics		
	Surge Bay[1]	Torch Bay[2]	Sitka Sound[3]	F	df	P
Understory kelp density (inds./0.25 m²)	9.12[A]	6.43[A]	0.01[B]	18.25	2, 59	<0.0001
Surface canopy kelp density (inds./0.25 m²)	0.63[B]	2.03[A]	0.00[B]	7.07	2, 59	<0.0001
Turf cover (%)	60.6[A]	8.0[B]	4.3[B]	52.52	2, 59	<0.0001
Suspension feeder cover (%)	5.0[A]	0.9[B]	0.3[B]	6.95	2, 59	0.0019
Sea urchin density (inds./0.25 m²)	0.01[B]	0.02[B]	1.52[A]	35.38	2, 55	<0.0001
Sea urchin biomass (g/0.25 m²)	0.2[B]	0.1[B]	369.9[A]	65.23	2, 55	<0.0001

[1] *Sea otters long established.*
[2] *Sea otters recently reestablished.*
[3] *Sea otters absent.*

Temporal Patterns of Community Structure

The persistence of deforested or kelp-dominated communities was evaluated at four locations over periods of time ranging from 3 to 15 yr during which the status of otter populations remained unchanged. The categories of data are not entirely consistent among regions or times because the information was obtained from two independent research programs in which we employed different methodologies.

Locations lacking sea otters.—Three sites at Torch Bay were sampled during the late summers of 1976, 1977, and 1978. Two sites at Shemya Island were sampled during the summers of 1972 and 1987.

Locations with sea otters.—Five sites were sampled at Surge Bay during the summers of 1978 and 1988. Four sites at Amchitka Island were sampled in the summers of 1972 and 1987.

The extent and rate of change in benthic community structure following sea otter recolonizations were determined by comparing census data from benthic communities at Attu Island and Torch Bay both before and after otters arrived. Nine sites at Attu were sampled in the summers of 1976, 1977, 1979, 1981, 1983, 1986, 1987, and 1990, during which time the growing sea otter population spread into these sites. At Torch Bay, four sites were sampled in 1978 before sea otters became reestablished, and again in May 1988, ≈2 yr after their reestablishment.

Sea Urchin Growth and Recruitment

A common feature of sea urchin populations in the western Aleutian Islands was that they contained large numbers of small individuals (<15 mm test diameter). In contrast, populations in southeast Alaska contained few small animals. Because it was unclear whether these small animals in the Aleutian Islands represented frequent recruitment or slow growth, we used tetracycline markers (following the methods of Kobayashi and Taki 1969, Ebert 1975, 1980, 1982, Pearse and Pearse 1975, and Russell 1987) to measure their growth rates in situ. This was done by injecting tetracycline hydrochloride (1 mg/100 mL seawater) through the peristomial membrane. Dosages varied with test diameter as follows: <16 mm, 0.02 mL; 17–22 mm, 0.05 mL; 23–28 mm, 0.10 mL; 29–32 mm, 0.15 mL; 33+ mm, 0.20 mL; as recommended by Kobayashi and Taki (1969) and Pearse and Pearse (1975).

We marked sea urchins (*S. polyacanthus*) at Attu Island in isolated tide pools near the sublittoral fringe to increase the likelihood of recovering the same animals 1 yr later. Growth rates of animals in these habitats probably were as high or higher than those in sublittoral habitats because 1) algal drift from the littoral zones and sublittoral fringe frequently washes into such pools, thus increasing food availability and growth rates (Vadas 1977), and 2) urchins with the largest gonads and test diameters that we have found in the western Aleutian Islands occurred in tide pools (Mayer 1980). We marked every sea urchin found in each of two tide pools on the west side of Murder Point on 23 July 1986 and collected them on 14 July 1987. The sea urchins in two additional tide pools on Pisa

Point were similarly marked on 18 July 1986 and collected on 15 July 1987. In total, 486 animals were marked and 496 collected (there was some recruitment or immigration into the pool); 274 of the collected animals had recognizable tetracycline marks. At the time of collection, the test diameter of each animal was measured and its Aristotle's lantern (a calcified feeding structure) removed. Later we dissolved the remaining soft tissue from Aristotle's lantern in bleach (a dilute NaOCl solution), removed the jaws, and examined them under ultraviolet light through a dissecting microscope with an ocular micrometer. Regions of the jaw in which tetracycline had been incorporated fluoresced. When such marks were present, we measured the growth increments between their proximal margins and the respective ends of the jaw. Jaw length when marked was estimated by subtracting these increments from jaw length at the time of collection. Since jaw length and test diameter are well correlated ($r^2 = 0.968$, Simenstad et al. 1978), we estimated test diameter from jaw length at the time of marking. Jaw length at the time an animal was marked was plotted against growth increment at the time it was collected. However, the Richards variable growth functions (Ebert 1980a) could not be parameterized so as to produce realistic growth rates for animals < 5 mm or > 45 mm test diameter at the time they were marked. Therefore average growth rates were estimated for 5 mm intervals in test diameter (i.e., 1– < 5, 5– < 10 mm, etc.) and the size–age relation determined by assuming the urchins settled at 1 mm test diameter and adding the appropriate growth increments at 0.1-yr intervals.

The frequency and intensity of urchin recruitment was assessed from population size distributions. While size is not a precise indicator of age in urchins, the virtual absence of small size classes at several locations was taken as an indicator of infrequent recruitment.

Size Selective Foraging by Sea Otters

Prior studies of sea urchin populations from Aleutian Islands with and without sea otters demonstrated that 1) large sea urchins (>35–40 mm test diameter) were absent from exploited populations and disappeared rapidly (<1–2 yr) in the wake of expanding sea otter populations, 2) sea urchins >60 mm test diameter were common in unexploited populations, and 3) even areas with sea otter populations at or near equilibrium density supported sea urchin populations with high densities of small individuals (<35–40 mm test diameter) (Estes 1978, Simenstad et al. 1978, this study). These observations, together with the common presumption that sea otters, like many other predators, forage so as to maximize net rate of energy intake (Krebs 1978), led us to hypothesize that otters preferentially consumed the largest available sea urchins. We tested this idea by collecting sea otter scats from Attu and Amchitka islands. The scats were softened in warm water, sea urchin remains separated from organic debris and the remains of other species, and the length of all jaws measured under a dissecting microscope. Size distributions of sea urchins eaten by sea otters were estimated from these data and compared with size distributions of living populations from the same areas.

Results

Spatial Variation in Community Structure
Kelp density.—Understory and surface-canopy kelp species (e.g., Dayton 1975, Duggins 1980) were analyzed separately because of their structural and ecological differences. Understory species seldom grow >2 m above the substratum whereas the surface-canopy species may grow from >30 m depth and often accumulate most of their biomass at the ocean surface. The understory kelps are typically competitive dominants over surface-canopy species in both the western Aleutian Islands (Dayton 1975) and southeast Alaska (Duggins 1980). Common species are listed in Table 4. *Laminaria* spp. composed most of the biomass and individuals in habitats that were shallow (<about 10 m depth) or exposed to moderate to heavy wave-generated surge in both regions. *Agarum* spp. (and *T. clathrus* in the Aleutian Islands) were more common in deeper and more protected habitats (Dayton 1975, Estes et al. 1978), especially where there was moderate grazing by sea urchins. The surface canopy was formed by a single species (*Alaria fistulosa*) in the Aleutian Islands and by three species (*A. fistulosa, Nereocystis leutkeana*, and rarely *Macrocystis integrifolia*) in southeast Alaska.

Understory kelp density in southeast Alaska varied strikingly between locations with and without sea otters (Fig. 2, Table 3). Understory kelp densities did not vary significantly between Surge and Torch bays (Table 3) even though sea otters had occupied these locations for 1–2 (Torch) and ≈20 yr (Surge). This suggests that kelp assemblages in southeast Alaska

TABLE 4: *Relative abundance (%) of species in kelp assemblages of the western Aleutian Islands and southeast Alaska. Calculations based on total plants counted from all quadrats sampled at each location.*

	Location			
	Aleutian Islands		Southeast Alaska	
Kelp species	Adak	Amchitka	Surge	Torch
Agarurm cribrosun	3.3	21.3	2.6	0.0
Alaria fistulosa	12.0	16.1	0.0	2.9
A. marginata	a*	a	6.0	14.5
Costaria costata	a	a	0.7	1.5
Cymathera triplicata	0.2	0.9	1.2	0.9
Laminaria spp.	83.9	58.0	75.5	61.5
L. yezoensis	0.0	2.4	5.4	2.4
Macrocystis intregrifoli	a	a	0.2	0.0
Nereocystis leutkeana	a	a	0.7	6.9
Pleurophycus gardneri	a	a	7.6	9.4
Thalassiophyllum clathrus	0.6	1.4	a	a

* a = *absent (outside geographic range of species).*

proliferate quickly following sea otter recolonization, progress rapidly through a successional sequence ending in the dominance of understory species, and remain largely unchanged thereafter.

Patterns of variation in understory kelp abundance between locations in the Aleutian Islands with and without sea otters were similar to those measured in southeast Alaska (Fig. 2). The abundance of understory kelps among quadrats sampled at Attu Island (Fig. 2) was intermediate to and significantly different from those measured at Alaid-Nizki-Shemya islands and Adak-Amchitka islands (Table 2), thus suggesting that kelp recovery following sea otter recolonizations occurs more slowly in the western Aleutian Islands than in southeast Alaska.

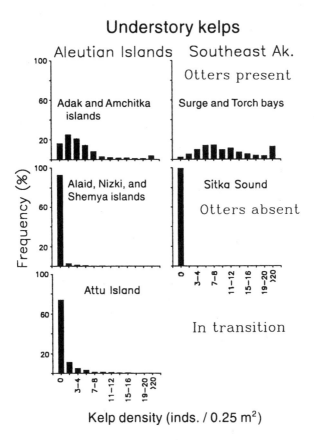

FIGURE 2: *Frequency distributions of understory kelp densities (inds./ 0.25 m²) from random samples taken at A) Torch and Surge bays; B) Sitka Sound: C) Adak and Amchitka Islands: D) Alaid, Nizki, and Shemya islands: and E) Attu Island.*

Less extreme differences were measured for densities of surface-canopy-forming kelp between locations with and without sea otters in southeast Alaska and the Aleutian Islands (Fig. 3). These differences probably resulted from competitive dominance (and exclusion) of surface canopy kelps by understory species and the longer time this interaction had been occurring in Surge Bay.

Urchin density and biomass.—Sea otters had predictable though different effects on sea urchin populations in southeast Alaska and the Aleutian Islands. In southeast Alaska, sites without sea otters (Torch Bay 1976–1978 and Sitka Sound 1988) supported dense urchin populations (Tables

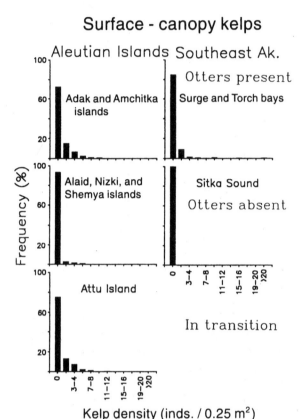

FIGURE 3: *Frequency distributions of surface-canopy forming kelps (inds./ 0.25 m²) from random samples taken at A) Torch and Surge bays; B) Sitka Sound; C) Adak and Amchitka islands; D) Alaid, Nizki, and Shemya Islands; and E) Attu Island.*

3 and 6). We observed numerous broken urchin tests in Surge Bay in 1978 and Torch Bay in 1988, and urchins were nearly absent from all sites at otter-dominated locations in southeast Alaska (Torch Bay 1988, Surge Bay 1978 and 1988) (Tables 3 and 6).

In contrast, both sites with and without sea otters in the Aleutian Islands supported dense urchin populations (Tables 2 and 5) of various size distributions. At sites lacking otters, urchin density and biomass was greatest just below the sublittoral fringe, whereas these measures increased with depth at sites with otters (Fig. 4). Sea urchin density was

TABLE 5: *Abundance and population characteristics of kelps and sea urchins at Amchitka and Shemya islands in 1972 and 1987 (shown as means ± 1 SE, across sites). The same four sites at Amchitka and two sites at Shemya were sampled in both years*. Sea otters were continuously abundant at Amchitka and absent from Shemya during the 15-yr period.*

	Amchitka Island		Shemya Island	
	1972	1987	1972	1987
Kelp species (inds./0.25 m^2)				
Alaria fistulosa	1.6 ± 1.30	0.3 ± 0.22	0	0.5
Laminaria spp.	2.3 ± 0.49	3.9 ± 0.95	0	0
Agarum cribrosum	1.2 ± 0.61	0.5 ± 0.42	0	0
Thalassiophyllum clathrus	0.1	0	0	0
Total kelps	5.1 ± 0.66	4.7 ± 1.15	0	0.5
Sea urchins				
Maximum test diameter (mm)	30.5 ± 1.34	27.3 ± 3.24	72.5 ± 0.71	70.5 ± 4.95
Biomass (g/0.25 m^2)	45.1 ± 16.9	36.7 ± 15.0	368.2 ± 151.7	369.3 ± 14.3
Density (inds./0.25 m^2)	27.9 ± 14.5	23.4 ± 7.5	50.0 ± 14.6	38.6 ± 1.4

*The 1972 data were obtained from 10 haphazardly placed 0.25-m^2 quadrats/site, the 1987 data from 20 randomly placed 0.25-m^2 quadrats/site.

greater at Attu Island than it was at either Adak-Amchitka or Alaid-Nizki-Shemya islands. However, sea urchin biomass at Attu was intermediate. This discrepancy resulted from the fact that foraging sea otters selectively removed the largest urchins at Attu (see following section on size-selective predation), thus eliminating the major contributors to biomass and perhaps releasing the population from intraspecitic competition, which in turn may have caused the number of small individuals to increase.

Sea urchin size structure.—The size distributions of sea urchin populations in the Aleutian Islands differed between locations with and without sea otters (Fig. 5). Maximum test diameter ranged from 45–85 mm among the 42 sites (297 quadrats, 6915 individuals) sampled at Alaid,

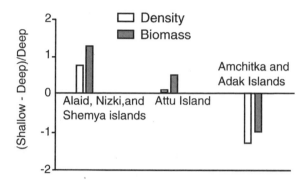

FIGURE 4: *Relative differences in sea urchin density and biomass between shallow (6–7-m depth) and deep (13–15-m depth) sites at locations where the population status of sea otters differed. Deep water values were arbitrarily chosen as standards for comparison (e.g., a quotient of +1 indicates a 2x greater value in shallow water, a quotient of -2 indicates a value 3x lower in shallow water, etc.). Alaid, Nizki, and Shemya islands (otters absent); Attu Island (otters recently reestablished); Adak and Amchitka Islands (otters long established, at or near population equilibria).*

Nizki, and Shemya islands and all but one of these sites contained individuals >56 mm. In contrast, the largest animal from 75 sites (483 quadrats, 7505 individuals) sampled at Adak and Amchitka islands was 39 mm and most (>99%) were <30 mm.

Sea urchins as small as 2–3 mm test diameter occurred at all six locations sampled in the Aleutian Islands, and individuals < 10 mm comprised 11–31% of the total (Fig. 6). Minimum size could not be shown to vary significantly (one-way ANOVA, $F_{5, 25} = 1.79$, $P = 0.12$) among sites within or among locations (Fig. 5).

Small sea urchins were sparse or absent at deforested sites in southeast Alaska (Fig. 5). Urchin size distributions were not determined at otter-dominated sites in southeast Alaska because they were virtually absent.

Kelp density vs. sea urchin biomass.—There was little (Aleutian Islands) or no (southeast Alaska) overlap in the distributions of kelp density vs. sea urchin biomass between locations with and without sea otters (Fig. 7). Kelp density was high and variable whereas sea urchin biomass was consistently low among sites with sea otters. In contrast, sites lacking

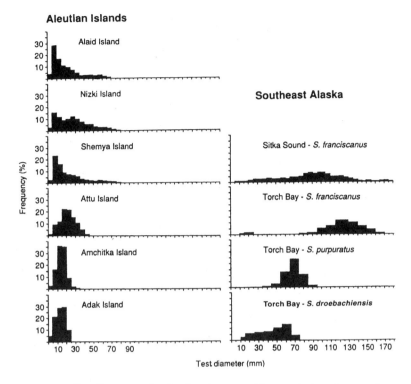

FIGURE 5: *Size-frequency distributions in sea urchin populations from study locations in the Aleutian Islands and southeast Alaska. Data from the Aleutian Islands are for the green urchin,* Strongylocentrotus polyacanthus, *and were obtained in 1987. Data from Torch Bay were obtained in 1978 and from Sitka Sound in 1988. Frequency distributions were obtained by pooling samples taken over all sites at the respective locations.*

sea otters had comparatively low and invariate kelp densities with high and variable sea urchin biomass (Tables 2 and 3).

Different distributions of kelp density vs. urchin biomass are evident between the Aleutian Islands and southeast Alaska. The most obvious of these was that sea urchin biomass was higher at sites where otters were present in the Aleutian Islands than at similar sites in southeast Alaska (Tables 2 and 3). Significant differences could not be demonstrated between Adak and Amchitka islands or Torch and Surge bays.

Algal turfs.—This group, which is comprised largely of red algae, contains numerous undescribed taxa in western Alaska, and field identification of

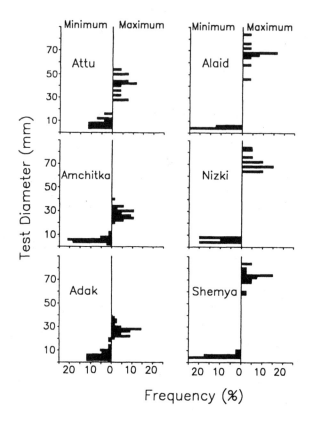

FIGURE 6: *Frequency distributions of the maximum and minimum sizes of sea urchins* (S. polyacanthus) *among sites sampled across study locations in the Aleutian Islands. The population status of sea otters at each study location is noted in Fig. 4 legend.*

many of the described taxa is difficult. Consequently, we have not differentiated the members of this assemblage. Except for crustose forms and erect coralline algae, all functional groups (sensu Littler and Littler 1980, Steneck and Watling 1982, Steneck and Dethier 1994) were included.

Patterns of algal turf abundance between locations with and without sea otters (Fig. 8) were similar to but less extreme than those for understory kelps. Mean turf cover differed significantly between Sitka Sound vs. Surge and Torch bays (one-way ANOVA on arcsine-transformed data, $P < 0.0005$). Similar patterns occurred in the Aleutian Islands (Fig. 8). Mean turf cover at Attu Island (10.8%) and the frequency distribution of

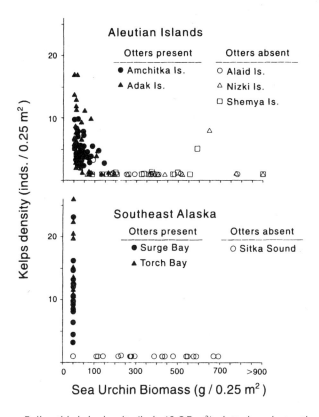

FIGURE 7: *Epibenthic kelp density (inds./0.25 m²) plotted against estimated sea urchin biomass (g/0.25 m²) for the Aleutian Islands and southeast Alaska. Points represent averages for sites within locations. Sea urchin biomass was estimated from samples of population density, size–frequency distribution, and the functional relation between test diameter and wet mass.*

estimated turf cover among quadrats (Fig. 8) were intermediate between those measured at Adak-Amchitka and Alaid-Nizki-Shemya, although they were more similar to the latter. Mean turf cover differed Adak Islands significantly between Adak-Amchitka and Alaid-Nizki-Shemya islands; turf cover at Attu Island also differed significantly from similar measures at Adak-Amchitka but did not differ significantly from those at Alaid-Nizki-Shemya (Table 2).

Suspension feeders.—Suspension feeders comprise a taxonomically diverse group of organisms including bivalves, barnacles, holothurians,

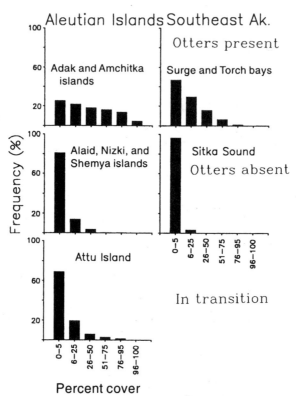

FIGURE 8: *Frequency distributions of estimated percent cover of red algal turf from random samples taken at A) Torch and Surge bays; B) Sitka Sound; C) Adak and Amchitka Islands; D) Alaid, Nizki, and Shemya Islands; and E) Attu Island.*

cnidarians, bryozoans, sponges, and tunicates. Members of these taxa consume particulate matter derived directly or indirectly from kelp and other macroalgae (Duggins et al. 1989), are consumed by foraging sea urchins (Vance 1979), and thus might be expected to differ between locations with and without sea otters. Alternatively, suspension feeders may compete with algae for space, an interaction that would decrease their abundance at locations with sea otters.

Differences in the percent cover of suspension feeders between locations with and without sea otters were apparent but not striking (Fig. 9).

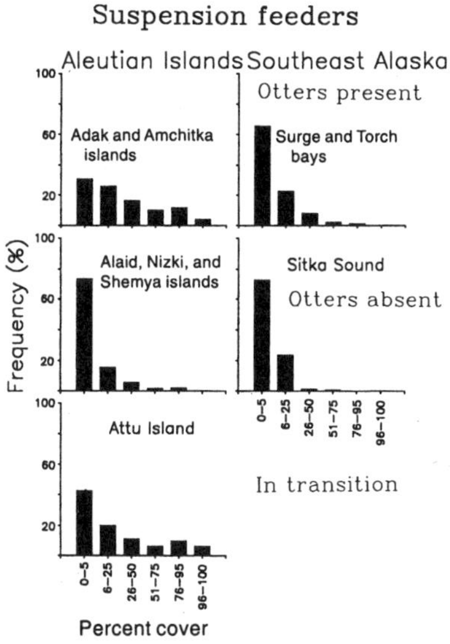

FIGURE 9: *Frequency distributions of estimated percent cover of suspension feeding invertebrates from random samples taken at A) Torch and Surge bays; B) Sitka Sound; C) Adak and Amchitka islands; D) Alaid, Nizki, and Shemya islands; and E) Attu Island.*

Mean percent cover of suspension feeders was greater, although not statistically significant (one-way ANOVA on arcsine-transformed data, P = 0.075) in Surge and Torch bays than in Sitka Sound. Similar patterns occurred in the Aleutians, although differences between locations with and without sea otters were larger and highly significant (P < 0.0001). As was true for kelp abundance and algal turf cover, the percent cover of suspension feeders at Attu was intermediate between Alaid-Nizki-Shemya islands and Adak-Amchitka islands (Fig. 9).

Temporal Patterns in Community Structure
Sites with no change in status of otters.—We surveyed varying numbers of sites at 2 locations in the Aleutian Islands (Amchitka and Shemya islands) and 2 locations in southeast Alaska (Surge and Torch bays) over periods ranging from 3 to 15 yrs. There was no change in the status of sea otters over the reported time periods at any of these locations; otters were continuously present at Amchitka Island and Surge Bay, and continuously absent at Shemya Island and Torch Bay.

TABLE 6: *Abundance and population characteristics of kelps and sea urchins at Torch Bay (1976–1978) and Surge Bay (1978 and 1988) shown as means ± 1 SE, across sites. Sea otters were continuously absent at Torch Bay and present at Surge Bay during these time periods.*

	Torch Bay			Surge Bay	
	1976	1977	1978	1978	1988
Kelps (inds./m^2)					
Annuals[1]	2.1 ± 1.39	0.2 ± 0.25	11.6 ± 6.69	2.1 ± 0.45	3.7 ± 2.34
Perennials[2]	0.1 ± 0.11	0	0.9 ± 1.14	48.4 ± 6.33	50.3 ± 7.46
Total	2.2	0.2	12.5 ± 5.56	50.5 ± 6.43	54.0 ± 9.33
Sea urchins (inds./m^2)					
S. franciscanus	3.6 ± 3.05	3.8 ± 2.55	4.9 ± 3.71	0	0
S. purpuratus	1.0 ± 0.75	2.3 ± 2.52	0.3 ± 0.41	0	0
S. droebachensis	3.4 ± 2.24	1.5 ± 0.95	0.2 ± 0.18	0.02	0.04
Total	8.0 ± 4.56	7.6 ± 5.78	5.4 ± 4.27	0.02	0.04

[1] *Primarily* Alaria fistulosa *and* Nereocystis leutkeana.
[2] *Primarily* Laminaria groenlandica.

The two locations with sea otters (Amchitka Island, Table 5; and Surge Bay, Table 6) were remarkably consistent among years for all taxa surveyed. Kelp density at Amchitka Island could not be shown to vary significantly among sites or among years (two-way ANOVA, $P = 0.70$ for sites and 0.69 for years). Urchin density, biomass, and maximum test diameter were also similar among years and sites (Density: $P = 0.79$ for sites and 0.75 for years; Biomass: $P = 0.67$ for sites and 0.65 for years; maximum test diameter: $P = 0.41$ for sites and 0.22 for years). In 1978, only one sea urchin was found in 80 1-m^2 quadrats sampled over the five sites at Surge Bay, and in 1988 only two urchins occurred in 100 0.25-m^2 quadrats sampled at each of the same five sites. Kelp abundance also did not vary significantly among sites ($P = 0.17$) or over time ($P = 0.70$, two-way ANOVA).

Patterns of variation were similar at the two locations without otters, with several notable exceptions. There was little temporal variation for any of the measured groups or taxa in the Aleutian Islands. Kelps were nearly absent in both 1972 and 1987 from the two sites at Shemya Island where we have data spanning the 15-yr period. Urchin densities, biomass, and maximum test diameter were consistently high at both sites in both years (Density: P = 0.44 for sites and 0.44 for years; Biomass: P = 0.44 for sites and 0.99 for years; maximum test diameter: P = 0.56 for sites and 0.25 for years).

A shorter time interval (1976–1978) is available for Torch Bay because otters had recolonized the location before we returned in 1988. However, the pattern over three consecutive years was considerably more variable than it was at any of the three locations described above. The patchiness of urchins in space, and the dynamic nature of these patches (Duggins 1983) led to high among-site and among-year variation (Table 6). Kelp density varied significantly among the three years ($F_{2,4}$ = 8.24, P = 0.038) but not among sites ($F_{2,4}$ = 2.18, P = 0.23). Urchin density varied significantly among sites ($F_{2,4}$ = 25.62, P = 0.005) but not among years ($F_{2,4}$ = 1.10, P = 0.42). In general, kelp density was low; however, annual species of kelp were abundant at two sites in 1978 because of anomalous physical and biological conditions (Duggins 1981). While kelp populations at Torch Bay in the absence of sea otters were unpredictable and variable (Table 6), the benthic community was nonetheless controlled by herbivory, as demonstrated by experimental urchin removals (Duggins 1980) and by changes that followed the eventual recolonization of sea otters.

Changes following sea otter recolonization. — In contrast with the generally low spatial and temporal variation in populations of herbivores and benthic algae at otter-free locations, changes that followed sea otter recolonization at each location were striking. We surveyed benthic species before and after the arrival of sea otters from nine sites at Attu Island and four sites at Torch Bay. Only data on kelp and sea urchin abundance are presented here.

Kelp density and urchin biomass at the pre-otter Torch Bay sites (1978: Table 6) were similar to those measured at Sitka Sound in 1988 (Table 3, Figs. 2 and 7). In 1988, 2 yr after sea otters recolonized Torch Bay, kelp density had increased 85x and urchin biomass had declined >10 000x. The post-otter kelp density and urchin biomass measured in Torch Bay were

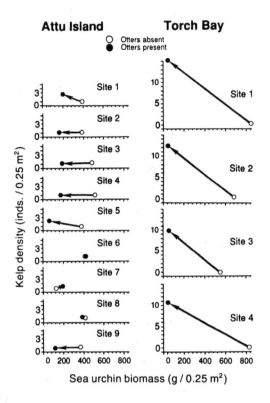

FIGURE 10: *Epibenthic kelp density (inds./ 0.25 m²) vs. estimated sea urchin biomass (g/ 0.25 m²) measured before and after the recolonization of sea otters from nine sites at Attu Island (1983, otters absent; 1987, otters present) and four sites at Torch Bay (1978, otters absent; 1988, otters present). Note: sea otters recolonized the Torch Bay sites in 1985–1986 and the Attu Island sites in 1984–1985, so both locations were sampled about 2 yr after being recolonized by sea otters.*

within the range of variation observed at other sites in southeast Alaska where sea otters were present (compare Figs. 7 and 10). Both kelp and urchin density at the post-otter sites in Torch Bay differed significantly from values obtained from these same sites before the arrival of otters (paired t tests: kelp, $t_7 = 3.626$, $P = 0.036$; urchins, $t_7 = 3.256$, $P = 0.047$) as well as from sites sampled in Sitka Sound (Table 3). Neither differed significantly from sites sampled in Surge Bay where sea otters were present (see previous section, Table 3).

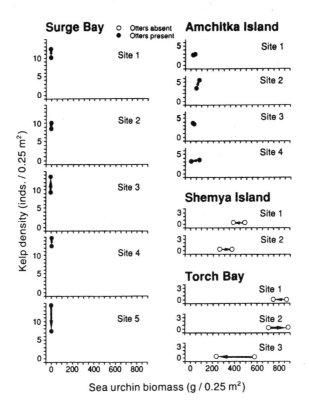

FIGURE 11: *Epibenthic kelp density (inds./0.25 m²) vs. and estimated sea biomass (g/0.25 m²) measured through time at locations where the status of sea otter populations did not change (Surge Bay 1978 and 1988, sea otters present; Amchitka Island 1972 and 1987, sea otters present; Shernya Island 1972 and 1987, sea otters absent; Torch Bay 1977 and 1979, sea otters absent).*

Samples taken from the five Surge Bay sites in 1978 and 1988 did not differ significantly in either kelp abundance or sea urchin biomass (Fig. 11; see previous section), thus indicating that the temporal changes observed at Torch Bay during this same time interval were caused by the arrival of sea otters rather than larger scale temporal changes in kelp forest community structure in southeast Alaska.

In southeast Alaska, sea urchin population changes in response to sea otter predation were rapid and extreme. Sea urchins were the most conspicuous macroinvertebrate in Torch Bay from 1975–1979, during which time population density and biomass averaged 4.5 individuals/0.25 m²

TABLE 7: *Density and biomass of sea urchin populations before and after reestablishment of sea otters via natural range expansion into Massacre Bay, Attu Island and into Torch Bay, southeast Alaska.*

	Population density (inds./0.25 m²)			Biomass (g/0.25 m²)		
Location/Species	Before	After	% diff.	Before	After	% diff.
Massacre Bay						
S. polyacanthus	31.5	39.6	+25.7	374.9	180.7	-51.8
Torch Bay						
S. francisanus	1.48	0.0	-100	982.8	0.0	-100
S. droebachiensis	1.22	0.0	-100	161.5	0.0	-100
S. purpuratus	1.45	0.05	-96.6	76.3	0.3	-99.6

TABLE 8: *Two-way ANOVA of the effects of location (Aleutian Islands vs. southeast Alaska) and the reestablishment of sea otters (before vs. 1–2 yr after) on kelp density and sea urchin biomass.*

Source	df	F	P
Kelp density			
Region	1, 22	44.78	0.0001
Time	1, 22	82.91	0.0001
R X T	1, 22	77.61	0.0001
Sea urchin biomass			
Region	1,22	11.76˙	0.0024
Time	1,22	32.23	0.0001
R X T	1,22	27.75	0.0001

and 1221 g/0.25 m², respectively (Table 6). By May 1988, ≈2 yr after sea otters had recolonized in Torch Bay, urchins were virtually absent from our four long-term sites as well as from 14 other sites sampled only in 1988. *Strongylocentrotus franciscanus* and *S. purpuratus* had disappeared from the four long-term sites and the density and biomass of *S. droebachiensis* had declined >96% and >99%, respectively (Table 7). In Surge Bay during May 1988, sea urchin density and biomass remained near zero at the five long-term sites, as well as at 15 other sites sampled only in 1988.

The short-term changes in kelp density and sea urchin biomass at nine Attu Island sites following the arrival of sea otters differed from those seen in southeast Alaska (Fig. 10). Kelp density at Attu did not change significantly over time within sites (paired t test, t_8 = 1.503, P = 0.17). Changes in urchin biomass at Attu Island also were small compared with those that occurred in southeast Alaska (Fig. 10).

The broad patterns of community structure, their changes in response to sea otter predation, and the manner in which these changes differ between the Aleutian Islands and southeast Alaska are captured in a two-way ANOVA of kelp density and sea urchin biomass (Table 8): Both measures differed significantly between the Aleutian Islands and southeast Alaska, and although sea otter predation ultimately drove the configuration of sublittoral reef communities from deforested habitats to kelp forests at both locations, the highly significant Region x Time interaction reflects the much more rapid effect in southeast Alaska.

Three main patterns in urchin demography and population structure are evident from our time series of data in the Aleutian Islands: (1) recruitment was frequent and broadly occurring; (2) population density and size structure were constant at locations where otter population status remained unchanged; and (3) population density and size structure changed at locations that were recolonized by otters.

Temporal variation in the Aleutian Islands in the rate and extent of response of sea urchin populations to sea otter predation is evident from samples taken during 1972–1990 at Attu Island. Two sites (Pisa Point, northeast coast, and Murder Point, southeast coast) were sampled intermittently during this period. Sea otters were present at Pisa Point in roughly constant numbers throughout the period. In contrast, otters were absent from Murder Point through the 1970s, began spreading into the area in the early 1980s, and were common at our study sites by the mid-1980s (Estes 1990).

Sea Urchin Size

1. *Recruitment.*—Relatively large numbers of urchins with test diameters <10 mm were present in all sample years at both sites (Fig. 12), thus indicating either strong annual recruitment or very slow growth. In situ growth measures argue against the latter explanation (see following section). This finding and the prevalence of small animals over a broad spatial scale in 1987 (Figs. 3 and 5) indicate that sea urchins recruited heavily to most areas in the western and central Aleutian Islands during all or most of the nearly 20 yr of our study.

2. *Size structure.*—Although the exact form of the sea urchin size-frequency distribution at Pisa Point varied among the six sample years (especially 1979 when individuals of test diameter 14–18 mm were relatively abundant), minimum and maximum sizes changed little (Fig. 12, Table 9). Maximum size declined from ≈48 to 40 mm during this period, perhaps because of continuing or increased intensity of size-selective predation by sea otters. Overall, the size distributions of sea urchins at Pisa Point were more similar to those measured at Adak and Amchitka islands than they were to those measured at islands without otters (Table 9).

   Maximum urchin size at Murder Point declined markedly following the spread of sea otters into this area (Table 9), and the size distribution eventually converged upon those measured at other locations in the Aleutian Islands with established sea otter populations (i.e., Pisa Point, Adak and Amchitka islands). In 1972 and 1976 respectively, prior to the arrival of otters in Massacre Bay, 26.2 and 21.9% of the urchins were >50 mm, and in both years individuals >80 mm were obtained. These population characteristics were similar to those from other areas lacking sea otters in the Aleutian Islands (i.e., Alaid, Nizki, and Shemya islands). A few otters had spread into Massacre Bay by 1983 (10 were counted in the west Massacre Bay-Murder Point survey areas in 1983 and seven more in areas to the west [Estes 1990]), and although some urchins with test diameters >80 mm still occurred at that time, those >50 mm had declined to comprise only 3.6% of the population. Sea otters were well established in the area by 1986 (29

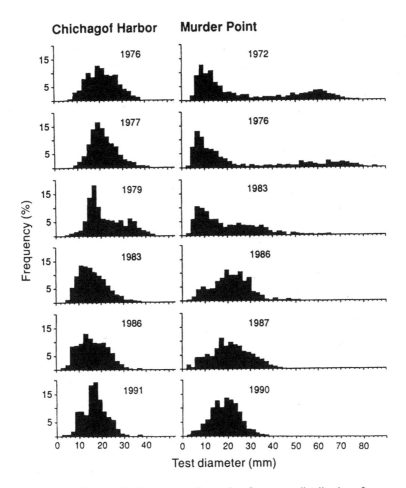

FIGURE 12: *Sea urchin (S.* polyacanthus*) size-frequency distributions from Attu Island (two sites at Chichagof Harbor and two sites at Murder Point) sampled at various times between 1972 and 1991. Sea otters were continuously present at Chichagof Harbor during the years shown; they recolonized the Murder Point area in the early 1980s.*

were counted in the west Massacre Bay–Murder Point survey area in 1986 and an additional 485 were seen in areas to the west [Estes 1990]). Sea urchins with test diameters >80 mm had disappeared from our study sites by 1986 and 1987, and only 0.2 and 0.05% of the individuals sampled were >50 mm during these respective years. The

TABLE 9: *Characteristics of sea urchin populations in the western and central Aleutian Islands, contrasting areas with and without sea otters and temporal changes at particular sites following the growth and spreading of otter populations into these sites.*

Location and Date	Test diameter (mm)		Percentage of population		
	Minimum	Maximum	>30 mm	>40 mm	>50 mm
Alaid Island[1]					
1987	3	83	23.9	21.1	14.7
Nizki Island[1]					
1987	3	84	42.1	25.1	14.7
Shemya Island[1]					
1987	3	84	30.6	19.7	12.9
Adak Island[3]					
1987	2	38	0.4	0	0
Amchitka Island[3]					
1987	3	39	0.5	0	0
Attu Island					
Chichagof Harbor					
1976[2]	5	44	10.5	0.2	0
1977[2]	5	48	7.8	0.2	0
1979[2]	5	44	21.1	2.1	0
1983[2]	4	44	2.9	0.1	0
1986[2]	2	40	1.5	0	0
Murder Point					
1972[1]	4	86	28.5	26.9	21.9
1983[2]	4	86	23.7	9.1	3.6
1986[2]	4	52	10.3	1.9	0.2
1987[2]	3	54	17.6	1.4	0.1
1990[2]	3	39	3.6	0	0

[1] *Sea otters absent.* [2] *Sea otters present but below equilibrium density.* [3] *Sea otters at equilibrium density.*

sea otter population at Attu continued to increase through the late 1980s, and in 1990 39 animals were counted in the west Massacre Bay–Murder Point survey area with an additional 649 seen in areas to the west (J. Estes, unpublished data). By 1990 the size structure of sea urchins at Murder Point had become similar to that of other exploited populations in the western and central Aleutian Islands (e.g., Pisa Point, Amchitka Island, Adak Island; Fig. 10 and Table 8).

The progressive loss of large individuals from the sea urchin population at Murder Point closely coincided with otters spreading into the area and preying on sea urchins in a size-selective fashion. Changes in urchin population structure between 1972–1976 and 1986–1987 reflect the nature and rate of this transition. Urchin biomass at Attu declined about 50% and density actually increased about 25% during this time (Table 7). These data contrast with our time series on sea urchin populations from Torch Bay (i.e., 1978 vs. 1988), where density declined by 99% and biomass by >99% following the recolonization of sea otters.

Sea Urchin Growth

Sea urchin abundance and size distribution were similar in the tide-pool populations marked with tetracycline for growth measurements between July 1986 (marked) and June 1987 (collected) at Attu Island. The total number of animals found in the four pools was 2% greater in 1987 than in 1986 (496 vs. 486 urchins), the smallest animals from each pool had test diameters of <10 mm in both years, and in both years the largest animals in pools at Pisa Point had test diameters of 40–50 mm whereas those in the Murder Point pools had test diameters of 60–70 mm.

Slopes of the linear regressions of 1n test diameter vs. 1n jaw length did not vary significantly among the four pools ($F_{3,496}$= 2.357), and since size-specific growth increments also appeared similar, the data were pooled (Fig. 13). Two hundred and seventy-four urchins had distinctive tetracycline marks, among which the estimated annual increase in test diameter ranged from 0 to 13.08 mm. Although highly variable among individuals, the average growth rate increased with size up to the 15–<20 mm diameter size class, declined somewhat for animals 20– <30 mm, and declined more abruptly for animals >30 mm (Table 11). Together with the size–frequency distributions of sea urchins (Figs. 5, 6, and 12), these data indicate that significant recruitment events occur at least every 2–3 yr. Despite the fact that we marked numerous individuals with

TABLE 10: *Percentage of total change in sea urchin biomass kelp density following the arrival of sea otters to long-unoccupied sites in the Aleutian Islands and southeast Alaska.*[1]

	Percentage of total change	
	Aleutian Islands	Southeast Alaska[2]
Sea urchin biomass	50% decline	100% decline
Epibenthic kelp density	1% increase	103% increase

[1] *Total change was estimated from measurements at sites with and without sea otters. Data are from Tables 2 and 3.*

[2] *Data from area occupied by equilibrium-density sea otter population not available for southeast Alaska. However, data from Surge Bay sites assumed to represent equilibrium situation, based on data in Fig. 2.*

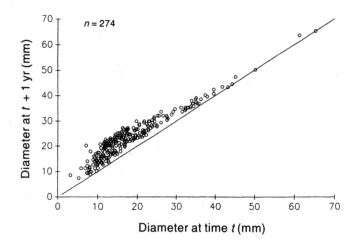

FIGURE 13: *Size-specific annual growth increments in test diameter (corrected to 365 d) for tetracycline-marked sea urchins (S. polyacanthus) from four tide pools at Attu Island. The solid line represents zero growth. Each point (0) represents the annual growth increment of one sea urchin.*

TABLE 11: *Size-specific growth increments from tetracycline-marked sea urchins at Attu Island, Alaska. ND = no data.*

Size range (mm)	n	Diameter growth (mm/yr)	
		X	SD
0–<5	1	5.41	...
5–<10	30	4.39	2.73
10–<15	93	6.39	2.70
15–<20	63	6.72	2.27
20–<25	35	5.02	1.74
25–<30	19	5.25	1.67
30–<35	9	3.40	1.22
35–<40	14	2.43	0.91
40–<45	6	1.93	0.79
45–<50	3	0.95	1.23
50–<55	0	ND	...
55–<60	0	ND	...
60–<65	0	ND	...
65–<70	1	2.48	...

test diameters >50 mm, few marked animals were recovered in this size range thus preventing us from reliably making age estimates of the larger sea urchins. A number of these larger animals appeared to have dim tetracycline marks at the terminal ends of their jaws, thus indicating that they had been marked but accrued little or no net growth in the ensuing year. These observations, together with the negative relationship between size and growth rate for animals of test diameter >20–30 mm, indicate that the large individuals (> about 50–60 mm test diameter and which characterize nearly all sea urchin populations in the western Aleutian Islands that are unexploited by sea otters) are several to many decades old (Fig. 14).

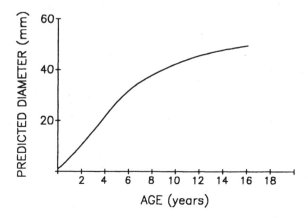

FIGURE 14: *Estimated size (test diameter) vs. age relation for sea urchins (S. polyacanthus) from Attu Island. Estimates are based on average growth rates for 5-mm size increments (Table 11).*

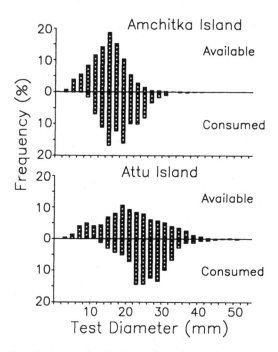

FIGURE 15: *Size-frequency distributions of sea urchins (S. polyncanthus) in natural populations (available) and those consumed by sea otters at Amchitka and Attu islands.*

Size-selective Predation

Urchin size distributions, estimated from jaw lengths in otter scats collected during 1986 and 1987 from Amchitka and Attu islands, demonstrated that otters selectively consumed the largest available prey (Fig. 15). An alternate interpretation is that the jaws of small urchins either were not consumed or, because of their relative fragility, were selective destroyed between consumption and excretion. These possibilities are unlikely. First, sea otters tend to consume small urchins in their entirety whereas the tests and other calcareous tissues of the larger ones are discarded. Second, larger calcified structures, simply by virtue of their comparatively large surface area, may be more prone to inadvertent damage by the grinding motion of an otter's heavy molars. In fact, our subjective impression was that many of the larger jaws extracted from scats were partially fractured whereas the smaller ones were less frequently damaged. In any case, because of their highly conserved form, it was easy to estimate lengths of most partially fractured jaws.

The mean respective urchin sizes (test diameter) available to and consumed by sea otters at Attu (based on otter scats and living populations sampled near Murder Point) were 21.9 and 26.7 mm. Similarly, mean respective urchin sizes available to and consumed by sea otters on the northeast coast of Amchitka Island were 14.9 and 19.0 mm. Size distributions of animals eaten differed significantly from those in living populations in both instances (Attu, $\chi^2 = 31.65$, df = 17, $P = 0.017$; Amchitka Island, $\chi^2 = 34.52$, df = 13, $P = 0.001$).

Sea otters did not eat the smallest available sea urchins in Aleutian Island populations. Only 0.3% of the animals consumed at Attu were <10 mm whereas about 11% of the available population was <10 mm; 1.9% of the urchins consumed at Amchitka Island were <10 mm whereas about 19% of the population were <10 mm. These data indicate that the smallest sea urchins in the Aleutian Islands have a refuge in size from sea otter predation and further imply an increased risk of predation with increased size.

Although we have no comparable data on the sizes of sea urchins consumed by otters in southeast Alaska, those sizes that appear to offer a refuge from sea otter predation in the Aleutian Islands (i.e., <30 mm) were sparse (red and green urchins) or absent (purple urchins) in southeast Alaska.

Discussion

Ecological communities, and the processes that structure them, typically vary in space and time (Dayton and Tegner 1984, Wiens 1986). Furthermore, such variation exists at a broad range of scales (Powell 1989). Therefore, the ability to generalize ecological patterns and processes depends on the magnitude and scale of spatial and temporal variation in nature. This point is substantiated by a rapidly growing list of published studies (Sutherland 1981, Williams 1983, Dayton and Tegner 1984, Hay 1984, Schneider and Piatt 1986, Wiens 1986, Wiens et al. 1986, Dethier and Duggins 1988, Paine 1988, Powell 1989, Foster 1990, Rose and Leggett 1990, Angel 1991, Witman and Sebens 1992).

Our study is perhaps the first rigorous attempt to define the generality and variation of a community ecological paradigm—in this case the claim that sea otter predation on sea urchin populations reduces intensity of herbivory, thereby enhancing abundance and production of kelps and other macroalgae. A number of previous studies (see Table 1) have provided evidence that the interactions exist, and there is reasonably strong evidence for a range of indirect consequences as well (Trapp 1979, Estes et al. 1982, Irons et al. 1986, Duggins et al. 1989, Estes et al. 1989). However, claims have also been made that these interactions occur under a limited range of circumstances, that previous studies were done under these limited circumstances or in habitats chosen to support the paradigm's predicted outcomes, and thus that the paradigm is of more limited importance than published studies imply (Foster and Schiel 1988, Foster 1990).

A rigorous evaluation of the extent to which any character or process is general requires that (1) the sample space or population of interest be defined, (2) sample units be appropriately defined, (3) a process for selecting those sample units be chosen that will provide unbiased (or at least representative) estimates of the character or process, and (4) a sufficiently large number of sample units be selected to provide reasonably precise estimates. These fundamental principles of probability sampling (Cochran 1963) raise problematic questions for community ecology. In contrast with many statistical populations (i.e., sample spaces, Hoel 1971), spatial and temporal boundaries of biological communities are usually either difficult to define or nonexistent. At a minimum it is necessary to know the geographical ranges of component species to properly define community boundaries, characteristics that often are poorly understood. This problem is exacerbated by the likelihood that species ranges vary discordantly (Curtis and McIntosh 1950, Whittaker 1975), making the precise spatiotemporal definition of a community necessarily arbitrary. Furthermore, in contrast with many biological populations for which individuals are the obvious sample units (those comprising clonal organisms are exceptions), the appropriate sample units in community ecological studies are unclear and usually arbitrarily selected. Because many natural populations vary extensively on small spatial and temporal scales, the structure of natural communities, however they are defined and measured, may vary at least as much, and on the same scales, as the populations that comprise them. The sources of such variation must be understood and accounted for in defining the sample space, which is probably possible only if large numbers of units can be sampled to provide a sufficiently robust generalization. It is for these reasons that rigorous attempts to generalize even the most well known and important of community ecological paradigms are rarely, if ever, done, and consequently, why community ecological paradigms, as they become popular or well known, often become contentious.

Generality and Variation in Alaskan Sea Otter–Kelp Forest Systems

We do not claim to have met all of the above stated requirements for evaluating the generality of a community ecological paradigm, although we were able to meet two of the more important ones. That is, our evaluation was based on rigorous probability sampling of sites within locations and plots within sites, thus providing unbiased estimates of the parameters we measured. Furthermore, our sample sizes were large. This helped assure that our measurements captured the range of natural variation, thus permitting realistic measurements of statistical confidence in the estimates. We believe these accomplishments are unprecedented in prior community ecological studies in which specific a priori hypotheses have been at issue.

Our findings showed that the influence of sea otter predation on kelp forest community structure is consistent in some ways and variable in others. For instance, the distributions of kelp density vs. sea urchin biomass among sites with and without sea otters (Fig. 6) were largely non-overlapping in the western Aleutian Islands and mutually exclusive in southeast Alaska, thus indicating that, based on these measurements, a random sample of 20 0.25-m^{-2} plots from a given site is sufficient to predict, with a high level of confidence, the presence or absence of sea otters in Alaskan kelp forest communities. This is not to say that the configurations of these systems are describable as precise point estimates. In communities lacking sea otters, urchin biomass was high and variable whereas kelp density was low and relatively invariant. In contrast, communities with sea otters had kelp densities that were relatively high and variable whereas sea urchin biomass was low and invariant. These patterns indicate that abundances of autotrophs and herbivores in this system both are precisely set when directly limited by top-down control. This precision declines when direct top-down control is relaxed, perhaps because a broader array of limiting processes (e.g., competition, physical disturbance) come into play. Thus, neither sea urchin biomass nor kelp density alone was a good predictor of the influence of sea otter predation although together they provided a predictable measure of the presence or absence of sea otters.

The most reliable indicator of the presence or absence of sea otters was the maximum size of sea urchins. In the Aleutian Islands, the frequency distributions of this measure did not overlap between islands lacking sea otters and those where otters were at or near equilibrium density (Fig. 5). Data from Attu Island indicate that the larger sea urchins (i.e., >35–40 mm test diameter) were consumed within 1–2 yr of otter recolonization. The speed of this change and the virtual population uniformity thereafter probably resulted from the size-selective nature of sea otter predation and the slow growth rate of sea urchins in the western Aleutian Islands. This pattern was not seen in southeast Alaska because sea urchin populations there were effectively eliminated by sea otters shortly following recolonization.

The abundance of red algae and benthic suspension feeders varied less between areas with and without sea otters than did the abundance of sea urchins and kelps. The abundance of red algae was greater at sites with sea otters than at sites without them, and although this difference was statistically significant, the abundance of fleshy red algae was not a predictable measure of the presence or absence of sea otters at any given site. The abundance of suspension feeders was an even poorer predictor of the presence or absence of sea otters, even though growth rates of several suspension feeding species are enhanced significantly by particulate organic carbon derived from kelp and other macrophytes (Duggins et al. 1989). The failure of these guilds of organisms to covary as strongly as urchins and kelp with the presence and absence of sea otters probably results from both positive and negative influences imposed on these groups in both ecological settings. For instance, the presence of sea otters should enhance the red algal turf via reduced herbivory and inhibit it via increased competition for light or space with the kelp overstory. Suspension feeding invertebrates should be subject to similar opposing forces, in this case enhancement from algal-derived detritus and reduced grazing and inhibition by competition with kelp and red algae.

Foster (1990) and Foster and Schiel (1988) argued that the importance of sea otters as a structuring element in California kelp forests has been greatly exaggerated, and that other factors, such as physical disturbance, are primarily responsible for patterns of kelp species abundance and distribution within the otter's current range. Foster's arguments are based upon the observation that outside the range of otters, sea urchin and kelp are highly variable in their abundance, with a relatively small proportion (≈20%) of benthic habitats qualifying as deforested areas. Unfortunately, no corresponding data are provided by Foster regarding kelp or urchin distribution within the sea otter's range. Such variation is in fact typical of habitats without otters at virtually all sites with which we are familiar along the northeast Pacific coast. Even in the Aleutian Islands, where expansive deforested undersea areas are ubiquitous, microhabitats exist where urchins are either ineffective or rare; especially at wave exposed, shallow sites where extreme turbulence creates conditions in which urchins cannot forage (Dayton et al. 1984, 1992). Particularly striking is how little variation there is in kelp and urchin abundance within the present range of otters (including sites in California) and how different measures of these parameters are from sites without otters.

The preponderance of observations and experimental studies in California (Table 1) support the sea otter paradigm. Additionally, in the only long term (historical) analysis of changes in kelp abundance following otter reintroduction in California, VanBlaricom (1984) presented compelling observations supporting the importance of otters. His pre-otter vs. post-otter comparisons suffer (as do most such comparisons) from the fact that only two points in time are compared. Kelp surface canopies (e.g., *Macrocystis*) are temporally variable in extent regardless of whether otters are present, and two-point comparisons must be viewed with caution given the number of co-variables which could contribute to kelp variation. However, VanBlaricom's observations came from three sites, with pre-otter assessments made in two years (1911 and 1912), and post-otter assessments (all of which showed dramatic increase in surface canopy cover) made at three different times spanning ≈40 yr.

If differences in the structure of kelp forests do exist between California and other regions, two life-history attributes may be responsible. In California, physical disturbance may be particularly important in that the competitively dominant kelp is a long-lived, surface-canopy species. Not only are these surface canopies more susceptible to damage by storm turbulence, but their removal has broad ramifications to other algal species. North of northern California, surface canopies are composed of competitively inferior annual species that exert little influence on other benthic plants. Additionally, in California (as in the Aleutian Islands), urchin recruitment may be considerably less variable than to the north. If our proposal for the interactive importance of urchin recruitment and size-selective predation by sea otters is correct, we would expect that California kelp forests would respond more slowly to otter reintroduction than their counterparts in Washington, British Columbia, and southeast Alaska.

While we agree completely with Foster and Schiel that sea otters are not the only important factor regulating kelp-dominated communities, we believe that it is unlikely that California represents a substantial exception to the sea otter paradigm. Even if rates of change in kelp forests following otter reintroduction were significantly slower in California, the end results (a highly significant increase in algal abundance, biomass, and distribution) are likely to be similar to those documented farther north.

Mechanisms of Persistence and Change in Plant-herbivore Interactions
The results of this study show that rocky reef communities in Alaska tend to persist as either kelp-dominated or deforested assemblages, and that intermediates between these community configurations are both rare and highly transitory. Two sources of information from our study support this conclusion. First, the bivariate plots of sea urchin biomass vs. kelp density (Fig. 7) were strongly hyperbolic in both the western Aleutian

Islands and southeast Alaska. If the community was not characterized by 2 domains of attraction separated by unstable intermediates, one would expect herbivore biomass and plant density to be either uncorrelated or more nearly correlated as a first-order function. Second, the transition from deforested to algal-dominated communities in response to the reestablishment and growth of sea otter populations occurs rapidly. In southeast Alaska, this transition occurred soon after sea otters had expanded into previously unoccupied habitats whereas in the western Aleutian Islands it was often delayed for years or decades. Nonetheless, the shift to a kelp-dominated community probably occurs abruptly in the latter region because we have seen no evidence for the gradual recovery of kelp forests at Attu in >20 yr of observation. Abrupt transitions between forested and deforested communities have been reported in kelp forests worldwide, especially in the northern hemisphere (Harrold and Pearse 1989). Furthermore, Steneck (1993) also reported a strongly hyperbolic pattern of association between the abundances of sea urchins (*Diadema antillarum*) and macroalgae on a Caribbean reef, thus suggesting that alternative stable equilibria in plant–herbivore interactions occur broadly in benthic marine systems.

Since algal-dominated or deforested communities in Alaska appear to be determined in large part by the presence or absence of sea otters, they are more properly characterized as boundary points (Lewontin 1969) than as alternate stable communities (Holling 1973, Sutherland 1974, 1981). The mechanisms responsible for maintaining these points are complex and varied, and the temporal patterns change differ remarkably between the western Aleutian Islands and southeast Alaska. Our findings together with those of other workers, suggest that three processes—sea urchin recruitment size-selective predation, and the dynamical properties of plant production and herbivore consumption—are important to the stability of and rates of transition between these community configurations.

Sea urchin recruitment.—The most striking difference in sea urchin population structure between the Aleutian Islands and southeast Alaska was that small animals (<15–20 mm test diameter) were abundant in the former region and virtually absent in the latter (Fig. 5). High densities of small sea urchins characterized most of the numerous locations we sampled in the central and western Aleutian Islands, and observations made during a cruise through the Aleutian archipelago in 1987 indicate that this situation persists eastward to at least the Islands of Four Mountains (Fig. 1; J. Estes, unpublished data).

Small sea urchins were also consistently abundant through time in Aleutian Islands populations. We found similar size–frequency distributions of sea urchins at Amchitka Island in the late 1960s and early 1970s (Estes et al. 1978) and again in 1986 and 1987 (this paper). Small individuals occurred at all sites sampled at Attu Island in 1975–1981, 1983, 1986, 1987, 1990, and 1991. These patterns could only be maintained by persistent recruitment or slow growth.

The relative importance of growth vs. recruitment to the abundance of small individuals in Aleutian Islands sea urchin populations is somewhat equivocal because growth was indeed slow and we have direct evidence for settlement. Nonetheless, given an average growth rate for tetracycline-marked animals of ≈5 mm/yr, major settlement events must have occurred at least every 2–3 yr to produce the observed patterns. In addition, such settlement patterns must be broadly occurring as we have yet discover an urchin population in the Aleutian Islands that lacked small individuals. Whatever the exact cause, its main consequence is that echinoid populations throughout the Aleutian Islands contain numerous individuals in the size range of 5–35 mm test diameter.

Although our records are less extensive for southeast Alaska, settlement there was episodic and apparently did not occur in most areas during most years. For instance, our samples of sea urchin populations (mostly red urchins) from 21 sites in Sitka Sound generally lacked individuals < about 35 mm test diameter, an often recurrent pattern for red and purple sea urchins from southeast Alaska to central California (Ebert 1983, Paine 1986, Harrold and Pearse 1987, Pearse and Hines 1987, Sloan et al. 1987). In 5 yr of study at Torch Bay, we observed a single settlement event for red urchins at one study site, and none for purple urchins. The main consequence of this recruitment pattern is that urchin populations in southeast Alaska contain few or no individuals between the sizes of 5–35 mm test diameter.

Processes responsible for this difference in echinoid population structure between the Aleutian Islands and southeast Alaska are uncertain. Biological differences among echinoid species represent one possible explanation. A second possibility is that small sea urchins in southeast Alaska occurred in cryptic habitats and entered the observable population only upon becoming larger. Third, comparable rates of urchin recruitment may have occurred in southeast Alaska and the Aleutian Islands, but with the small recruits lost to some other predator unique to southeast Alaska. The sunflower star, *Pycnopodia helianthoides*, an urchin predator, is in fact abundant in southeast Alaska and rare or absent in the central and western Aleutian Islands. These and other predators are known to influence the distribution and abundance of sea urchin populations. However, this species is not known to selectively exploit small sea urchins. The fact that a single strong recruitment pulse of red urchins, and periodic weak recruitment by green sea urchins, did occur during the time of our studies in southeast Alaska, argues against all these possibilities, especially the latter two. A final possibility is that physical oceanographic processes responsible for transporting larvae differ between the Aleutian Islands and southeast Alaska. Even in regions where recruitment is generally more predictable (e.g., California and Oregon, Ebert and Russell 1989) it is patchy at meso-scales, probably because of offshore larval transport. Recent studies have demonstrated that meso-scale patterns of larval and spore transport influence adult populations of many marine organisms including fish (Cowen 1985), invertebrates (Gaines and Roughgarden 1985, Roughgarden et al. 1988, Ebert and Russell 1989), and kelps (Reed et al. 1988). Little is known at present about meso-scale current patterns and larval transport in the Aleutian Islands, although presently available evidence makes us favor this last explanation

Size-selective predation.—Sea otters selectively exploited the largest available sea urchins from populations in the Aleutian Islands, avoiding prey <15-20 mm test diameter almost entirely (Fig. 15). The likely reason for such selective feeding is that the differential costs of locating, capturing, and consuming different sized sea urchins are probably negligible whereas nutritional benefits, which scale as a cubic function of test diameter, are obtained from larger prey. Because of frequent recruitment coupled with slow growth rates, sea urchin populations in the Aleutian Islands contain numerous individuals that apparently are too small to be eaten profitably. Thus, the largest sea urchins are selectively eaten and eliminated by expanding sea otter populations in the Aleutian Islands whereas individuals below the optimal to minimal sizes consumed by otters are reproductively mature and sufficiently abundant to prevent the recovery of kelp populations. Consequently, deforested habitats persisted in the Aleutian Islands. Although we have no data on either growth rates or the sizes of sea urchins consumed by predators in southeast Alaska, there is no reason to expect that the size-specific costs and benefits to foraging on sea urchins should differ between the regions. Thus, even the smallest sea urchins available in typical southeast Alaska populations probably are readily captured and consumed by sea otters, leaving no size refuge from predation. Consequently, abundant sea urchins disappeared and kelp forests developed quickly following the reestablishment of sea otters in southeastern Alaska.

A Dynamical Model for Plant–Herbivore Interactions

A conceptual model is used to help envision how the processes discussed above might act to preserve or disrupt a stable equilibrium between plant growth and herbivore consumption in the face of recovering sea otter populations. Several of the model's features require clarification. First, we do not present it as a strict portrayal of empirical reality in our systems, but rather as a heuristic aid for discussing how dynamical properties and natural histories of those systems and their key players might be linked to explain observed patterns. Second, although rates of net production and consumption are characterized as independent functions of plant biomass, these processes in fact may interact. Finally, our system probably is atypical in that extremely high herbivore abundance is somehow maintained even after their fleshy macroalgal food resources have been eliminated. Whether the maintenance of high herbivore abundance results because of a unique ability by sea urchins to persist under starvation conditions or because food resources are subsidized from elsewhere is uncertain.

The existence of two stable equilibria in a plant–herbivore system, one with low plant biomass and the other with plant biomass near the maximum attainable, is a recurrent feature of kelp forest and some other natural communities (Noy-Mier 1975, May 1977, Walker et al. 1980). The proposed explanation for these alternate stable equilibria stems from several assumptions about the dynamical interaction between biomass-dependent plant production and herbivore consumption. The first assumption is that biomass-dependent plant production (dV/dt) is a standard yield curve in which net production rate is zero when plant biomass (V) is both zero and maximum, and maximum at some intermediate value (Fig. 16). A second assumption is that herbivores satiate with increased food availability. Equilibria occur at values of plant biomass where net production and consumption rates coincide. These equilibria are stable within the contiguous range of low plant abundances for which production exceeds consumption (thus driving plant abundance upward toward the equilibrium point) and the contiguous range of high plant abundances for which consumption exceeds production (thus driving plant abundance downward toward the equilibrium point). The equilibria are unstable when the opposite conditions apply. Plant biomasses that define stable equilibria depend on the shapes of the yield curve and the herbivore consumption rate at satiation. Although the forms of these curves are unmeasured in kelp forests, stable equilibria should occur near V_0 and V_{max} under many realistic circumstances. Algal–herbivore dynamics in kelp forest communities frequently conform to the predictions of this model. That is, fleshy algal stands typically are either deforested or largely ungrazed (Harrold and Pearse 1987), organizational states that we suggest represent the two stable equilibria. Intermediate conditions usually are transitory, and we suggest these correspond with transition intervals or unstable equilibria depicted in Fig. 16.

When these general conditions apply, changes in the abundance of sea otters (i.e., intensity of predation on sea urchins) drives the herbivore satiation plateau (i.e., consumption rate vs. algal biomass) upward or downward. For systems lacking sea otters, herbivore consumption exceeds net algal production at all values of algal biomass except V_0 which thus defines a single stable equilibrium. Similarly, for systems in which sea otters are at or near equilibrium density, production exceeds consumption for all values of V_x except V_{max}. V_0 is unstable and the system has a single stable equilibrium at V_{max}. Hence the system supports dense algal stands. These extreme conditions appear to pertain in the Aleutian Islands and southeast Alaska. That is, we found communities lacking sea otters to be generally deforested, and those with otter populations at or near equilibrium density generally to support dense algal stands with little evidence of grazing damage. However, the reestablishment of low density otter populations has profoundly different short-term effects between the Aleutian Islands and southeast Alaska, apparently in large measure due to the otter's size-selective foraging behavior and the pres-

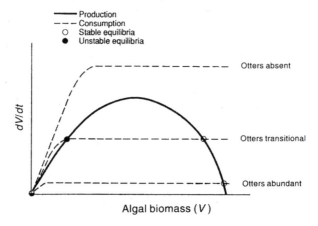

FIGURE 16: *Graphical model for the maintenance of stable equilibria between the production and consumption of kelp as a function of sea otter population status in the Aleutian Islands. Note that the equilibrium where algal biomass = 0 is stable when sea otters are absent or transitional but unstable when otters are abundant. This discrepancy is an arbitrary consequence of the way we have drawn the satiation curves.*

ence or absence of high densities of sea urchins smaller than the otter's size preference threshold. Thus, the plateau of the herbivore consumption curve in southeast Alaska is driven effectively to zero by recolonizing sea otters whereas in the Aleutian Islands a similar intensity of otter predation has less of a depressing influence on the consumption rate plateau. The magnitude of reduction in herbivore consumption may be further compromised in the Aleutian Islands by an increased density of small sea urchins in response to the selective removal by sea otters of large individuals (Fig. 4, Tables 7 and 10), the result being that herbivore consumption rate in the Aleutian Islands is only modestly influenced in the short term by the recolonization of an area by sea otters. Similar in situ kelp consumption rates and survival rates of whole kelp plants (J. Estes, unpublished data) in deforested habitats with and without sea otters at Attu Island support these conclusions.

Conclusions

The findings of this study leave little doubt that Alaskan kelp forests are broadly dependent on sea otter predation for protection against

destructive grazing. That is, the predicted outcomes of sea otter predation are broadly recurrent, which is not to say they occur invariably,
but rather that they occur at most times, places, and under most natural circumstances. This notwithstanding, the patterns of variation are
also intriguing and broadly relevant. On relatively small spatial scales
(metres to kilometres), sampling variation and the variation among
sites within locations help define the "domains of attraction" (Holling
1973: 4) between stable boundary points of community organization.
The fact that these points were essentially non-overlapping indicates
that the proposed equilibrium states are mutually exclusive and that
transitional or intermediate states of community organization are
rare events. Nonetheless, the extent to which small-scale (among site)
measurements varied in space and time indicate that both kelp- and
urchin-dominated communities are highly dynamic, and thus variable,
within their respective domains of attraction.

Variation at the regional scale was both significant and informative,
in particular the vastly different frequencies and intensities of urchin
recruitment between the Aleutian Islands and southeast Alaska. This
finding provides insight into the coupling of meso-scale oceanographic
processes that influence larval life history stages and microscale processes that influence adult life history stages. In the context of adult
life history phases, Alaskan kelp forests are dominated by top-down
forces, i.e., carnivores limit herbivores, and when carnivores are absent,
herbivores limit plants. However, in the broader context of larval life
histories, interregional comparisons between the Aleutian Islands and
southeast Alaska indicate subtle but important donor-controlled or bottom-up forces, namely the supply of planktotrophic larvae to the rocky
benthos. The coupling of scales (micro and meso), systems (rocky reefs
and demersal), and life history stages (propagule and adult), when inter-
faced with a strongly connected food web and an understanding of the
natural history of consumer choice, provides the conceptual elements
that are both necessary and sufficient to explain strikingly different
interregional rates of transition between deforested conditions and the
development of kelp forests with the reintroduction of sea otters.

Acknowledgments

Numerous people have assisted us over the past quarter-century. Our thanks to all for hard work, friendship, and for sharing countless adventures with us in the majestic wilderness of coastal Alaska. We are especially grateful to our life-long friends and colleagues, D. B. Irons and C. A. Simenstad. Earlier drafts were critiqued by M. Dethier, M. S. Foster, S. Gaines, D. Reed. P. D. Steinberg, R. S. Steneck, and J. D. Witman. This study would not have been possible without the cooperation and support of several agencies. In particular, we thank the Alaska Maritime National Wildlife Refuge for their continuing interest, encouragement, and assistance with our work in the Aleutian Islands, the National Park Service for providing vessel and field support for work in southeast Alaska, the United States Air Force and Navy for providing access to their facilities in the Aleutian Islands, and the United States Coast Guard for providing air transport of personnel and equipment to the western Aleutian Islands, even after we broke a 4-L bottle of raw formaldehyde in the cargo bay of one of their HC-130 aircraft. Special thanks to C. Jones who served as Director of the National Fish and Wildlife Laboratory and Denver Wildlife Research Center during the early phases of our research. The project would never have been started without his vision, leadership. and support. Funds were provided by the United States Fish and Wildlife Service, National Park Service, and National Science Foundation Grant No. DPP-8421362.

Literature Cited

Anderson, G. R. V., A. H. Ehrlich, P. R. Ehrlich, J. D. Roughgarden, B. C. Russell, and F. H. Talbot. 1981. The community structure of coral reef fishes. American Naturalist 117:476–495.

Angel, M. V. 1991. Variations in time and space: is biogeography relevant to studies of long-term scale change? Journal of the Marine Biological Association of the United Kingdom 71:191–206.

Belsky. A. J. 1986. Does herbivory benefit plants'? A review of the evidence. American Naturalist 127:870–892.

Benech, S. V. 1977. Preliminary investigations of the giant red sea urchin resources of San Luis Obispo County, California, *Strongylocentrotus franciscanus* (Agassiz). Thesis. California Polytechnic University. San Luis Obispo, California. USA.

Bowlby, C. E., B. L. Troutman, and S. J. Jeffries. 1988. Sea otters in Washington: distribution, abundance, and activity patterns. Washington Department of Wildlife. Olympia, Washington, USA. Final Report to Coastal Resources Research and Development Institute, Newport, Oregon, USA.

Breen, P. A., T. A. Carson, J. B. Foster, and E. A. Stewart. 1982. Changes in subtidal community structure associated with British Columbia sea otter transplants. Marine Ecology Progress Series 7:13–20.

Chapman, D. G. 1981. Evaluation of marine mammal population models. Pages 277–296 in C. W. Fowler and T. D. Smith, editors. Dynamics of large mammal populations. John Wiley and Sons, New York, New York, USA.

Cochran, W. G. 1963. Sampling techniques. John Wiley and Sons, New York. New York, USA.

Cowen, R. K. 1985. Large scale pattern of recruitment by the labrid, *Semicossyphus pulcher*: causes and implications. Journal of Marine Research 43:719–742.

Cowen, R. K., C. R. Agegian, and M. S. Foster. 1982. The maintenance of community structure in a central California giant kelp forest. Journal of Experimental Marine Biology and Ecology 64:189–201.

Curtis, J. T., and R. P. McIntosh. 1950. The interrelations of certain analytic and synthetic phytosociological characters. Ecology 31:434–455.

Dayton, P. K. 1975. Experimental studies of algal canopy interactions in a sea otter-dominated kelp community at Amchitka Island, Alaska. Fishery Bulletin 73:230–237.

———. 1984. Processes structuring some marine communities: are they general? Pages 181–197 in D. R. Strong Jr., D. Simberloff. L. G. Abele, and A. B. Thistle, editors. Ecological communities: conceptual issues and their evidence. Princeton University Press, Princeton, New Jersey, USA.

Dayton, P. K., V. Currie, T. Gerrodette, B. D. Keller, R. Rosenthal, and D. Ven Tresca. 1984. Patch dynamics and stability of some California kelp communities. Ecological Monographs 54:253–289.

Dayton, P. K., and M. J. Tegner. 1984. Catastrophic storms, El Niño, and patch stability in a southern California kelp community. Science 224:283–285.

Dayton, P. K., M. J. Tegner, P. E. Parnell, and P. B. Edwards. 1992. Temporal and spatial patterns of disturbance and recovery in a kelp forest community. Ecological Monographs 62:421–445.

Dean, T. A,. S. C. Schroeter, and J. Dixon. 1984. Grazing by red and white sea urchins and its effect on the recruitment and survival of kelp. Marine Biology 78:301–313.

Dethier, M. N. 1984. Disturbance and recovery in intertidal pools: maintenance of mosaic patterns. Ecological Monographs 54:99–118.

Dethier, M. N., and D. O. Duggins. 1988. Variation in strong interactions in the intertidal zone along a geographical gradient: a Washington–Alaska comparison. Marine Ecology Progress Series 50:97–105.

Dethier, M. N., E. S. Graham, S. Cohen, and L. M. Tear. 1993. Visual random-point percent cover estimations: "objective" is not always better. Oecologia 96:93–100.

Druehl, L. D. 1970. The pattern of Laminariales distribution in the northeast Pacific. Phycologia 9:237–247.

Duggins, D. O. 1980. Kelp beds and sea otters: an experimental approach. Ecology 61:447–453.

———. 1981. Sea urchins and kelp: the effects of short-term changes in urchin diet. Limnology and Oceanography 26: 391–394.

———. 1988. The effects of kelp forests on nearshore environments: biomass, detritus, and altered flow. Pages 192–201 in G. R. Van Blaricom and J. A. Estes, editors. The community ecology of sea otters. Springer-Verlag. Berlin, Germany.

Duggins, D. O., S. A. Simenstad, and J. A. Estes. 1989. Magnification of secondary production by kelp detritus in coastal marine ecosystems. Science 245:170–173.

Ebeling, A. W., D. R. Laur, and R. J. Rowley. 1985. Severe storm disturbances and reversal of community structure in a southern California kelp forest. Marine Biology 84:287–294.

Ebert, E. E. 1968a. A food-habits study of the southern sea otter. *Enhydra lutris nereis.* California Department of Fish and Game 54:33–42.

———. 1968b. California sea otter—census and habitat survey. Underwater Naturalist 1968:20–23.

Ebert, T. A. 1975. Growth and mortality of post-larval echinoids. American Zoologist 15:755–775.

———. 1980a. Estimating parameters in a flexible growth equation, the Richards function. Canadian Journal of Fisheries and Aquatic Sciences 30:467–474.

———. 1980b. Relative growth of sea urchin jaws: an example of plastic resource allocation. Bulletin of Marine Science 30:467–474.

———. 1982. Longevity, life history and relative body wall size in sea urchins. Ecological Monographs 52:353–394.

———. 1983. Recruitment in echinoderms. Pages 169–203 in M. Jangoux and J. M. Lawrence, editors. Echinoderm studies. A. A. Palkema. Rotterdam, The Netherlands.

Ebert, T. A., and M. P. Russell. 1988. Latitudinal variation in size structure of the west coast purple sea urchin: a correlation with headlands. Limnology and Oceanography 33:286–294.

Estes, J. A. 1990. Growth and equilibrium in sea otter populations. Journal of Animal Ecology 59:385–401.

———. 1991. Status of sea otter (*Enhydra lutris*) populations. Pages 27–35 in C. Reuther and R. Röchert, editors. Proceedings of the Fifth International Otter Colloquium. Habitat 6.

Estes, J. A., R. J. Jameson, and E. B. Rhode. 1982. Activity and prey selection in the sea otter: influence of population status on community structure. American Naturalist 120:242–258.

Estes, J. A., and J. F. Palmisano. 1974. Sea otters: their role in structuring nearshore communities. Science 185:1058–1060.

Estes, J. A., N. S. Smith, and J. E Palmisano. 1978. Sea otter predation and community organization in the western Aleutian Islands, Alaska. Ecology 59:822–833.

Foster, M. S. 1990. Organization of macroalgal assemblages in the northeast Pacific: the assumption of homogeneity and the illusion of generality. Hydrobiologia 192:21–33.

———. 1991. Rammed by the Exxon Valdez: a reply to Paine. Oikos 62:93–96.

Foster, M. S., C. Harrold, and D. D. Hardin. 1991. Point vs. photo quadrat estimates of the cover of sessile marine organisms. Journal of Experimental Marine Biology and Ecology 146:193–203.

Foster, M. S., and D. R. Schiel. 1988. Kelp communities and sea otters: keystone species or just another brick in the wall? Pages 92–108 in G. R. VanBlaricom and J. A. Estes. editors. The community ecology of sea otters. Springer-Verlag, Berlin, Germany.

Gaines, S., and J. Roughgarden. 1985. Larval settlement rate: a leading determinant of structure in an ecological community of the marine intertidal zone. Proceedings of the National Academy of Sciences 82:3707–3711.

Gotchall, D. W., L. L. Laurent and F. E. Wendell. 1976. Diablo Canyon power plant site ecology study annual report July 1 1974–June 30 1975 and quarterly report no. 8, April 1 1975–June 30 1975. California Department of Fish and Game, Marine Resources Administrative Report 76–8.

Harrold, C., and J. S. Pearse. 1987. The ecological role of echinoderms in kelp forests. Pages 137–233 in M. Jangoux and J. M. Lawrence. editors. Echinoderm studies. A. A. Balkema, Rotterdam.

Harrold, C., and D. C. Reed. 1985. Food availability, sea urchin grazing, and kelp forest community structure. Ecology 66:1160–1169.

Hay, M. E. 1984. Patterns of fish and urchin grazing on Caribbean coral reefs: are previous results typical? Ecology 65:446–454.

Hoel, P. G. 1971. Introduction to mathematical statistics. John Wiley and Sons. New York, New York, USA.

Holling. C. S. 1973. Resilience and stability of ecological systems. Annual Review of Ecology and Systematics 4:1–23.

Hunter, M. D., and P. W. Price. 1992. Playing chutes and ladders: heterogeneity and the relative roles of bottom-up and top-down forces in natural communities. Ecology 73:724–732.

Hurlbert, S. H. 1984. Pseudoreplication and the design of ecological field experiments. Ecological Monographs 54:187–211.

Irons, D. B., R. G. Anthony, and J. A. Estes. 1986. Foraging strategies of glaucous-winged gulls in a rocky intertidal community. Ecology 67:1460–1474.

Jameson, R. J., K. W. Kenyon, A. M. Johnson, and H. M. Wight. 1982. History and status of translocated sea otter populations in North America. Wildlife Society Bulletin 10:100–107.

Jones, R. D., Jr. 1965. Sea otters in the Near Islands, Alaska. Journal of Mammalogy 46:702.

Kenyon, K. W. 1969. The sea otter in the eastern Pacific Ocean. North American Fauna 68:1–352.

Kobayashi, S., and J. Taki. 1969. Calcification in sea urchins. I. A tetracycline investigation of growth of the mature test in *Strongylocentrotus intermedius*. Calcified Tissue Research 4:210–223.

Krebs, J. R. 1978. Optimal foraging: decision rules for predators. Pages 23–63 in J. R. Krebs and N. B. Davies, editors. Behavioral ecology: an evolutionary approach. Blackwell, London, England.

Kvitek, R. G., and J. S. Oliver. 1992. Influence of sea otters on soft-bottom prey communities in southeast Alaska. Marine Ecology Progress Series 82:103–113.

Kvitek, R. G., J. S. Oliver, A. R. DeGange, and B. S. Anderson. 1992. Changes in Alaskan soft-bottom prey communities along a gradient in sea otter predation. Ecology 73:413–428.

Kvitik, R. G., D. Shull, D. Canestro, E. C. Bowlby, and B. L. Troutman. 1989. Sea otters and benthic prey communities in Washington State. Marine Mammal Science 5:266–280.

Laur, D. R., A. W. Ebeling, and D. A. Coon. 1988. Effects of sea otter foraging on subtidal reef communities off central California. Pages 151–167 in G. R. VanBlaricom and J. A. Estes, editors. The community ecology of sea otters. Springer-Verlag, Berlin, Germany.

Laurent, L. L., and S. V. Benech. 1977. The effects of foraging by sea otter (*Enhydra lutris*) along their southern frontier in California from 1973 to 1977. 58th Annual Meeting, Western Society of Naturalists, December 1977, Santa Cruz, California, USA.

Lebednik, P. A., and J. F. Palmisano. 1977. Ecology of marine algae. Pages 353–393 in M. L. Merritt and R. G. Fuller, editors. The environment of Amchitka Island, Alaska. United States Energy Research and Development Administration, Springfield, Virginia, USA.

Lensink, C. J. 1960. Status and distribution of sea otters in Alaska. Journal of Mammalogy 41:172–182.

———, C. J. 1962. The history and status of sea otters in Alaska. Dissertation. Purdue University, Lafayette, Indiana, USA.

Lewontin, R. C. 1969. The meaning of stability. Brookhaven Symposia in Biology 22: 13–24.

Littler, M. M., and D. S. Littler. 1980. The evolution of thallus form and survival strategies in benthic marine algae: field and laboratory tests of a functional form model. American Naturalist 116:25–44.

Lowry. L. F., and J. S. Pearse. 1973. Abalones and sea urchins in an area inhabited by sea otters. Marine Biology 23:213–219.

McNaughton, S. J. 1983. Serengeti grassland ecology: the role of composite environmental factors and contingency in community organization. Ecological Monographs 53:291–320.

Mapstone, B. D., and A. J. Fowler. 1988. Recruitment and the structure of assemblages of fish on coral reefs. Trends In Ecology and Evolution 3:72–77.

Matson, P. A., and M. D. Hunter. 1992. The relative contributions of top-down and bottom-up forces in population and community ecology. Ecology 73:723.

Maurer, B. A. 1985. Avian community dynamics in desert grasslands: observational scale and hierarchial structure. Ecological Monographs 55:295–312.

May, R. M. 1977. Thresholds and breakpoints in ecosystems with a multiplicity of stable states. Nature 269:471–477.

Mayer, R. M. 1980. A study of the population ecology of three nearshore communities of the sea urchin *Strongylocentrotus polyancanthus*. Thesis. University of Washington, Seattle, Washington, USA.

Menge, B. A. 1992. Community regulation: under what conditions are bottom-up factors important on rocky shores? Ecology 73:755–765.

McLean, J. H. 1962. Sublittoral ecology of kelp beds of the open coast near Carmel, California. Biological Bulletin 122:213–219.

Noy-Meir, I. 1975. Stability in grazing systems: an application of predator–prey graphs. Journal of Ecology 63:459–481.

O'Clair, C. E. 1977. Marine invertebrates in rocky intertidal communities. Pages 395–449 in M. L. Merritt and R. G. Fuller, editors. The Environment of Amchitka Island, Alaska. United States Energy Research and Development Administration, Springfield, Virginia, USA.

O'Neill, R. V. 1989. Perspectives in hierarchy and scale. Pages 140–156 in J. Roughgarden, R. May, and S. Levin, editors. Perspectives in ecological theory. Princeton University Press, Princeton, New Jersey, USA.

Ogden, J. C., and J. P. Ebersold. 1981. Scale and community structure of coral reef fishes: a long-term study of a large artificial reef. Marine Ecology Progress Series 4:97–103.

Orians, G. H. 1975. Diversity, stability and maturity in natural ecosystems. Pages 139–150 in W. H. van Dobben and R. H. Lowe-McConnell, editors. Unifying concepts in ecology. Dr. W. Junk, The Hague, Netherlands.

Ortega, S. 1986. Fish predation on gastropods on the Pacific coast of Costa Rica. Journal of Experimental Marine Ecology 97:181–191.

Oshurkov, V. V., A. G. Bazhin, V. I. Lukin, and V. E Sevost'yanov. 1988. Sea otter predation and the benthic community structure of Commander Islands. Biologia Morya (Vladivostok) 6:50–60.

Ostfeld. R. S. 1982. Foraging strategies and prey switching in the California sea otter. Oecologia 53:170–178.

Paine, R. T. 1980. Food webs: linkage, interaction strength, and commnnity infrastructure. Journal of Animal Ecology 46:667–685.

———. 1986. Benthic community–water column coupling during the 1982–1983 El Niño: are community changes at high latitudes attributable to cause or coincidence? Limnology and Oceanography 31:351–360.

———. 1988. Food webs: road maps of interactions or grist for theoretical development? Ecology 69:1648–1654.

———. 1991. Between Scylla and Charybdis: do some kinds of criticism merit a response? Oikos 62:90–92.

Palmer, A. R. 1979. Fish predation and the evolution of gastropod shell structure: experimental and geographic evidence. Evolution 33:697–713.

Pearse, J. S., and A. H. Hines. 1979. Expansion of a central California kelp forest following the mass mortality of sea urchins. Marine Biology 51:83–91.

Pearse, J. S., and A. H. Hines. 1987. Long term population dynamics of sea urchins in a central California kelp forest: rare recruitment and rapid decline. Marine Ecology Progress Series 39:275–283.

Pearse, J. S., and V. B. Pearse. 1975. Growth zones in the echinoid skeleton. American Zoologist 15:731–753.

Powell, T. M. 1989. Physical and biological scales of variability in lakes, estuaries, the coastal ocean. Pages 157–176 in J. Roughgarden, R. May, and S. Levin, editors. Perspectives in ecological theory. Princeton University Press, Princeton, New Jersey, USA.

Reed, D. C., D. R. Laur, and A. W. Ebeling. 1988. Variation in algal dispersal and recruitment: the importance of episodic events. Ecological Monographs 58:321–335.

Rose, G. A., and W. C. Leggett. 1990. The importance of scale to predator–prey spatial
correlations: an example of Atlantic fishes. Ecology 71:33–43.

Rotterman, L. M., and T. Simon-Jackson. 1988. Sea otter. Pages 237–275 in J. W. Lentfer,
editor. Selected marine mammals of Alaska. National Technical Information
Service. PB88-178462, Springfield, Virginia, USA.

Roughgarden, J. 1983. Competition and theory in community ecology. American
Naturalist 122:583–601.

Roughgarden, J., S. D. Gaines, and H. P. Possingham. 1988. Recruitment dynamics in
complex life cycles. Science 241:1460–1466.

Russell, M. P. 1987. Life history traits and resource allocation in the purple sea urchin
Strongylocentrotus purpuratus (Stimson). Journal of Experimental Marine Biology
and Ecology 108:199–216.

Sale, P. F. 1978. Coexistence of coral reef fishes—a lottery for living space.
Environmental Biology of Fishes 3:85–102.

Schneider, D. C., and J. F. Piatt. 1986. Scale dependent correlation of seabirds with
schooling fish in a coastal ecosystem. Marine Ecology Progress Series 32:237–246.

Simenstad, C. A., J. A. Estes, and K. W. Kenyon. 1978. Aleuts, sea otters, and alternate
stable-state communities. Science 200:403–411.

Sloan, N. A., C. P. Lanridsen, and R. M. Harbo. 1987. Recruitment characteristics of the
commercially harvested red sea urchins *Strongylocentrotus franciscanus* in southern
British Columbia. Fisheries Research 5:55–69.

Steneck, R. S. 1993. Is herbivore loss more damaging to reefs than hurricanes? case studies from two Caribbean reef systems (1978–1988). Pages 220–226 in R. N.
Ginsburg, editor. Proceedings of the colloquium on global aspects of coral reefs:
health hazards and history. Rosenstiel School of Marine and Atmospheric Science,
University of Miami, Miami, Florida, USA.

Steneck, R. S., and M. N. Dethier. 1994. A functional group approach to the structure of
algal-dominated communities. Oikos 69:476–498.

Steneck, R. S., and L. Watling. 1982. Feeding capabilities and limitations of herbivorous
molluscs: a functional group approach. Marine Biology 68:299–319.

Strong, D. R. 1992. Are trophic cascades all wet? Differentiation and donor-control in
speciose ecosystems. Ecology 73:747–754.

Sutherland, J. P. 1974. Multiple stable points in natural communities. American
Naturalist 108:859–873.

———. 1981. The fouling community at Beaufort, North Carolina: a study in stability.
American Naturalist 118:499–519.

Trapp, J. L. 1979. Variation in summer diet of glaucous winged gulls in the western
Aleutian Islands: an ecological interpretation. Wilson Bulletin 91:412–419.

Underwood. A. J., and E. J. Denley. 1984. Paradigms, explanations, and generalizations in
models for the structure of intertidal communities on rocky shores. Pages 151–197
in D. R. Strong. Jr., D. Simberloff, L. G. Abele, and A. B. Thistle, editors. Ecological

communities: conceptual issues and the evidence. Princeton University Press, Princeton, New Jersey, USA.

Vadas, R. L. 1977. Preferential feeding: an optimization strategy in sea urchins. Ecological Monographs 47:337–371.

VanBlaricom, G. R. 1984. Relationships of sea otters to living marine resources in California: a new perspective. Pages 361–381 in V. Lyle, editor. 7–10 November. Ocean Studies Symposium, Asilomar, California. California Coastal Commission and California Department of Fish and Game, Sacramento, California, USA.

Vance, R. R. 1979. Effects of grazing by the sea urchin *Centrostephanus coronatus* on prey commnnity composition. Ecology 60:537–546.

Vequist, G. W. 1987. Sea otter re-colonization of ancestral range in Glacier Bay National Park. Unpublished National Park Service Publication, National Park Service, Gustavus, Alaska, USA.

Walker, B. H., D. Ludwig, C. S. Holling, and R. M. Peterman. 1981. Stability of semi-arid savanna grazing systems. Journal of Ecology 69:473–498.

Watanabe, J. M., and C. Harrold. 1991. Destructive grazing by sea urchins *Strongylocentrotus* spp. in a central California kelp forest: potential roles of recruitment, depth, and predation. Marine Ecology Progress Series 71:125–141.

Watson, J. C. 1993. Effects of sea otter *Enhydra lutris* foraging on rocky sublittoral communities off northwestern Vancouver Island. British Columbia. Dissertation. University of California. Santa Cruz, California, USA.

Whittaker, R. H. 1975. Communities and ecosystems. Second edition. Macmillan, New York, New York, USA.

Wiens, J. A. 1986. Spatial scale and temporal variation in studies of shrubsteppe birds. Pages 154–172 in J. Diamond and T. J. Case, editors. Community ecology. Harper and Row, New York, New York, USA.

Wiens, J. A., J. F, Addicott, T. J. Case, and J. Diamond. 1986. Overview: the importance of spatial and temporal scale in ecological investigations. Pages 145–153 in J. Diamond and T. J. Case, editors. Community ecology. Harper and Row, New York, New York, USA.

Wild, P. W., and J. A. Ames. 1974. A report on the sea otter, *Enhydra lutris* L., in California. California Department of Fish and Game. Marine Resources Technical Report Number 20.

Williams, D. M. 1983. Daily, monthly, and yearly variability in recruitment of a guild of coral reef fishes. Marine Ecology Progress Series 10:231–237.

Witman, J. D., and K. P. Sebens. 1992. Regional variation in fish predation intensity—a historical perspective. Oecologia 90:305–315.

Body Mass Patterns Predict Invasions and Extinctions in Transforming Landscapes

CRAIG R. ALLEN, ELIZABETH A. FORYS, AND C. S. HOLLING

CRAIG R. ALLEN and C. S. HOLLING, *Department of Zoology, 110 Bartram Hall, University of Florida, Gainesville, Florida 32611.*

ELIZABETH A. FORYS, *Department of Environmental Science, 4200 54th Avenue South, Eckerd College, St. Petersburg, Florida 33711.*

CRAIG R. ALLEN *current address, South Carolina Cooperative Fish and Wildlife Research Unit, Room G27, Lehotsky Hall, Clemson University, Clemson, South Carolina 29634, USA.*
e-mail: Allencr@Clemson.edu

With kind permission from Springer Science+Business Media: *Ecosystems,* Body Mass Patterns Predict Invasions and Extinctions in Transforming Landscapes (1999) 2:114–121, Craig R. Allen, Elizabeth A. Forys, and C. S. Holling.

Abstract

SCALE-SPECIFIC PATTERNS OF RESOURCE distribution on landscapes entrain attributes of resident animal communities such that species body-mass distributions are organized into distinct aggregations. Species within each aggregation respond to resources over the same range of scale. This discontinuous pattern has predictive power: invasive species and extinct or declining species in landscapes subject to human transformation tend to be located at the edge of body-mass aggregations (P < 0.01), which may be transition zones between distinct ranges of scale. Location at scale breaks affords species great opportunity, but also potential crisis.

Key words: cross-scale; ecosystem structure; endangered species; Everglades ecosystem; extinctions; invasions.

Introduction

Landscape pattern is scale dependent (O'Neill and others 1991; Milne and others 1992), and differently sized animals living upon the same landscape perceive their environment at different scales (Milne and others 1989; Holling 1992; Peterson and others 1998). Increasing evidence suggests that ecosystems are structured by relatively few key processes operating at specific temporal and spatial scales (Carpenter and Leavitt 1991; Levin 1992; Holling and others 1995). The distinct temporal frequencies and spatial scale characterizing these key processes create hierarchical landscape structures with scale-specific pattern. The scale-specific effect of key processes leads to a discontinuous distribution of ecological structure and pattern (Burrough 1981), which in turn entrains attributes of animals residing on the landscape (Holling 1992). This entrainment reflects adaptations to a discontinuous pattern of resource distribution acting on animal community assembly and evolution both by sorting species and by providing a specific set of evolutionary opportunities and constraints. On the animal community level, this is expressed by an aggregated pattern of species body masses (Holling 1992). Animals within a particular body-mass aggregation perceive and exploit the environment at the same range of scale.

The discontinuous, aggregated pattern in animal communities may result from the interaction between key self-organizing processes, landscapes, and animals (Holling 1992; Perry 1994). Human development of landscapes changes some of the self-organizing processes so that landscape patterns begin to be transformed. If animal body-mass aggregations are linked to scale-specific structures, such perturbations should reveal themselves by changes in species turnover that affect body-mass aggregation patterns. We tested five competing hypotheses by determining whether invasive species and extinct or declining species were nonrandomly distributed in terms of vertebrate community body-mass patterns. In addition to testing those hypotheses, we also established that discontinuous body-mass distributions exist in ecosystems other than the boreal forest, and in taxa other than birds and mammals (Holling 1992).

Methods and Analysis

Species Lists

We used data from the Everglades ecoregion of south Florida to investigate the relationship between body-mass patterns in animal communities and biological invasions and extinctions. This region has experienced large-scale landscape transformations, and a large percent of its vertebrate fauna is threatened, declining, extinct, or nonnative. Three general vertebrate taxonomic groups were used as replicates: herpetofauna, birds, and mammals. Species distributions were determined from museum records, published accounts, and the *Florida Breeding Bird Atlas* (Kale and others forthcoming). Only species that had established breeding populations in the Everglades ecoregion (Bailey 1983) were included in the analysis. Because species distributions were available at the county level, we used records from the seven southernmost Florida counties within the Everglades ecoregion. Oceanic and deep-water aquatic species were excluded from the analysis because they interact with their environment differently than do terrestrial species (Holling 1992) and may be trophically compartmentalized from terrestrial systems (Pimm and Lawton 1980).

A species was considered to be a biological invader if it became established after it was introduced to south Florida by humans or if it was a nonindigenous species that naturally expanded its range following anthropogenic landscape transformations associated with European colonization. A species was considered to be endangered if it was listed by the state of Florida (Florida Game and Freshwater Fish Commission 1994) as being extinct, endangered, threatened, or a species of special concern (hereafter, *declining species*). Listed subspecies were not included unless they were the only subspecies occurring in south Florida, or if all subspecies were listed.

In most cases, data on vertebrate body masses were collected from published sources. For a portion of the herpetofauna, body mass was determined from unpublished field data or by weighing a sample (*n*= 10) of preserved museum specimens. Although some weight changes occur during the preservation process, these changes tend to be less than 10% (Haighton 1956; Mount 1963). In all cases, adult male and female weights were averaged.

Analysis

The goals of our analysis were outwardly simple. First, we sought to determine whether body masses of the three taxonomic groups analyzed were distributed discontinuously and, if so, we sought to determine where discontinuities occurred in the data (that is, determine the body-mass patterns of the preinvasion faunas). Second, we sought to determine whether invasive, and extinct and declining, species were randomly or nonrandomly distributed in relation to the body-mass pattern of each vertebrate group.

All native species within each vertebrate class (including recently extinct species) were ranked in order of body mass to determine whether discontinuities existed within the ranked distribution of the recent historical fauna (Figure 1). Invasive species were not included in the faunal list when determining body-mass patterns in order to reconstitute preinvasion communities. Body-mass distributions were analyzed by using simulations that compared actual data with a null distribution established by estimating a continuous unimodal kernal distribution of the log-transformed data (Silverman 1981). Significance of discontinuities in the data was determined by calculating the probability that the observed discontinuities were chance events by comparing observed values with the output of 1000 simulations from the null set [see Restrepo and others (1997)]. Because *n* in our three data sets varied from 35 (mammals) to 106 (birds), and because we were most interested in determining community structure, we maintained a constant

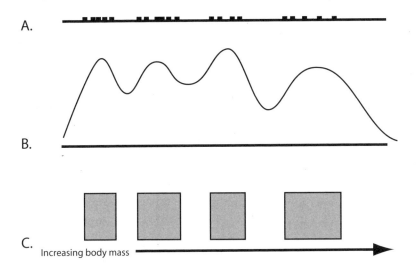

FIGURE 1: *The rank-ordered distribution of animal body masses are discontinuous. A) Location of species in a hypothetical community along a log-transformed body-mass axis. Species form distinct aggregations, separated by gaps. B) Density plot of data from A. C) Stylized portrayal of the body-mass pattern of the hypothetical animal community in A.*

statistical power of approximately 0.50 when setting alpha for detecting discontinuities. Although we believe that the application of a null model is the best method for determining body-mass aggregations, we confirmed our results with a form of the split moving-window boundary analysis [SMW; Webster (1978) and Ludwig and Cornelius (1987)] and hierarchical cluster analysis (SAS Institute 1985).

A gap was defined as an area between successive body masses that significantly exceeded the discontinuities generated by the continuous null distribution. A species aggregation was a grouping of three or more species with body masses not exceeding the expectation of the null distribution. Body-mass aggregations were defined by the two end-point species that defined either the upper or the lower extremes of the aggregation.

Invasive and declining species could be distributed in the body-mass patterns of the animal communities in a number of possible ways (Figure 2). (a) A random distribution would indicate that the scale-specific structure

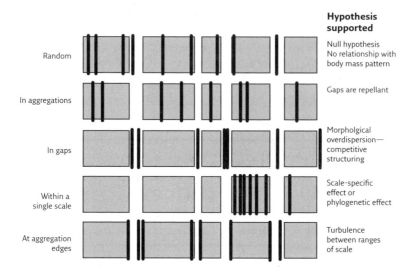

FIGURE 2: *Possible pattern of invasions or deletions in the context of animal community body-mass structure. A different pattern, or a lack of a pattern, supports a different hypotheses.*

of animal communities was not related to invasions or declines. (b) Invasive species could be distributed only within body-mass aggregations, well separated from the gaps, which would indicate that gaps are "forbidden zones," but indicate little else. (c) Invasive and declining species could be restricted to a limited range of body masses, which would indicate that it is at the ecological scale corresponding to these species that perturbation has had its greatest impact (Morton 1990) or that phylogeny had a predominant influence. (d) Invasive species could occur only in gaps, which we would interpret as indicative of competitive processes unrelated to scale-specific animal community structure (Moulton and Pimm 1986). (e) Invasive and declining species could occur at the edge of aggregations, which would indicate that the areas between distinct ranges of scale are most susceptible to changes in ecological process and landscape structure.

After determining the body-mass pattern of the historical faunas, we visually inspected the distributions for the aforementioned patterns. Adding invasive species to the extant community structure may be likened to throwing darts at the distribution and determining whether

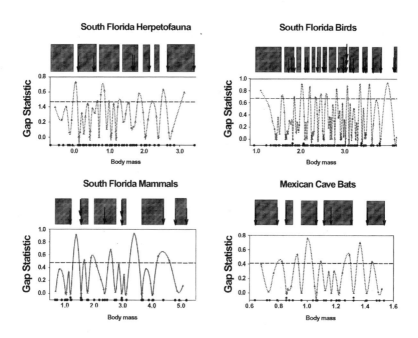

FIGURE 3: *Gap statistic, body-mass pattern, and occurrence of listed species for A) Everglades herpetofauna, B) birds and C) mammals, and D) Mexican cave bats. All data are presented in the* bottom graph, *whereas the* top graph *displays a stylized version of the body-mass pattern and location of listed species (arrows). Power for detecting gaps was kept constant, resulting in a criterion line (gap statistic, dotted line) that varied from 0.43 (bats, n = 28) to 0.68 (birds, n = 106). Aggregations (shaded) were defined as groups of 3 or more species bordered by significant gaps; this criteria led to us to disregard some high values of the gap statistic. Note, however, that changes in body-mass patterns due to the above make no difference in the overall patterns detected.*

there was pattern to where the darts stuck. For endangered species, we quantified the pattern of listed species within the overall community. Chi-squared analysis was used to compare observed with expected frequencies, with taxonomic groups used as replicates.

Results
The south Florida herpetofauna, bird, and mammal body-mass distributions were discontinuous. Distinct aggregations of species were detected

TABLE 1: *Location of Endangered, Declining, and Invasive Species in the Everglades Ecosystem Vertebrate Fauna in Terms of Body Mass Pattern*

		Listed			Invasive		
	Native	at edge (%)			at edge (%)		
	n	*n*	obs	exp	*n*	obs	exp
Herpetofauna	50	7	57	28	21	29	16
Avifauna	106	21	38	24	36	25	13
Mammals	35	9	78	34	11	55	19
Average % at scale breaks (exp)			58	29		36	16

A large percentage (obs) of both declining and invasive species occur at the edge of body mass aggregations. Expected (exp, as a percent) declining and invasive fauna occurring at aggregation edges are given in brackets.

in each taxa, by all methods. The results of simulations of the null model, SMW analysis, and cluster analysis converged. We observed 6–13 discontinuities in our datasets. In contrast, random draws of the same n from a unimodal null revealed that 91% of the randomly generated datasets were unimodal or bimodal (Sendzimir 1998). Fewer than 1% had over five discontinuities. Also, mean gap size exceeded the variation inherent in mean body size 100% of the time in tested mammal datasets and 97% of the time in birds (Sendzimir 1998); that is, normal variation is unlikely to mask the gaps that we identified.

Invasive and declining species were nonrandomly distributed in relation to animal community body-mass patterns. Both invasive and declining species were concentrated at the edge of aggregations for each taxonomic group (Figure 3 and Table 1). For both invasive (χ^2 = 12.94, 2 df; P = 0.002) and declining (χ^2 = 8.61, 2 df; P = 0.01) species, across all taxonomic replicates, approximately twice as many species as expected occurred at aggregation edges (Table 1). No other patterns in the data were detected. Phylogeny was not responsible for observed body-mass pattern (that is, aggregations do not tend to consist of a single family or order) or for endangered status (that is, the largest member of a family did not tend to be listed). Other traditional simple, single species-based hypotheses were not supported (Forys and Allen 1998).

Three additional analyses confirmed those conclusions. First, to provide an alternative analysis of our invasion data, we used the same data for successful invasive bird species in the Everglades, but also incorporated unsuccessful invaders to test the prediction that the body masses of unsuccessful invaders were *not* associated with aggregation edges. We eliminated the effect of small propagule size by including only those species that were known to have been introduced as multiples, known or suspected to have bred, and persisted for at least 5 years. This conservative cut yielded data for 36 successful introductions and invasions and 46 unsuccessful introductions. We then compared the distributions of distances to body-mass aggregation edges (distances were determined by calculating the distance of all species, in log body-mass terms, to the nearest body-mass-edge defining species) between the successful and unsuccessful groups. The distance to aggregation edge for successful species was significantly less than for unsuccessful species (Mann-Whitney rank-sum test, $T = 1217.5$ and $P = 0.027$).

As an additional independent test of endangered species association with aggregation edges, we analyzed the cave bats of Mexico. The main cave-roosting bats of Mexico comprise 28 species, nine of which have been categorized as fragile or vulnerable (Arita-Watanabe 1992). Seven of the nine species categorized as fragile or vulnerable had body masses located at the edge of body-mass aggregations (Figure 3D). If that distribution was random, we would have expected fewer than four fragile or vulnerable species to occur at aggregation edges. The distance to aggregation edge for fragile or vulnerable species was significantly less than for the other native species (Mann-Whitney rank sum test, $T = 85.0$ and $P = 0.014$).

As an independent test of invasive species association with aggregation edges, we also analyzed the birds and mammals from Mediterranean-climate Australia. There are 141 native bird species associated with the Mediterranean ecotype of southwestern Australia (Schodde 1981) and 31 native mammals in the Mediterranean ecotype of south-central Australia (Strahan 1995). Nine nonindigenous birds and 10 mammals have established successfully in those regions. Four times as many nonindigenous birds were found at aggregation edges than would be expected by chance, and three times as many mammals, accounting for 50% of the invasions in both cases. Again, our previous results were confirmed: in transforming landscapes, both endangered and invasive species have body masses close to the edge of body-mass aggregations.

Discussion

Discontinuous body-mass distributions were found in all ecosystems and taxa that we examined. Such discontinuous or "lumpy" distributions also have been demonstrated for boreal region mammals and birds (Holling 1992), for tropical forest birds (Restrepo and others 1997), and for pre- and postextinction Pleistocene mammal faunas (Lambert and Holling 1998). The present analysis extends that conclusion to include animal communities in additional ecotypes (Everglades wet savanna and Australian Mediterranean) and taxa (herpetofauna and bats).

Although we are discovering that discontinuous patterns in body-mass distributions seem ubiquitous, it is easy to ignore them. They represent patterns of departure of body masses from some central tendency that typically is represented by a simple continuous distribution. Traditional approaches seek a unimodal distribution that best describes the data [for example, see May (1978) and Schoener and Jansen (1968)], and the residuals are ignored. It is the patterns of residuals around such a unimodal distribution that form the "lumps" or aggregations of similar body masses. As a result, traditional statistical tests designed to minimize type I error can miss identifying real aggregations. Manly (1996), for example, applied an elegant but particularly conservative test to the original boreal animal datasets [in Holling (1992)], and concluded that, at the most, two lumps or aggregations of body masses were significant, rather than the eight and more that Holling identified. The data presented here similarly would show few or no significant aggregations by using such a conservative method.

Conservative tests, of course, reduce the chance of being wrong (type I error)—but they also reduce the ability to detect real pattern. If that is done too early in an investigation, potentials for novel discovery are sharply reduced. Hence, in the tests of discontinuities presented here, we attempt to minimize, initially, type II error; that is, to reduce the chance of missing a real pattern. That sets the stage for a sequence of tests of increasing breadth and rigor in order to develop multiple lines of evidence that might converge on a robust demonstration and explanation.

In this case, we are following four steps to prove and expand the concept:

1. *Tests for pattern in distributions using tests that, initially, minimize type II error.* These focus on examining patterns of residuals—that is, on sequences of departures from a unimodal null distribution—but it is easy to miss the pattern. For example, tests of probabilities of individual departures of various degrees from a null distribution alone are inappropriate, because they ignore the pattern caused by sequences of departures. Similarly, tests using uniform distributions as a null, as is common in the community ecology literature, are inadequate because such distributions create false body-mass gaps when fitted to distributions that have (as these do) a prominent mode or modes. The Silverman (1981) kernal-density-estimate method, as used here, is an excellent way to identify an unbiased unimodal null distribution.

2. *Expanding the comparative evidence.* There is a need for evidence of discontinuous patterns revealed by using the same objective methods in different ecotypes, different taxa, and for different attributes. So far, there is evidence published for body-mass data from boreal prairies, boreal forests, wet savannas, tropical lowland and highland forests, and Mediterranean-type ecosystems; and for birds, mammals, herpetofauna, and bats.

3. *Establishing causation.* Competing hypotheses for explaining pattern need to be established and multiple sets of different data used to separate among the hypotheses. Body-mass aggregations and the gaps between them could be caused by founder effects, by phenological organization, by trophic or competitive interrelations, by locomotory constraints, and/or by entrainment to landscape/vegetation patterns. A number of causes are likely, but evidence so far converges on the latter as a dominating explanation [here and see Holling (1992)].

4. *Relating to existing theory.* The original idea came from a set of regional ecosystem studies [summarized in Holling (1992)] where hierarchy theory (Allen and Starr 1982; O'Neill and others 1986) combined with theories of ecosystem dynamics (Carpenter and Leavitt 1991; Levin 1992; Holling and others 1995) led to the proposition that ecosystem and animal community attributes should be distributed discontinuously across scales in time and space.

5. *Developing independent datasets to test causation.* Strong tests that minimize type I error begin to be appropriate when entirely independent datasets are compared. Examples include comparison of body-mass patterns in similar and dissimilar ecotypes on different continents, or of patterns in different taxa or trophic groups in the same ecosystem. This report presents one of the strongest tests to date. The animal body-mass data presented here are entirely independent of the datasets designating endangered and invasive species. The body-mass aggregations and gaps between them are identified by the methods in step 1, where type II error is minimized. The body-mass aggregations and gaps so identified show unambiguously that more endangered and invasive species exist at or near the edge of the body-mass lumps than could be expected by chance alone. This is consistently the case for four different taxa (birds, mammals, herpetofauna, and bats) in three different ecotypes.

In short, body-mass aggregations identified with the methods used here are powerful predictors of endangerment and invasiveness in transforming landscapes. If we could demonstrate that the relationship breaks down, over some scale ranges, in stable landscapes, the test would be even more convincing. Unfortunately, it is difficult to find convincing examples of ecosystems and landscapes that are not transformed by human activity.

The strong correspondence between the independent attributes of population status and body-mass pattern in three different taxa confirms the existence of discontinuous body-mass distributions. It may seem initially surprising that both invasive and declining species are located at the edge of body-mass aggregations. These results suggest that

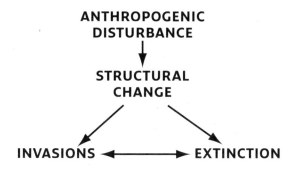

FIGURE 4: *The relationship between disturbance, ecological structure, invasions, and extinctions. Disturbance exceeding the resilience of a system affects ecological structure, which in turn leads to changes in animal communities. Additionally, changes in animal community membership may in turn lead to further changes: invasions, especially by predators, may lead to further declines in native species (Simberloff 1981), and extinctions may lead to further invasions (Diamond and Veitch 1981).*

something similar must be shared by the two extreme biological conditions represented by invasive species and declining species.

Note that our investigation has used definitions of endangerment and decline derived by an administrative process of government. Such definitions are conservative and tend to lag behind actual fluctuations in animal populations. For invasive species, historical contingency has played a large role in introductions. For instance, small animals such as mice and rats often are inadvertently transported, but antelope have likely never been stowaways on a ship or plane. It is not useful to describe historical happenstance; rather, the relevant question is, given that a number of species are introduced, which will succeed? We believe that the scale of environmental use by a species and landscape-level ecological transformation are important in predicting species' success or failure.

We hypothesize that the expected results of disturbance that exceeds the resilience of an ecological system (Holling 1973) are simultaneous events of invasion and deletion (Figure 4). Intense or extensive anthropogenic disturbance and land-use change may exceed the resilience of a system, disrupting the key processes that structure ecosystems and entrain biotic variables. When the resilience of a system is exceeded, new processes begin to control the system, and landscape structure proceeds through a transition to a new dynamic stable state. Because animal communities in an ecosystem are entrained by ecological structure, perturbation affects the structure of animal communities. Changes in key processes cause a transition from one type of self-organizing landscape structure to another. This results in changes in the animal community, including successful invasions and the decline or extinction of susceptible native species. Species with body masses near aggregation edges are the first to encounter changes occurring as landscape patterns start to transform.

At a global scale, rapid anthropogenic disturbance is affecting the processes that structure most ecosystems. Feedback from variables adapted to the altered ecosystem structure (for example, invasive species) prevents return to the original state. As more invasive species become established, they may further alter the environment and promote or entrench structural change. In south Florida, continuing invasions and declines in native species indicate that the community is still in a state of flux, responding to changes in landscape structures that lag behind changes in disturbance regimes. Even if original key processes are reestablished, the original animal community is not likely to be reassembled (Case 1990; Drake and others 1996).

Unusual characteristics are associated with the edge of body-mass aggregations. The edges of aggregations may be considered zones of crisis and opportunity, depending on the way a given species at these scales exploits resources and interacts with its environment, and may be analogous to phase transitions. In perturbed systems, we documented that biological invasions and extinctions tend to occur at aggregation edges. However, we suspect that variability in species composition and population status is higher at scale breaks (the edge of body-mass aggregations) whether or not the system has been perturbed. Human landscape transformations simply heighten the inherent variability. Highly variable behavior such as this has been demonstrated for the area between domains of scale in physical systems (Nittmann and others 1985; O'Neill and others 1986; Grebogi and others 1987) and postulated for biological communities (Wiens 1989). This investigation is the first to document unusual characteristics associated with scale breaks in animal communities.

Acknowledgments

We thank P. Marples, J. Sendzimir, and other colleagues in the Arthur Marshall Ecology Lab for adding to the conceptual foundation upon which this article was based. We thank D. Auth for providing access to the Florida Museum of Natural History herpetological collection, and K. Dodd for providing unpublished data on herpetofauna body mass. The manuscript was improved by comments from T. Allen, L. Gunderson, M. Moulton, G. Peterson, and an anonymous reviewer. This work was partially supported by a NASA/EOS grant and a NASA grant.

References

Allen TFH, Starr TB. 1982. Hierarchy: perspectives for ecological complexity. Chicago: University of Chicago Press.

Arita-Watanabe H. 1992. Ecology and conservation of cave bat communities in Yucatan, Mexico [PhD dissertation]. Gainesville: University of Florida.

Bailey RG. 1983. Delineation of ecosystem regions. Environ Manage 7:365–73.

Burrough PA. 1981. Fractal dimensions of landscapes and other environmental data. Nature 294:240–2.

Carpenter SR, Leavitt PR. 1991. Temporal variation in paleolimnological record arising from a trophic cascade. Ecology 72: 277–85.

Case TJ. 1990. Invasion resistance arises in strongly interacting species-rich model competition communities. Proc Natl Acad Sci USA 87:9610–4.

Diamond JM, Veitch CR. 1981. Extinctions and introductions in the New Zealand avifauna: cause and effect? Science 211:499–501.

Drake JA, Huxel GR, Hewitt CL. 1996. Microcosms as models for generating and testing community theory. Ecology 77:670–7.

Florida Game and Freshwater Fish Commission. 1994. Official lists of endangered and potentially endangered fauna and flora in Florida. Tallahassee: Florida Game and Freshwater Fish Commission.

Forys EA, Allen CR. 1998. Biological invasions and deletions: community change in south Florida. Biol Conserv 87:341–347.

Grebogi C, Ott E, Yorke JA. 1987. Chaos, strange attractors, and fractal basin boundaries in nonlinear dynamics. Science 238:632–8.

Haighton R. 1956. The life history of the slimy salamander, *Plethodon glutinosus*, in Florida. Copeia 1956:75–93.

Holling CS. 1973. Resilience and the stability of ecological systems. Annu Rev Ecol Syst 4:1–23.

———. 1992. Cross-scale morphology, geometry, and dynamics of ecosystems. Ecol Monogr 62:447–502.

Holling CS, Schindler DW, Walker BW, Roughgarden J. 1995. Biodiversity in the functioning of ecosystems: an ecological synthesis. In: Perrings C, Mäler CKG, Folke C, Holling CS, Jansson BO, editors. Biodiversity loss: economic and ecological issues. New York: Cambridge University Press. pp. 44–83.

Kale HW II, Pranty B, Stith BM, Biggs CW. The atlas of the breeding birds of Florida. Forthcoming.

Lambert WD, Holling CS. 1998. Causes of ecosystem transformation at the end of the Pleistocene: evidence from mammal body-mass distributions. Ecosystems 1:157–75.

Levin SA. 1992. The problem of pattern and scale in ecology. Ecology 73:1943–67.

Ludwig JA, Cornelius JM. 1987. Locating discontinuities along ecological gradients. Ecology 68:448–50.

Manly BF. 1996. Are there clumps in body-size distributions. Ecology 77:81–6.

May RM. 1978. The dynamics and diversity of insect faunas. In: Mound LA, Waloff N, editors. Diversity of insect faunas. Oxford: Blackwell Scientific. pp. 188–204.

Milne BT, Johnston K, Forman RTT. 1989. Scale dependent proximity of wildlife habitat in a spatially-neutral Bayesian model. Landscape Ecol 2:101–10.

Milne BT, Turner MG, Wiens JA, Johnson AR. 1992. Interactions between the fractal geometry of landscapes and allometric herbivory. Theor Popul Biol 41:337–53.

Morton SR. 1990. The impact of European settlement on the vertebrate animals of arid Australia: a conceptual model. Proc Ecol Soc Aust 16:201–13.

Moulton MP, Pimm SL. 1986. The extent of competition in shaping an introduced avifauna. In: Diamond J, Case TJ, editors. Community ecology. New York: Harper and Row. pp. 80–97.

Mount RH. 1963. The natural history of the red-tailed skink (*Eumeces egregius* Baird). Am Midl Nat 70:356–85.

Nittmann J, Daccord G, Stanley HE. 1985. Fractal growth of viscous fingers: quantitative characterization of a fluid instability phenomenon. Nature 314:141–4.

O'Neill RV, DeAngelis DL, Waide JB, Allen TFH. 1986. A hierarchical concept of ecosystems. Princeton: Princeton University Press.

O'Neill RV, Turner SJ, Cullinam VI, Coffin DP, Cook T, Conley W, Brunt J, Thomas JM, Conley MR, Gosz J. 1991. Multiple landscape scales: an intersite comparison. Landscape Ecol 5:137–44.

Perry DA. 1994. Self-organizing systems across scales. Trends Ecol Evol 10:241–4.

Peterson G, Allen CR, Holling CS. 1998. Ecological resilience, biodiversity and scale. Ecosystems 1:6–18.

Pimm SL, Lawton JH. 1980. Are food webs compartmented? J Anim Ecol 49:879–98.

Restrepo C, Renjifo LM, Marples P. 1997. Frugivorous birds in fragmented neotropical montane forests: landscape pattern and body mass distribution. In: Laurance WF, Bierregaard RO, editors. Tropical forest remnants: ecology, management and conservation of fragmented communities. Chicago: University of Chicago Press. pp. 171–89.

SAS Institute Inc. 1985. SAS user's guide: statistics. Cary (NC): SAS Institute.

Schodde R. 1981. Bird communities of the Australian mallee: composition, derivation, distribution, structure and seasonal cycles. In: di Castri F, Goodall DW, Specht RL, editors. Mediterranean-type shrublands. New York: Elsevier Scientific. pp. 387–415.

Schoener TW, Jansen DH. 1968. Notes on environmental determinants of tropical versus temperate insect size patterns. Am Nat 102:207–24.

Sendzimir J. 1998. Patterns of animal size and landscape complexity: correspondence within and across scales [PhD dissertation]. Gainesville: University of Florida.

Silverman BW. 1981. Using kernel density estimates to investigate multimodality. J R Stat Soc [B] 43:97–9.

Simberloff D. 1981. Community effects of introduced species. In: Nitecki MH, editor. Biotic crisis in ecological and evolutionary time. New York: Academic Press. pp. 53–81.

Strahan R. 1995. Mammals of Australia. Washington (DC): Smithsonian Institution Press. p. 53–81.

Webster R. 1978. Optimally partitioning soil transects. J Soil Sci 29:388–402.

Wiens JA. 1989. Spatial scaling in ecology. Funct Ecol 3:385–97.

Empirics and Models

Commentary on
Part Three Articles

CRAIG R. ALLEN, LANCE H. GUNDERSON, AND C. S. HOLLING

THE THIRD SECTION OF THIS BOOK presents three groundbreaking articles that developed a set of methods for understanding and diagnosing ecological resilience. Moreover, they blended an elegant set of models with real-world applications and were the first to demonstrate how resilience can be lost because of management activities that focus on an optimal control strategy of a single target variable.

The first article (Holling and Chambers 1973) also introduced the use of models to help understand the complex dynamics exhibited by such resource systems. This work preceded the first adaptive management volumes by half a dozen years, yet the kernels of adaptive management are presented in article 10. Article 10 is also rich because it explicitly includes the human dimensions of complex resource issues by depicting six caricatures of personality types that are still common at adaptive management workshops. The final two articles (articles 11 and 12) depict aspects of modeling spruce budworm outbreaks in eastern Canadian forests. The models were initially developed in conjunction with programs at the International Institute for Applied Systems Analysis in Austria and the University of British Columbia in Canada. Much of the art of modeling—the parsimonious selection of a few key variables, the need for data, and the search for understanding complex dynamics—is presented in these seminal articles.

By the early 1970s, it was apparent that anthropogenic alterations of the earth's land cover and the effluent associated with increasing affluence were creating novel changes in ecosystems for which existing theory was insufficient. This strikingly parallels the current state of the world; only the particular challenges are different. Holling and Chambers (1973) wrote: "Suddenly [we] could act as if the world was unlimited and the only constraints were economic.... We structured our institutions, our disciplines, and our behaviour in response to aspirations for long life and material possessions. And within a few years we faced a new set of problems involving the environment."

Holling and Chambers recognized the inherently cross-disciplinary nature of these new environmental problems, and the fact that in many cases ecosystems were on the brink of collapse due to accumulating human impacts. They also recognized that the ecosystems' ability to absorb more insults—that is, their resilience—was in danger of being exceeded.

An increasing threat and incidence of surprise results from anthropogenic influences that slowly accumulate; increasing global connectivity increases the threats we face (Holling 1994). Nature, people, and economies are now coevolving and may be characterized as a complex adaptive system. Nature, people, and economies are now affecting one another in novel ways and at such large scales that traditional modes of governance and management are challenged and systems of people and nature may be pushed beyond their adaptive capacity.

The universality of nonlinearities in systems, such as thresholds and limits, is increasingly recognized in ecology. Multiple stable states are the rule when complex behaviors are incorporated into relatively simple models. The behaviors of systems are controlled by the interactions of a small set of key processes, both biotic and abiotic (Holling 1992, Allen and Holling 2002), with each set operating at distinct spatial and temporal scales. Those scales may be very large—far larger than the scales commonly used by humans for management and planning. These sets of processes are robust and resilient because of the large degree of spatial heterogeneity and functional diversity present both within and across scales (Peterson et al. 1998, Folke et al. 2004).

Holling (1994) describes a pathology whereby early exploitation of the environment and early management efforts are often successful (because, we add, they have survived and evolved over time to states of high resilience), but exploitation and management efforts erode resilience by mechanisms such as species loss, the loss of genetic diversity, and an increase in spatial homogeneity. Correspondingly, management agencies become more efficient but also more rigid and less responsive, reducing the available management options over time (Meffe and Holling 1996). This leads to an overall pathology whereby management becomes more rigid, society becomes more dependent on those rigidities and stability, and the systems as a whole—the interconnected social-ecological-economic system—becomes less resilient, inevitably leading to eventual surprise and collapse.

The pathology described can be broken by adaptive change and readjustment that create new opportunities and novel institutions and institutional responses. In the past few decades, however, the loss of species (and, thus, functional diversity and cross-scale redundancy) and spatial heterogeneity and connectivity has become a global phenomenon for the first time. The consequences of this have yet to manifest.

Holling and Chambers (1973) decry the lack of connection between theory and practice. Thus, their paper foreshadowed many paradigmatic changes that have occurred during the past thirty years: the concept and formalization of resilience; adaptive resource management as a method to test theory and policy with management practice; the formation of the Resilience Alliance as a loose experimental organization with cross-disciplinary representation; and models to explore alternative scenarios and generate "better questions" rather than provide answers.

The boreal forest system, especially in exploited forests vulnerable to spruce budworm outbreaks, provided rich data for the understanding of resilience. Clark et al. (1979) attempt to translate that understanding into policy. This represents one of the first attempts to suggest managing social-economic-ecological systems for resilience. In contemporary times, resilience is considered a component of complex systems of both people and nature (Walker et al. 2004, Walker and Salt 2006), but such thinking took decades to develop.

This paper represents a response to management that maintained large areas of Canadian boreal forest at a stage extremely vulnerable to spruce budworm. Incipient outbreaks were controlled by applying pesticides over large areas, and for brief periods variability in the system was reduced to maximize the variables of interest to the managers (a high potential volume of wood available coupled with low populations of spruce budworm). But management had the undesired and unanticipated consequence of making the system ever more vulnerable to widespread spruce budworm outbreaks. Management of the forest lowered the resilience of the forest ecosystem.

Clark et al. (1979) highlight what are now, three decades later, frontiers in ecology: systems with nonlinear dynamics, spatial components, high dimensionality, high uncertainty, and links among the ecological systems, economic systems, and social systems. All of these conspire to make complex systems of people and nature highly unpredictable. This paper is contemporary with the proposition of adaptive resource management (Walters and Hilborn 1976, Holling 1978), and the challenges described call for an adaptive management approach. The adaptive management approach flowed from the recognition of multiple stable states in nature. Given that unexpected or very rare events are expected, the shift of management should be from avoiding extreme events toward minimizing the effects of such events. This is the key insight made by Clark et al. (1979) three decades ago. Management's emphasis should be on maintaining and enhancing resilience. Management has traditionally sought to optimize output of some form; managing for resilience seeks to guarantee output in the face of change.

Despite the complexity of the spruce budworm–boreal forest dynamics, a handful of variables drives the system. Recognizing these key drivers, and the scales on which each operates, is all that is needed to produce the observed dynamics. This observation led to the eventual proposition of the Extended Keystone Hypothesis (Holling 1992). The relatively simple models described produce complex dynamics with multiple equilibria.

Clark et al. (1979) provide an exhaustive examination of the relationship between ecological states and management options, focusing on the ecological, economic, and social variables and indicators. They provide an early examination of the pitfalls of command-and-control approaches to natural resource management, whereby a policy is seemingly effective in the short term—as were insecticide applications to control budworm populations—but the short-term success inevitably leads to catastrophic failure. Similar examples include policies of fire control in the United

States that eventually led to the 1988 Yellowstone National Park fire. As is often the case with reactive policies, these disasters were required before an entrenched policy was changed. Unluckily, such failed policies tend to lock in so that there is no easy way out when years of pest or fire suppression lead to an enormous volume of vulnerable biomass, other than to continue more strenuously the policies that led to that trap in the first place.

Human desire for certitude often blinds us to reality. Humanity's unintentional natural experiments—represented at local scales by, for example, altering hydrology in watersheds via dams and irrigation, and at the global scale by, for example, carbon dioxide-mediated climate change—often fail because the potential for being wrong is present and being wrong can be extremely costly. Humans' perceptions of the environment are simply anthropogenic myths based on our personal viewpoints (Holling 1982).

The cost of management failures increases as the probability of failure decreases. Surprises lead to human institutions that attempt to reduce the probability of more surprises, which in turn reduces future options. Adaptation is a better option than either regulation or ever more precise predictions. The latter option is reasonable for responding to known threats, but adaptation is the best option for surviving unknown threats.

Most of our institutions focus on increasing stability and minimizing disturbance, and they tend to react to surprises as crises. Alternatively, institutions that develop policies to embrace change and encourage novelty respond to surprise in an adaptive manner and are strengthened by it. Institutions that embrace change recognize the explicit link between resilience and adaptive resource management.

As with Clark et al. (1979), described above, the boreal forest–spruce budworm interaction provided the insight for the paper by Ludwig et al. (1978). Here, simple qualitative models were created that explained a wide range of the dynamics observed in the forest system. The key for the success of these qualitative models was the separation of the variables based on their scale. Variable "scales" were not explicitly considered (in the sense of both space and time domains); rather, variables were segregated by their relative speeds—fast versus slow. This presaged more sophisticated understanding of cross-scale dynamics in ecosystems.

Implicitly and partially explicitly, slow and fast variables also differ in the spatial extent of response. Fast variables respond quickly but only over small areas, whereas slow variables respond over very large extents. The simple equations and their response dynamics as described by Ludwig et al. (1978) eventually gave rise to cross-scale ecology, resilience theory, adaptive management, and panarchy theory.

The model presented readily captures the dynamics of managed boreal forest–spruce budworm dynamics, characterized by two stable states separated by an unstable state. The attraction of the system to one or another of the potential equilibria is determined by budworm limitation via predation and food availability (forest volume).

Insights from these simple models have direct applications for managing complex resource systems. Management can increase the possibility of outbreak, reduce it, or maintain it endemically. The boreal forest was the model system, but the authors contend that data sufficient to reveal complex dynamics are at hand for most ecological systems.

Literature Cited

Allen, C. R., and C. S. Holling. 2002. Cross-scale structure and scale breaks in ecosystems and other complex systems. *Ecosystems* 5:315–18.

Clark, W. C., D. D. Jones, and C. S. Holling. 1979. Lessons for ecological policy design: A case study of ecosystem management. *Ecological Modeling* 7:1–52.

Folke, C., S. Carpenter, B. Walker, M. Scheffer, T. Elmqvist, L. Gunderson, and C. S. Holling. 2004. Regime shifts, resilience, and biodiversity in ecosystem management. *Annual Review of Ecology and Systematics* 35:557–81.

Holling, C. S. 1978. *Adaptive environmental assessment and management.* Caldwell, NJ: Blackburn.

———. 1982. Resilience in an unforgiving society. In *Sakherets-politik och system analysis*, ed. A. Andersson and A. Erikson, 15–30. Umea, Sweden: Umeå universitet, Avdelning för regional ekonomi.

———. 1992. Cross-scale morphology, geometry, and dynamics of ecosystems. *Ecological Monographs* 62:447–502.

———. 1994. An ecologist view of the Malthusian conflict. In *Population, economic development, and the environment*, ed. K. Lindahl-Kiessling and H. Landberg, 79–104. New York: Oxford University Press.

Holling, C. S., and A. D. Chambers. 1973. Resource science: The nurture of an infant. *BioScience* 23 (1):13–20.

Ludwig, D., D. D. Jones, and C. S. Holling. 1978. Qualitative analysis of insect outbreak systems: The spruce budworm and forest. *Journal of Animal Ecology* 47:315–32.

Meffe, G. K., and C. S. Holling. 1996. Command and control and the pathology of natural resource management. *Conservation Biology* 10:328–37.

Peterson, G., C. R. Allen, and C. S. Holling. 1998. Ecological resilience, biodiversity and scale. *Ecosystems* 1:6–18.

Walker, B., C. S. Holling, S. R. Carpenter, and A. Kinzig. 2004. Resilience, adaptability and transformability in social–ecological systems. *Ecology and Society* 9 (2):5. http://www.ecologyandsociety.org/vol9/iss2/art5/.

Walker, B., and D. Salt. 2006. *Resilience thinking: Sustaining ecosystems and people in a changing world.* Washington, DC: Island Press.

Walters, C. J., and R. Hilborn. 1976. Adaptive control of fishing systems. *Journal of the Fisheries Research Board of Canada* 33:145–59.

Resource Science
The Nurture of an Infant

C. S. HOLLING AND A. D. CHAMBERS

C. S. HOLLING, an ecologist, is the Director of the Institute of Resource Ecology, University of British Columbia, Vancouver 8, B.C., Canada.

A. D. CHAMBERS, a resource scientist, is a Science Adviser, Science Council of Canada, Ottawa, Canada.

BioScience, Vol. 23, No. 1. (1973), pp. 13–20.
Originally published in BioScience, copyright American Institute Biological Sciences.

IT IS A TEMPTING AND SAFE ACADEMIC DEVICE to approach any problem from a traditional viewpoint. By so doing we assume that the twenty or so civilizations of man and the few thousand years of recorded history are sufficient to have faced all problems and devised all solutions. Society now seems to be facing problems of resources and environment, however that are more intensive and extensive than those experienced in the past. A critical new ingredient was added to man's history with the Industrial Revolution. Suddenly men could act as if the world was unlimited and the only constraints were economic. Responding to a past filled with crises of pestilence and poverty, we used our new powers to dispatch these past tormentors. We structured our institutions, our disciplines, and our behaviour in response to aspirations for long life and material possessions. And within a few years we faced a new set of problems involving the environment.

Now Mother Nature is not in a delicate state of balance—community wisdom to the contrary. Natural systems have evolved over a very long time and have experienced traumas and shocks long before man came on the scene. Unexpected catastrophes are part of the natural order, and selection operates so that those organisms, populations, and communities that survive are explicitly those that are able to take the unexpected trauma in their stride. They have developed an internal resilience to absorb punishment, and it is this resilience that has absorbed the consequences of man's intrusions. But the resilience is not infinite; we now face the first signs of the consequences of stretching the bounds of resilience of ecological and social systems. The three hundred years of ignoring these limits has left us with a baggage of approaches and solutions that are only admirable as instruments for resolving fragments of problems. Wherever we look there are gaps—gaps between methods, disciplines, and institutions.

Our analytical tools and training emphasize rigor at the expense of holism. We can, therefore, do an admirable job of learning more and more about less and less. There are, as a consequence, serious gaps within research activities themselves, with hypothesis and theory often having rare and sporadic contacts with empirical testing and reality. Just as there are gaps between methodologies, so also between disciplines. Environmental problems have economic, social, ecological, and physical dimensions, and no one discipline can possibly encompass all the necessary concepts and techniques.

But even if an ideal interdisciplinary research activity could be mobilized to produce a better mousetrap, no one would beat a path to its door. It would be destined only for display and preservation as an historical oddity since there are serious institutional gaps that inhibit the flow of new knowledge of these large-scale problems into policy formulation, testing and implementation. A university can be an effective environment for research, but it is weak on the pragmatic experience required to implement it. On the other hand, institutions like government agencies, that have experience in policy formulation and implementation, are so fragmented in their charge—e.g., highways or agriculture—that they are forced to concentrate on the fragments and not the whole. Neither the university nor the government agencies alone can bridge this gap between abstraction and rigor on the one hand, and policy formulation and implementation on the other.

Finally, even if links can be forged between methods, disciplines, and institutions, a gap still exists that inhibits the formulation and implementation of policies that are responsive to people's needs. In this world of increasing complexity and with the apparent need for technological expertise, it becomes massively difficult for the citizen and his political spokesman to communicate well enough so that humane controls can be applied to technology and policy. It is this gap in communication between constituencies that must be bridged if the new approaches are not to yield social as well as ecological DDT's.

But at least now there are signals that the world is limited and that resilience of ecological and social systems is not infinite. There is a groping awareness of the problem and a growing desire for new ways. But the historical cards may be so stacked against us that magnificent steps to resolve our problems will fail. How many universities have relabeled old boxes with "Faculty of Environmental Studies?" How many governments have recycled existing activities into "Departments of Environment?" Such steps might lead to new panaceas that are more disastrous than the old because they are more global. Would it not be better to face the reality of the historical momentum that perpetuates the myth of "no limits" and proposes narrow solutions for narrow problems? By recognizing this reality we will be forced to develop a newly disciplined step-like approach in which each step is made digestible enough to be successful, and can lead to future steps which cover progressively more ground. We need to find directions towards solutions and not the Utopian solutions themselves.

It was against this background that the Resource Science Workshops were established. Each was to be one of a series of steps that would hopefully begin bridging the gaps. The first two workshops concentrated on bridging gaps between methods and disciplines. This had the advantage of concentrating the initial steps within one institution—the university. Moreover, the university represents, in microcosm, the extremes in society. There is no place where one can find more competent people—nor more incompetent; more responsible—nor more irresponsible; more selfless—nor more egocentric. Each of the departments within a university has developed its own methods and jargon, and if there has been communication between disciplines at all, it has been as much territorial posturing as anything else. Hence, if techniques can be developed to cross disciplines within a university, it seemed likely that the same techniques could be useful in bridging gaps anywhere.

The first of the three workshops concentrated on an overall conceptualization of a resource problem in an open forum. The second expanded on the results of the first and concentrated on improvement and extension of the details. The first workshop succeeded admirably. The second convinced us that workshops were inappropriate as an environment for the hard technical detail, data collection, and analysis necessary. This requires the persistent commitment that only an individual, not a group, can develop. The third workshop attempted to apply the lessons learned from the first two to a major project involving 15 departments of the university, five departments of a city government, and a regional planning agency. It, therefore, attempted to bridge not only gaps between disciplines but also between institutions and social roles—between academics and bureaucrats, for example. The success of this last workshop has allowed us now to take one more step to begin approaching gaps between constituencies. This will involve citizens and politicians in a workshop kind of activity.

In what follows, we will describe some strategies, tricks and gimmicks. All were useful, but we sense that the real success of the workshops rests as much with its style of operation as with its substance. The open workshop environment is a social game in which the rules and the players of the game jointly evolve as the game progresses.

Choosing the Game

A working focus is essential for the workshop, and to do justice to the ultimate goal, the problem chosen must have at least two or more of the economic, ecological, social, and physical dimensions that now characterize environmental problems. But recognizing the unlikelihood of success, the first problem chosen should be digestible, even if it is trivial. Our first two workshops chose a study of land use in the Gulf Islands off the coast of British Columbia. The problem was chosen for its simplicity:

the area is largely undeveloped and clearly seems to have a dominant recreational use now and in the future. Demand for land is largely generated outside the islands, mostly by population pressures from the urban centers of Victoria, Vancouver, and Seattle. Moreover, it did not at the time seem to be a political issue. This was important because we wished to concern ourselves with the very difficult technical problems before facing even more difficult and relevant political ones. But the islands have many of the same resource and environmental problems facing large regions. The economic, ecological, and social forces are producing escalating land prices, fragmentation of land into smaller and smaller parcels, and the erosion of those values that originally attracted people to the area. Most important for the purpose of the workshop, however, the components of the system could be hazily identified from the outset, and a vague form of the model imagined.

The primary purpose of the first two workshops was to discover how to develop and nurture interdisciplinary research, and, hopefully, to provide some answers to the problems of the region. The first is not a trivial problem. The successful examples of interdisciplinary groups are those in which the goal has been so specifically and narrowly defined that a disciplined, highly organized team approach could achieve it. In resource and environmental problems, however, the goals of society are as much an unknown as the mechanism to achieve those goals. A much more open environment is necessary to achieve progress that is not at the expense of individual creativity.

These first two workshops did satisfy their intent, and the successful strategies and tactics that evolved are described in the remainder of this paper. Their success led to another step in the third workshop, in which the approaches could be focused on a much larger problem—a study of man and environmental interaction in a large region dominated by a major urban center—Vancouver. Moreover, this problem became the focus for a cooperative effort between institutions—a university, a city, and an emerging regional government. By combining such diverse institutions as equal partners, we hoped to develop a more effective flow between research and the formation and testing of policies. We are now about to initiate the fourth workshop using the lessons learned to involve not just academics and bureaucrats but politicians and citizens in a still broader probing of the gaps in the total decision process.

FIGURE 1: *Flow Diagram of the Gulf Islands Recreational Land Simulator.*

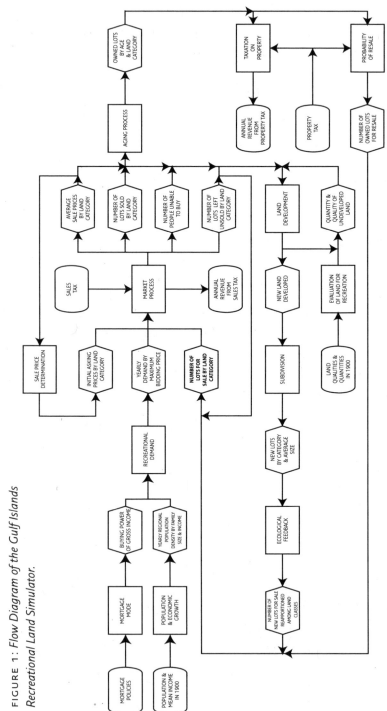

Each of these workshops, whether modest or grand, in a real sense became a game with highly flexible and evolving rules. The only immutable rule has been to stay in the game. To do so requires a highly resilient structure that can absorb the inevitable mistakes that are made.

Organizing the Game

Practically any kind of visible organization will work. What is much more important is the invisible organization that evolves. Nevertheless, it has become very obvious in each of the workshops that many people need the security of visible organization. In providing it, however, it is essential to maintain flexibility and responsiveness to change. As an example, on paper the workshop projects can often seem highly structured, with a leader, a core group having management and budget responsibilities, a research group whose activities are integrated through a synthesis group, and a staff of programmers and program assistants. But the reality can be very different: the real leadership always emerges to form a committed "Camarilla" whose individuals come from any one of the groups from core to staff. It is these people who have a sense of purpose and philosophy that makes them commit the time and energy necessary to make the workshop a success. They act primarily as trouble-shooters. Integration of the activity very often does not occur in a synthesis group, particularly if it is at all large. The integration occurs as the major investment of one to three of the Camarilla who thereby crystallizes the process of systems integration by periodically providing specific proposals or solutions. Without this activity the project becomes simply a constellation of unrelated parts.

The Players of the Game

Our first workshop had 15 people; the second, 35, and the third, 70. There is no question that the smaller number is a very effective and manageable size. It is about the best compromise between, on the one hand, wanting a large number of people for an adequate array of disciplines and constructive workers, and, on the other hand, a small number for ease of communication. But even the larger group is quite feasible and effective. All the larger group requires are additional devices to blunt and absorb the less constructive suggestions and diversions. As a first effort, however, a large group would be disastrous.

Benevolent Despot   Peerless Leader   Snively Whiplash

The choice of individuals should be very open: the selection process should not be an exercise of philately of disciplines. It would be ideal to have one of that, and one of that, and one of that, but any combination of more than two disciplines is a major advance. Much more important is to have people who have a degree of commitment and sympathy irrespective (or perhaps in spite) of their training. One of the major values of a workshop is as a device to identify that small proportion of individuals whose different talents in combination can make an effective project.

But however many individuals play the game, in each case they will begin to assume very distinct roles. And all roles are important. The leader of the workshop, for example, can best be played as a Benevolent Despot who balances delicately between humane omniscience and programmed (or real) stupidity. Peerless Leaders will emerge who, with astonishing and real sacrifices, take onto themselves onerous duties for the greater good. Inevitably too, there will be Snively Whiplashes who clearly detest the whole exercise, wish it to collapse, and stay for unknown reasons. But they are invaluable, for managed hostility inevitably provides a focus that can crystallize an admirable esprit. In fact, if you don't have one, invent one.

In a more positive way, the Compleat Amanuensis provides the essential day-by-day organized efficiency that is so foreign to other key members of our workshops. Even more important, if you are lucky, he or she will have the wisdom, judgment, and experience to gently deflate the more unrealistic exuberances. Some of these exuberances, but not all, come from The Utopians, who dream the impossible dreams. These are perhaps most difficult of all. At the beginning of a workshop, their ideas are absolutely essential to broaden perspectives. Moreover, the Utopians themselves benefit since other participants of the workshop can provide the filter, generally missing in Utopians, that separates imaginative ideas from fantasy. Once this is done, however, their inability to separate what is ideal from what is possible can continually divert and frustrate. It is easy for some to develop an arrogant contempt for them, but we are convinced that their elimination, for the sake of efficiency, can defeat the real value of a workshop. After all, who can better stimulate the articulation of goals and values than an Utopian?

Finally, if you are very lucky indeed, you might find one out of a hundred or more participants to be a Blunt Scot. He is the one person whose bluntness and sincerity of purpose transcend the mischievous irresponsibility that most of the rest of us succumb to occasionally. He identifies difficult personality clashes and intervenes in a way that the Benevolent Despot cannot.

Irrespective of the roles people play, it is, of course, essential to have a minimum set of people. There has to be at least one person who spends the majority of his time organizing and conceptualizing the project. In a project of any magnitude or consequence, it ideally would be two people, since the tensions can only be handled if occasionally one person spells another. Secondly, our experience suggests that the internal Camarilla of closely communicating individuals should be from three to six. Thirdly, there must be an efficient support staff, particularly secretarial and programming. The size of the project and the duration of the workshop determine very much the size of the staff. There was, for example, only one in our first workshop involving 15 people; two in the second workshop involving 35, and five in the third workshop involving 70.

The Tools of the Game

Every good game has attractive geegaws—spinning wheels, bouncing balls, clickety-clack. In our games the computer serves the same admirable purpose as the roulette wheel. It provides valuable glitter as well as random numbers. The danger is that the tool is so glamorous that it can become an end instead of a means. We only tempt this fate because the computer has unique potential to help bridge the gaps noted earlier.

The computer is capable of handling the complexity found in resource and environmental problems. But more important than this, it can be a powerful device for communication. It is first of all an idiot, so that there has to be a highly organized and simple way of communicating with it. Since all disciplines have to develop the same organized communication with the idiot, they are encouraged to begin communicating with each other, independent of disciplinary jargon. Secondly, an interactive computer system that is open shop can turn the machine into a powerful doodling pad to explore the unknown and allow an individual's intuition full play. Thirdly, since the complex technical trivia can be handled by the machine, people from many backgrounds can be freed to do the important thing—i.e., exploring goals and values and the consequences of different policies in a simulated world. It has the potential, therefore, of placing citizen, politician, technocrat, and academic on the same competitive footing. The exercise of playing the game with a simulation model enables people from very different backgrounds to develop rapidly a feeling for the systems character of environmental problems.

For our workshop we emphasized the use of a visual display terminal—a TV screen controlled by a large computer. These terminals allow people to act independently or together as they intervene in the simulated world. Considerable emphasis is placed on output in a visual form—maps, graphs, movies—since people relate quickly to visual information. Such an interactive computer center, if emphasized as a useful tool, not a panacea, can make for the best game in town.

The Rules of the Game
Beyond general statements of motherhood and the need for communication and unity of purpose, there is only one set of rules, and these have been designed to adjudicate arguments. There are three classes of arguments that emerged in our workshops.

Arguments of Principle
The first touches very closely on principle and must be faced and resolved early in the activity. If it does not emerge naturally within the workshop setting, it should be made to do so. Usually the point of principle is identified around a rather specific tactical issue, and this must be crisply articulated. In our case, the principle has been that we must not presume we know man's goals or the values he assigns to them. Quite the opposite, we argue that the goals are multifaceted and each individual has a unique weighting that he assigns to these goals. Our ignorance is so

sublime it would be enormously dangerous to presume otherwise. A large number of specific issues flow from this principle.

First, any model that we attempt to develop should not be an optimizing model in the strict analytic sense, for such models requires clear and usually simple identification of a unique goal and a unique path to that goal. Rather, the model must be a non-optimizing simulation model—a kind of dynamic Meccano set in which goals and values are explicitly assigned outside the model.

Second, it means that the models cannot be value-free but can be value-variable. That is, explicit opportunity is provided for each individual to weigh the information generated in the way he feels best reflects his unique value system. It is, therefore, extremely important to develop devices external to the model to allow individuals with various interests to apply their own view of the world.

Third, rather than confining use of the model to specific "legitimate" vested interests, it has to be open for use by all vested interests, even where that requires subsidization.

Fourth, truly open access to information and devices to use information means a change in political and institutional structure that can threaten but also improve the political process.

Arguments of Dogma

Other arguments do not touch on principle and can be resolved much more indirectly. In each of our workshops most individuals had a certain mind set about what the problems of the world were. The planner might argue that there is not and never will be a population problem and that what is wrong is that resources and people are not distributed in an optimal way. A classical economist might insist that Adam Smith lives, that the free market model is correct, and the world is wrong. If any problem does arise, however, the solution is to redistribute income. The conservationist might insist that population growth is the only problem and major parts of the world should be locked up (except, perhaps, for his socioeconomic group).

These are, of course, presented here as exaggerated caricatures, but, nevertheless, the viewpoints that emerge in a more rational form are totally resistant to resolution through discussion. They emerge because of the person's background and training, and attempts at objective discussion simply lead to polarization. We have found, in these cases, that each viewpoint can be coupled with a policy. Hence, the planner might argue that, given Olympian control of transportation planning, he could distribute people and resources to resolve any major regional problem. We record the point of view and the policy and assure each individual that the model will be programmed in such a way to allow intervention to test the specific policy. Hence, in simulation runs with the model, the planner is given total powers to develop transportation systems in the way he wishes. The economists can apply a number of monetary and fiscal policies, and the environmentalist land bank policies, zoning policies and park policies. Once each individual realizes he will be given the opportunity to expose the recalcitrance of the rest, these important policy issues are reserved for later resolution during simulation runs.

In fact, we discover that none of the policies works; they all have "unexpected consequences." The first reaction is to argue that the model is wrong, and indeed at times it is wrong, thereby triggering a useful cycle of improvements in the model. Eventually, however, each individual is forced to face his basic assumptions. The basic trap is that people conceive of a small fragment of a whole and the very best policy for that fragment can produce the reverse effect through interaction with other parts of the system.

The growing realization that the same thing happens in the real world (the more successful the freeway, the more congestion, the more sprawl) leads to major shifts of view and new and better kinds of questions.

Arguments of Detail
The third kind of argument is technical. It usually occurs in arguments over specific hypotheses of the effect of one variable upon another. When such conflicts emerge, the individuals must ask a series of questions.

They were first to ask whether they were talking about the same thing, since very often arguments can develop if hypotheses are not stated crisply. Very often the problem could be resolved at this point, but if an issue persisted, the next question to ask was "Is it important for the aims we have in mind or is it an important topic that is tangential to our aims?" This question, in essence, says we must not only do interesting things, but relevant things—a rather heretical notion in a university. If the answer is "No," the antagonists were asked to discuss it over beer on their own time, but if it was relevant, they were asked if there were data available to allow choice between the two hypotheses. If there were data easily available, then an individual was assigned to collect them. If the data were not readily available — as often they are not — then both hypotheses were preserved in the model, and in simulation each hypothesis was tried in turn.

Often, however important the points seem to be intuitively, even radically different hypotheses can have little effect on the output of the model. The effect is absorbed by other parts of the model. There is little point in spending time and resources when this is the case. There are, however, strategic points in the model to which the output is very sensitive. When this is discovered through simulating the effects of each alternative, further research effort becomes essential. Not only can technical arguments be resolved by this series of questions, at the same time the results can give direction to future activity by assigning research priorities.

The Game

Management Strategy

The only management strategy is to evolve an open flexible system that can accommodate conflicting views and the inevitable errors and traumas that will develop. In a sense, therefore, the workshop should develop the same kind of resilience that has evolved in natural systems. It is this resilience that can accommodate the unknown. The price paid is to reduce efficiency, and, initially at least, precision. But we would argue that efficiency can only be attained if the goal is trivial. We would rather have an inefficient management system that can accommodate creativity and independence than an efficient team with PERT charts dictating goals. To repeat, the best we can hope for is to identify towards a solution and not the best solution.

The Systems Tactic

In each of our workshops there has been an initial emphasis upon systems approaches and simulation techniques. Although such approaches do justice to the complexity of resource systems, and do provide a *lingua franca* for communication, the emphasis is only a tactic to initiate the program. Nevertheless, the emphasis does give an organization to the activities that have to be part of a workshop. They do so by providing a way to break a complex problem into constituent components with their interactions. The exercise of identifying these components and the processes that interrelate them is an important interdisciplinary exercise requiring information from a large number of disciplines. It leads, ultimately, to the identification of fragments that are sufficiently small that their analysis is not threatening to established patterns of thought. At this level, one can act in a traditional disciplinary way, without realizing that when the small fragments are put together something different in kind emerges.

The second feature of systems approaches is that insight does not have to be initially elegant in the sense that mathematicians would consider elegance. It frees the scientist from the classical trap of going deeper and deeper into a problem with more and more precision and less and less relevance. Not only does it free the scientist from his bias, but it also opens to those both quantitatively and qualitatively trained, a way to organize their knowledge in compatible forms.

The result is a model. A model that can be probed and explored in a simulated world and becomes an evolving device of self-instruction. Its value is not so much to give answers as to generate better questions; not to define a unique policy but to expose some of the consequences of alternate policies. It therefore becomes an essential but nevertheless small part of a larger process that requires radical reappraisal of decisions in our society.

What is really important is not what is put into the model but what is kept out of it. Not how to provide a mechanism, like a model, to handle the quantitative, but how to design a decision-making framework to handle the qualitative. Not whether the model is useful, but how it is used. The systems tactic provides one part of the definition of these alternatives.

Keeping it going

A large number of specific tasks have been developed by trial and error that seem to be useful in maintaining the momentum of the game. Since we only hazily understand why they work, the best we can do is to present them as a list.

1. Since our workshops run for a six-month period, everyone is
   expected to attend a regular meeting one afternoon each week.

It is a measure of commitment that they are willing to make this sacrifice. It is at this time that principles and concepts are evolved, and progress reviewed and criticized.

2. Deadlines and goals have to be established and one person (in our case the Compleat Amanuensis) should be given the responsibility to follow up on these.

3. A diary should be maintained with one person responsible for summarizing the events at each workshop meeting. These diary items should be circulated to all participants within 2 or 3 days of the meeting in order to maintain a sense of momentum and a continuing record of triumphs and traumas.

4. Frequent pay-offs are valuable as a reinforcement of the activity. We have never been terribly successful at this, tending to set up too large pay-offs too far in advance. For example, in the first workshop, the first major pay-off was a rough working model, and this required 14 weeks. It is much better to have some models available right at the start that can be used in a game-playing situation periodically. Although this will slow progress, it will keep people involved and aware of the pay-off they might expect if they continue detailed work.

5. There are times to be interdisciplinary and times to be disciplinary in a workshop. We have found that the interdisciplinary arguments and discussions are essential to establish principles, to develop an overall conceptualization, to identify interactions between components and to identify subgroups. In short, whenever new ground is broken.

The result is usually a flow diagram of the system that gives a qualitative representation of the process. An example is shown in Figure 1.

The development of such a flow diagram brings together knowledge of ecological, economic, social, and physical processes in a way that makes people aware of interactions. The flow diagram also provides a device to identify subgroups which focus on one process: In the Gulf Islands study, for example, a mortgage process, population growth, recreational demand or ecological feedback. We found, through trial and error, that such groups had to be largely composed of one discipline, otherwise the activity foundered in futile arguments. All

Compleat Amanuensis   The Utopians   Blunt Scot

they were asked to do was to recognize the inputs they would receive from the rest of the model and to generate outputs, as indicated on the flow diagram, using the techniques listed in item 9 below.

6. There is great value in dramatic emphasis because, after all, it is a game. The emphasis chosen is totally dependent on the project and the individuals involved. In our case, the acronym for the Gulf Islands study was an admirable example—i.e., GIRLS (Gulf Islands Recreational Land Simulator). It allowed one full opportunity for sophomoric humor and the light touch. Another was the analogy of land speculation to the predation process. There are very real analogies, but the main purpose of bringing them out was to give dramatic emphasis to the project.

7. Progress through frustration is an unhappy reality. Individuals of strong will and ability inevitably have to go through a period of probing tangential issues. This leads to frustration, as people work themselves into traps that make dignified retreat difficult. But let it happen, for the resulting frustration can be used to crystallize advances. But care must be taken, since if the frustration is diffused too soon the advance will not be accepted, and if left too late, the group will self-destruct. But at the right

moment major and rewarding progress can take place as the only alternative to mental breakdown, if nothing else.

8. At each session, at least one of the internal Camarilla should have in reserve a solution to the question being considered. It is only necessary to present this solution if progress seems impossible.

9. There are a number of specific techniques that we found extremely useful. The method of adjudicating technical arguments we described above, and others follow:

    a) *Some Useful Subgroup Guidelines*
       i. Specifically identify the problem.
       ii. List all the variables that might be present.
       iii.Organize these variables into classes identified by their common effect or source.
       iv. Specify hypotheses concerning the response to each class of variable and demonstrate graphically. Some thought should be given to the form of the independent and dependent variable in order to facilitate interfacing with the rest of the model.
       v. Identify, for each response, all reasonable alternate hypotheses and give guesstimates for the maxima, minima, and thresholds. Preserve alternate hypotheses for subsequent sensitivity tests in simulation.

    b) *Degree of Precision Table*
       With separate subgroups working, it is important at the start to have approximately the same degree of precision recognized by each subgroup. We found the best way to do this was to have each subgroup draw up a hierarchy of precision in their first meeting, identifying inputs, model detail and outputs for each level of precision. An example of one developed by a Transportation Subgroup is shown in Table 1.

       In most instances, the hierarchy is so structured that the group decides on level 2 or 3. This decision, in discussion with other subgroups, provides a useful way to ensure a degree of conformity of precision.

TABLE 1: *Transportation Submodel Hierarchy of Precision*

	Input	*Model Detail*	*Output*
Level 1	*Regional Totals* • Population • Households • Vehicle ownership	Tables Graphs	*Regional Totals* • Total trips, by mode • Vehicle-miles or vehicle-hours of travel • Person-miles or person-hours of travel • Total accidents by severity
Level 2	*By Traffic Districts* (20) • Level 1 inputs and income • Trip length character	Tables, graphs, regression, gravity, opportunity and probability distribution models	*By and Between Traffic Districts* (20) • Level 1 outputs and trips by 0-D, by mode • Major corridor flows
Level 3	*By Traffic Zones* (100+) Level 2 inputs and network character	Same as level 2 except in greater detail and minimum path tree building program	*By and Between Traffic Zones* (100+) Level 2 outputs and Arterial corridor flows
Level 4	*By Traffic Zones* Level 3 inputs and greater network detail (turn penalties, etc.)	Same as level 3 except in greater detail for minimum path program	*By and Between Traffic Zones* Level 3 outputs and network link flows

c) *Flow Diagram*

We have tried a variety of flow diagrams drawn from control system theory, cybernetics and information theory. The best seems to be the simplest in which one symbol designates an input or output, another an intervention and a third, a process. (See Figure 1 for example). These standard symbols are used throughout in describing both the model and its constituent submodels.

d) *Interaction/Interfacing*

With subgroups working independently, one of the most difficult tasks is to ensure effective interaction between groups and interfacing between the submodels they produce. We have tried many devices but the one that seems to work best is an interaction matrix.

This is designed to be used after each subgroup has developed a fairly clear understanding of the general form of their submodel, the inputs they expect to receive, and the outputs they expect to generate. This information is contained in subgroup reports and is used to make up the matrix. The matrix identifies the inputs each group expects to receive. After the table is drawn up, each group is then asked to see if the output expected of them is the kind of output they intend to produce. Very often it is not, and the group chairman is charged with undertaking bilateral negotiations with other subgroups. In many instances, the disagreement arises because of the different words people use to describe different inputs and outputs. In other cases, a group did not anticipate the need for one kind of information that could, in fact, be provided. Finally, there are some cases where a group will decide that it is not possible to generate certain kinds of information. In every instance there has been ultimate resolution of these problems and in the process, a growing interaction between subgroups. It leads, through a series of revisions, to a final interaction matrix acceptable to all.

Progression of the Game

Each of the three workshops has followed a remarkably similar pattern and timing. Although the pattern of phases is probably fairly general, the exact timing of the phases will probably vary with different durations of workshops.

The First Week Start. The stage is set and a series of steps predicted for the next 6 months. The workshop is placed in the context of Resource Science problems, emphasizing the kind of gaps described in the introduction. It ends with a discussion of a specific task—or defining a problem or a system really anything—to begin to get people to work. In the next 3 weeks a variety of written suggestions are solicited whose purpose is specifically to identify the interests and abilities of the workshop participants.

The Fourth Week Catharsis. During the first 3 or 4 weeks there is a growing level of frustration as strong-willed people attempt to dominate and determine the direction of the project. The level of frustration builds rapidly, tensions mount and a catharsis occurs in the fourth or fifth week which crystallizes future directions. At the end of the catharsis, defined aims are established and people set to work on projects of meaning to them.

In some instances, a flow diagram will have been developed by this time or will have emerged as a result of the catharsis, and this allows subgroups to be established. Individuals who have been identified as leaders are asked to lead each subgroup and each individual joins a group relevant to his interests, abilities, and expertise.

The Sixth Week Goal. By the sixth week, each subgroup is working independently and periodically returns to the full workshop to summarize progress and solicit recommendations and criticisms. Results begin to appear in the form of flow diagrams of submodels. Every device should be used to achieve some defined product by this time. These may be written reports, lists of concerns and variables, identification of temporal or spatial scale, or flow diagrams. Some of these attempts will fail, and the group should quickly probe other devices until a useful one is discovered.

The Tenth Week Triumph. By the tenth week, a global flow chart has been developed backed up by detailed flow charts of each submodel. Hypotheses concerning functional relationships have been tentatively presented. In addition, we have always attempted to provide on the tenth week, a Tantalizing Tidbit—an exciting theory, construct or method that has relevance to the issue, and has been developed independently by one of the group. Finally, in the tenth week, there should be a review of the progression that was predicted and a re-writing of the progression expected for the future.

Two Week Break. Within the university, a break automatically comes around Christmas time. It is a happy event. Certainly, by that time the directors of the workshop have their reserves almost exhausted and need to heal. But it also provides the time for a small inner core—the Camarilla—to do some hard work in order to produce a product for the group when they return at the end of the break. In the first workshop, this was an actual working model, and the two-week break was spent by two people programming the whole model and debugging it so that it would be available for the workshop in the first session.

The Post-break Three Months. The remaining 3 months exploit the advances and progress made in the first three. There is a review and a new projected scenario, and if a model is available, as ideally it should be, considerable time is spent in simulation runs. Out of the simulation runs logically flow the exploration of policies and the gradual articulation of more and more effective ones. This process is accompanied by revision and extensions to the model, and sensitive points are identified through simulation so that a research strategy can be designed for future work. Each of our three workshops has been followed by a significant research effort that capitalized on the research priorities that emerged.

Conclusion

The workshop provides an admirable setting to explore the unknown and to build the essential bridges between methods, disciplines, and institutions. Its very flexibility and openness avoid the dangers of institutionalization. But the workshop is not a place for the total research effort required. It is excellent at the cutting-edge of new developments and must always remain so. The hard back-up work required can parallel and be associated with the workshop activity but should operate independently. In summary, although the tricks and devices of our game are useful, it is the philosophy of openness and style that is important.

Acknowledgements

This paper is dedicated to the late Walter Jeffrey—the Blunt Scot of our first workshop. We are indebted to Gordon Harrison and William E. Felling for the continuous encouragement, advice and support they provided throughout the development of the three workshops. Phyl Norris gave unstintingly of her experience and ideas in the writing of the paper, and Peter Larkin and Ilan Vertinsky provided the critique it needed. And, most important, we wish to thank the participants of the workshops who all gave so freely of their talents.

ARTICLE 11

Lessons for Ecological Policy Design
A Case Study of Ecosystem Management

WILLIAM C. CLARK, DIXON D. JONES, AND C.S. HOLLING*

* Order of authorship was selected by lot.

Reprinted from Ecological Modeling, 7, William C. Clark, Dixon D. Jones and C. S. Holling, Lessons for ecological policy design: A case study of ecosystem management, 1–53, Copyright (1979) with permission from Elsevier.

Abstract

THIS PAPER EXPLORES the prospects for combining elements of the ecological and policy sciences to form a substantive and effective science of ecological policy design. This exploration is made through a case study whose specific focus is the management problem posed by competition between man and an insect (spruce budworm, Choristoneura fumiferana) for utilization of coniferous forests in the Canadian Province of New Brunswick. We used this case study as a practical testing ground in which we examined the relative strengths, weaknesses, and complementarities of various aspects of the policy design process. Where existing approaches proved wanting, we sought to develop alternatives and to test them in turn. In particular, we used a combination of simulation modeling and topological approaches to analyze the space–time dynamics of this ecosystem under a variety of natural and managed conditions. Explicit consideration was given to the development of invalidation tests for establishing the limits of model credibility. An array of economic, social, and environmental indicators were generated by the model, enabling managers and policy makers to evaluate meaningfully the performance of the system under a variety of management proposals. Simplified versions of the models were constructed to accommodate several optimization procedures, including dynamic programming, which produced trial policies for a range of possible objectives. These trial policies were tested in the more complex model versions and heuristically modified in dialogue with New Brunswick's forest managers. We explored the role of utility functions for simplifying and contrasting policy performance measures, paying special attention to questions of time preferences and discounting. Finally the study was shaped by a commitment to transfer the various models and policy design capabilities from their original academic setting to the desks and minds of the practicing managers and politicians. An array of workshops, model gaming sessions and nontraditional communication formats was developed and tested in pursuit of this goal.

This paper reports some specific management policies developed, and some general lessons for ecological policy designed learned in the course of the study.

Introduction

This paper reports steps towards the development of a science of ecological policy design. First, we show how a number of mathematical tools can be used in the effective dynamic description of specific ecological systems. These tools range from simulation models to differential equations to topological representations. Second, we show how these descriptive methods can be combined with prescriptive techniques from the policy sciences—techniques of optimization, utility analysis and decision theory. Our central argument is that these various elements, developed in separate fields, can now be combined, amplified and tested as a rigorous science of ecological policy design.

Such a science requires two essential ingredients: a conceptual framework and a coherent methodology. Concepts alone are not sufficient, for if not illuminated and evaluated through a rigorous methodology applied to specific problems, they inevitably lose touch with reality. But even the best of methodologies, if not provided with a framework for identifying key conceptual issues, can lead at best to an aimless proliferation of numbers and, at worst, to more intractable problems created more quickly and efficiently.

The key conceptual issue of ecological policy design is how to cope with the unknown and unexpected. Unexpected events bear on the future of every complex system. Our understanding is always incomplete; substantial ignorance is always guaranteed. The aim of sound ecological policy is not to predict and eliminate future surprises, but rather to design resilient systems which can absorb, survive, and capitalize on unexpected events when they do occur. The appropriate paradigm is not that of fail-safe design, but rather of design which is safe (or "soft") in the inevitable event of its failure (Holling and Clark, 1975; Branscomb, 1977).

The history of resource management, and indeed of the applied sciences in general, has been one of trial-and-error approaches to the unknown. Existing information is mobilized and organized to suggest a trial and the errors, when they are detected, provide additional information for modification of subsequent efforts. Such "failures" provide essential probes into the unknown—probes generating the experience and information upon which new knowledge grows. But the increasingly extensive and intensive nature of our trials now threatens errors larger and more costly than society can afford. This is the dilemma of "hypotheticality" posed by Haefele (1974), who argues that the design of policy is locked in a world of hypothesis because we dare not conduct the trials necessary to test and refine our understanding.

The heart of the policy design problem lies in the way systems we manage respond to unexpected events. This response is directly related to the stability properties of the systems.

A system which is globally stable is admirable for blind trial-and-error experimentation: it will always recover from any perturbation. It is this paradigm of an infinitely forgiving Nature that has been assumed implicitly in the past, but if a system has multiple regions of stability, then Nature can seem to play the practical joker rather than the forgiving benefactor. Policies, trials, and management will seem to operate effectively as long as the system remains within known stability domains. However, if the system moves close to a stability boundary, incremental perturbations can precipitate radically altered behavior (Holling, 1973).

Even more troublesome, the stability boundaries themselves may contract in response to management activities, again generating sudden changes in behavior (Clark, 1976; Peterman, 1977a). In either event, the real danger is that a past history of policy "success" will often result in sufficient institutional inflexibility to make timely management response to the new condition impossible.

Policy design therefore requires a clear understanding of the resilience and stability properties of ecological systems and the institutional and social systems with which they are linked. Two lines of relevant evidence are now accumulating. The first comes from recent efforts to develop structurally simple differential equation models of complex systems, emphasizing the qualitative form of the functional relationships. Such models have been proposed for ecological systems (Bazykin, 1974; Ludwig et al., 1978), institutional systems (Holling et al., 1976a) and social systems (Haefele and Buerk, 1976). Even these simple structures exhibit extremely rich topologies with multiple stability regions a dominant feature. Different regions of parameter space exhibit different numbers and configurations of these stability regions, suggesting that biological, cultural, or managerial "evolution" of the parameters can indeed cause the stability properties of a managed system to change in quite unanticipated ways.

The second line of evidence comes from empirical studies of specific systems. Preliminary findings have been reported in Holling and Goldberg (1971) and Holling (1973, 1976) and a detailed review will be published elsewhere. To summarize, a great variety of examples illustrating instances of multi-equilibria structure and of behavioral shifts among equilibria are documented in the ecological, water resource, engineering and anthropological literature. Among the ecological studies, there are cases concerning freshwater and oceanic fisheries, terrestrial grazing, insect pest, and tropical and temperate forest ecosystems (e.g. Ricker, 1963; Holling, 1973; Noy-Meir, 1975; Southwood, 1976). There is, in addition, a larger range of more anecdotal evidence that is part of the community wisdom of the resource manager. Typical examples are the effectively irreversible development of the Scottish moors after deforestation, the desertification of the Middle East, and the loss of productive land in tropical terrestrial systems as a consequence of extensive and intensive agricultural practices.

In brief, the manager would be prudent to view Mother Nature as less benignly forgiving than deviously mischievous. The ecological systems we seek to manage will more than likely exhibit complex multi-equilibria behavior. The combined effects of an uncertain world and an incomplete understanding of system structure guarantee eventual excursions of the managed world into regions near and beyond its local stability boundaries. A recognition of this problem leads us to seek a science of resilient policy design that explicitly articulates the qualitative stability properties of managed systems and develops and evaluates alternative management approaches which respect those properties.

This paper is one of a series that explores related problems of ecological policy design. Our aim here is to provide an overview of the methodological issues involved. A fuller treatment of the conceptual arguments outlined above can be found in Holling (1973, 1976), Holling and Clark (1975), Walters (1975a) and Hilborn et al. (1976). Detailed consideration of certain problems posed by the unknown in descriptive modeling is provided in, Ludwig et al. (1978). Finally, specific procedures for the design and evaluation of "safe-failure" policies in an uncertain world are a major focus of Holling (1978) and Yorque et al. (1978).

The Case Study Approach

A conceptual framework for policy design is meaningless unless a cohesive methodology links it to the constraints and realities of actual management practice. To develop such a methodology and to test the applicability and practical relevance of the resilience concept in policy design, we have chosen to analyze specific case studies typical of large classes of ecological problems.

Several case studies of policy design for ecological management already exist. Those of Conway et al. (1975), Gutierrez et al. (1977) and Kiritani (1977) are representative of recent work, but most of these studies have chosen to focus upon the development of specific solutions to specific problems. They have not been concerned with the critical evaluation of design approaches per se, nor with questions of generality or transferability. Precisely because these broader strategic issues have received so little attention in the past, we have made them the loci of our case studies.

The first requirement for the case studies was to represent a common class of problems not specific to any country, creature or resource. To ensure a realistic confrontation with the constraints imposed by feasibility and implementation considerations, we also looked for problems with an active and troubled management history. We selected cases in which existing management agencies could be involved in the analysis from the beginning, emphasizing policy design with, rather than for, the user. Furthermore, we selected problems in which the ecological issues do not altogether dominate the economic and social ones.

The case study analyses are both descriptive and prescriptive. For descriptive purposes, we wish to combine detailed understanding of some key ecological systems with the more promising developments in modeling—simulation, simplified differential equations and topological approaches. The goal is to capture the essential behavior of the system in a number of different but complementary forms so that questions relating to the existence and form of multiple equilibria can be specifically defined and explored.

Yet we are equally concerned with prescription. The descriptive models provide laboratory versions of the real world within which alternative policy prescriptions can be developed and evaluated. Just as our central conceptual interest in multiple equilibria has descriptive relevance, so it has prescriptive significance as well. In a policy sense, our goal is to design resilient or robust policies less sensitive to the unexpecteds and unknowns in every system's future. This has led to the application and testing of a variety of prescriptive methodologies from operations research and management science, including optimization techniques, utility analysis and decision theory. The full combination of the concepts and the descriptive and prescriptive techniques provides the essential building blocks for a new science of ecological policy design. Much of this experience is drawn together in Holling (1978). We will focus here on one of the most fully developed case studies, a forest/insect management problem that involves the spruce budworm and the boreal forests of North America.

The spruce budworm (Choristoneura fumiferana) is the most widely spread destructive forest insect of North America. It ranges from Virginia to Labrador and west across Canada into the Northwest Territories (Davidson and Prentice, 1967). Particularly in the northeastern part of its range the budworm periodically undergoes severe and extensive outbreaks, imposing heavy defoliation and mortality on its preferred hosts, balsam fir (Abies balsamea) and white spruce (Picea glauca). The outbreaks result in major social and economic disruptions and have been the object of intensive research and management efforts since the late 1940's (Morris, 1963; Belyea et al., 1975; Prebble, 1975). The present case study was undertaken as a cooperative venture with the scientists and managers of the Canadian Forest Service's Maritimes Forest Research Centre. A detailed report of the work is forthcoming (Yorque et al., 1978). Rather than reiterate that material, we shall concentrate here on the major lessons learned as we attempted to develop, test, and transfer the methodologies and concepts outlined above. These lessons tended to destroy many of our most treasured myths of ecological analysis and policy design—myths which we, our collaborators and the ecological modeling community have often accepted in the past. We have preserved the central myths to remind us of our errors and provide a convenient focus for the discussion which follows.

Prescription and Description

Myth: Policy design should begin with an analysis of the decision making environment.

If institutions were immutable, if notions of political and technical feasibility never changed, if ultimate goals were known and universally agreed upon, and if we were designing policy for a specific time and place, the decision making environment might be a good place to begin the analysis. However, these conditions are simply not met in reality.

First, our own goals concern not the specific but the general. We emphasize the transferability of concepts and methods to a constellation of problems occurring in various regions. It is true that to give focus we initially concentrate on a specific problem with the name budworm/forest and on one particular region—the Province of New Brunswick in Canada. As the steps of transfer began to take place, it ultimately did become necessary to examine a number of specific institutional settings, but these were pragmatic concessions to give specificity and allow for testing in real world situations. To make transfer a reality, the initial emphasis must be on those elements of the problem which are truly general.

This focus on generality is not possible in an analysis of institutional or decision behavior. The state of knowledge in these fields is still primitively rooted in specific examples. In contrast, the state of knowledge of ecological systems and of ecological processes allows for well-tested analyses that have generality beyond the specific focus. With the need to facilitate transfer, the first requirement is to develop an effective and validated dynamic description of the ecological constraints of the problem. In practice, this means a simulation model which can be used as a kind of "laboratory world" with some confidence that it will be responsive to the exploration of a variety of different policies and their consequences.

There is, however, an even more compelling argument for initially structuring the analysis around the descriptive ecological problem rather than the prescriptive decision problem. Decision systems — whether states or individuals — are by nature and necessity fluid creatures. Perceptions and objectives sporadically change in ways that are scarcely understood and are wholly unpredictable. Such shifts occurred repeatedly in the 3 years of our budworm/forest studies. New findings on insecticide spraying side effects (Crocker et al., 1976) and unprecedented court decisions on the liabilities and responsibilities of the spraying operators have changed insecticides from a weapon grudgingly accepted for forest protection to a political liability. Detailed consideration of policy alternatives has consequently moved from its former position among middle and senior level civil servants to the provincial and federal cabinets themselves. Unforeseen developments in the international wood processing industry have likewise posed novel technical opportunities and political problems. Because of the descriptive foundation of our analysis in budworm/forest ecology, we could readily respond to these unexpected changes in the decision environment. An analysis based upon a particular version of the decision structure would have been constantly changing, hopelessly out of date, or both.

Finally, it should be pointed out that good policy analysis will almost invariably cause changes in the decision environment and therefore cannot, even in principle, use that environment as a foundation for its efforts. When we began our budworm study, one major goal set by the decision makers was to find a policy for reducing forest inventories to a more economically productive level. An early result of the analysis was to show that inventories were already drastically depleted, and there were *no* long term policies which did not include a reduction of harvest rate and/or the institution of yield-increasing management. The perceived goals and constraints of the decision makers shifted accordingly and the analysis continued. A similar but happier shift occurred when the analysis showed that certain viruses, which inflict only low mortality on budworm and are therefore largely ignored as possible management tools, might have dramatic potential if applied in coordination with other recognized control measures. The decision response here was an expansion of the perceived feasibility region for policy options.

An ecological policy design program must be based on a generalizable description of the underlying biology if it is to be usefully responsive to inevitable but unpredictable changes in the decision environment and if it is to meet the requirements of transferability across a wide range of decision problems.

The Modeling Problem

Myth: The descriptive model should be as comprehensive as possible.

Any model represents an abstraction of reality. The problem is not whether, but what to leave out.

Ecosystem management problems are comprised of an immense array of interacting variables, conflicting objectives and competing actions. Attempts to comprehensively model such complexities are futile. At best they produce models as intricate and unfathomable as the real world. More likely, they founder in a limbo of unending data requirements, impossible "debugging" problems, nonexistent validation criteria, and general ineffectiveness (e.g. Cline, 1961; Shubik and Brewer, 1972; Brewer, 1973; Mar, 1974; Ackerman et al., 1974; Holcomb Research Institute, 1976; Mitchell et al., 1976). Our experience has suggested the opposite course: to be as ruthlessly parsimonious and economical as possible while still retaining responsiveness to the management objectives and actions appropriate for the problem. The variables selected for system description must be the minimum that will capture the system's essential qualitative behavior in time and space.

The initial steps of bounding the problem determine whether the abstract model will usefully represent that portion of reality relevant to policy design. Key decisions must be made regarding the policy domain, the ecosystem variables, the temporal horizon and resolution, and the spatial extent and resolution to be modeled.

Policy Domain

The policy domain can be defined, or bounded, by specifying the range of acts and indicators to which the dynamic model will be responsive. Particular policies and objectives can later be assembled from these components.

It is important to clarify what we will mean by the terms "policy", "acts", "indicators", and "objectives". *Objectives* are descriptions of desired system behavior. A New Brunswick policy maker might well (albeit wishfully) declare his objectives to be "full employment with low cost harvest even if it requires high quantities of insecticide". That objective could be unique to the specific policy maker but its component *indicators* are probably not. It is easy to imagine an ardent environmentalist using the same indicators—employment, harvest cost and insecticide quantity—but describing his desired combination of them (i.e. his objective) quite differently. *Acts* are the physical weapons in the manager's arsenal. They are actual things which he can do, such as cutting trees, killing budworm, building mills, and so on. Finally, *policies* are the rules or plans by which acts are applied to the system in order to obtain its desired behavior. A policy is something of the form: "when there are so-and-so many budworms, spray insecticide," or "harvest such-and-such a quantity of wood each year". As we use the terms, policies are prescriptions for action and, as such, wholly distinct from descriptions either of intent (objectives composed of indicators) or capability (acts).

The range of acts which have been or potentially could be applied to budworm/forest management is enormous, including the use of insecticides, biological control agents, genetic manipulation, and tree harvesting and planting schemes. Moreover, the acts which now seem to be economically impractical might, with future developments, become feasible. However, the whole range of actions feasible now and imagined for the future fall into three qualitative classes—control of insect numbers, harvest of the trees, and the manipulation of the forest through planting (Baskerville, 1975b). To be responsive to management questions, the descriptive model must allow intervention with any of these classes of acts at any moment in any place. The tactical details of how any specific act is implemented can be dealt with on an ad hoc basis, often outside the simulation model. Similarly, although a large number of performance indicators can be imagined, certain indicators of policy performance are independent of place and time and are of universal interest. These include employment, costs and quantities of wood harvested, timber losses to budworm, and a number of environmental indicators (Bell, 1977b, and below under the discussions of evaluation).

The policy domain resulting from these choices is one bounded between the macro level of provincial economics and the micro level of single stand forest management. By explicitly modeling the provincial forest/insect management system we are capable of interfacing with either of these levels.

Variables

Even the simplest ecosystem contains thousands of species and potential variables. However, existing knowledge, much of it summarized in Morris (1963), has allowed us to capture the strategically relevant part of the system's behavior with a limited subset of these variables.

The principal tree species of the system are birch (*Betula* sp.), white and black spruce (*Picea glauca, P. mariana*), balsam fir (*Abies balsamea*), and a variety of hardwoods (Loucks, 1962). These have a dynamic interaction of their own which is dependent on the influence of budworm. Fir is highly susceptible to damage, white spruce moderately so, black spruce only slightly, and birch and hardwoods not at all. Our rule of parsimony and our strategic level of interest dictate that we lump together the principal budworm hosts, fir and white spruce, into a single dynamic variable expressing the density of susceptible forest and eliminate non-host species from dynamic consideration.

The extensive measure of forest density must be coupled with an intensive measure of tree condition. This is closely linked with present foliage condition that, in conifers, can serve to retain the "memory" of past defoliation stress. Budworm's differential preference for fresh rather than aged needles dictates that this qualitative property be split into two variables, called new and old foliage in the model.

Host tree density and susceptibility to budworm are highly dependent upon forest age structure, as are the economic properties of trees as a crop for man. Consequently, we build a dynamic age class structure into the forest model. Repeated early attempts to treat age structure implicitly and so avoid the extra variables were dismal failures, yielding a model completely incapable of describing observed budworm/forest dynamics.

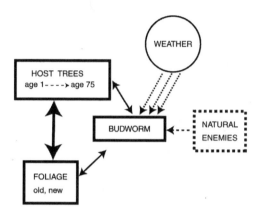

FIGURE 1 : *Key variables included in the model of the natural ecosystem.*

Between outbreaks the budworm is rare but present, its numbers being controlled by natural enemies such as insectivorous vertebrate predators and parasites. A key feature of this control is that there exists an upper threshold of budworm numbers which, once exceeded, allows the budworm to "escape" predation and multiply unchecked (Takahashi, 1964; Southwood, 1976). The response of vertebrate and parasite numbers to changes in budworm density is slow compared to the rate of interaction between the budworm and its host trees. As a first simple approximation it therefore seemed justified to model the effects of natural enemies implicitly, without resort to additional state variables.

Finally, weather is a key factor affecting budworm survival and dispersal, and is included as a stochastic driving variable in the model.

From the thousands of potential candidates, we abstract the structure of variables shown in Fig. 1 to model the local dynamics of the budworm/forest system.

Time

An analysis of tree rings (Blais, 1965, 1968) covering eight regions of eastern North America and extending as far back as 1704 provides valuable data on the long term temporal pattern of outbreaks. These data, together with more detailed information on recent outbreaks summarized in Brown (1970), indicate a distinctive 30–45 year cycle (Fig. 2). During the inter-outbreak periods the budworm is present in barely detect-

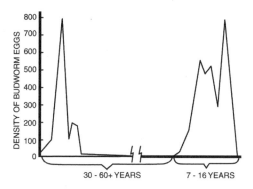

FIGURE 2: *Temporal outbreak pattern. Schematic portrayal of a typical outbreak cycle. Density is measured as the number per standard 10 ft² balsam fir branch (Morris, 1963). (10 ft² = 0.93 m² ≈ 1 m².)*

able numbers that, when appropriate conditions occur, can increase by three orders of magnitude during a 3–4-year period. Once the outbreak is initiated in a sufficiently large area it spreads over thousands of square kilometers, finally collapsing after 7–16 years, often with attendant high mortality to the forest. Because of the pattern of outbreaks shown in Fig. 2, the minimum time horizon required is one which can completely contain two outbreak cycles—that is, between 80 and 160 years.

The time resolution which will capture the dynamics of the system is 1 year—the generation time of the budworm. Seasonal events within the year can be implicitly represented. Equally important, the single year resolution is close to the operational time scale of the management agencies.

Space
The characteristic pattern in time is complemented by one in space. Typically, outbreaks spread from small regions of initiation and contaminate progressively larger areas (Brown, 1970). Collapse of the outbreaks occurs in the original centers of infestation, often in conjunction with severe tree mortality. The result is a high degree of natural spatial heterogeneity in forest age and species composition (Baskerville, 1975a).

The choice of spatial extent for the modeling effort was dictated by the dispersal properties of budworm, data availability, and the concerns of management. As with many pest species, the budworm has very strong

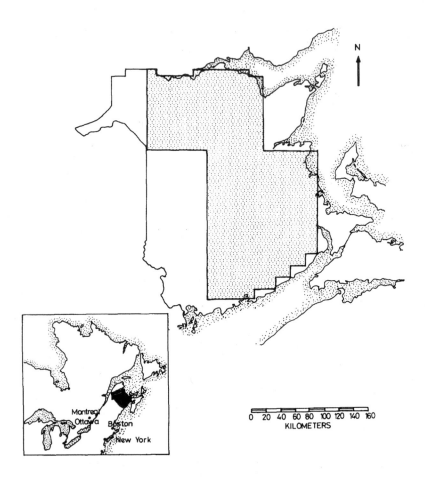

FIGURE 3: *The modeled area is a 4.5 x 10⁶-ha region comprising most of the softwood forest in the Province of New Brunswick, Canada (see insert).*

dispersal abilities. The modal distance of dispersal is about 40 km, but distances of several times this figure have been recorded (Greenbank, 1973). To study the implications of dispersal it was thought essential to model a total area with dimensions of the order of five times this modal distance. The particular area chosen was a 4.5 x 10⁶-ha region containing much of the softwood forest of Canada's Province of New Brunswick (Fig. 3). The peculiar shape is a pragmatic concession to the local management agencies but it does include the majority of the area for which validation and initialization data were available. Analysis efforts subsequent to those reported here have expanded the study area to include most of New Brunswick and substantially reduce the edge to volume ratio. A buffer zone approximately 80 km in width around this area compensates for edge effects.

The spatial resolution of the model is defined by the dispersal capabilities of budworm, the scale of spatial heterogeneity in the forest, and the available data base. The modal dispersal figure of 40 km dictates a minimum spatial resolution of, at most, half that distance. Since the standard management data unit consists of rectangular grids approximately 11 x 15 km, the overall region of Fig. 3 was divided into 265 of these biologically arbitrary but convenient spatial units, each containing just over 17,000 ha.

In summary, the decisions on bounding the problem are as follows:

Policy domain. — Responsive to the management acts of insect control, tree harvest and silviculture; generating indicators of employment, costs, harvest, insecticide distribution, environmental quality and timber losses to budworm.

Key variables. — Host tree species (with age structure), foliage condition, budworm density and weather.

Time horizon. — 80–160 years.

Time resolution. — 1 year with implicit seasonal causation.

Spatial area. — 4.5×10^6 ha.

Spatial resolution. — 265 subregions of 17,000 ha.

The number of state variables set by this bounding of the problem determines whether or not subsequent prescriptive steps, such as optimization, are feasible. Table 1 summarizes the final decisions made on the number of state variables required. Even though the previous steps of bounding may seem to have led to a highly simplified representation, the number of state variables generated is still very large.

The 79 variables in each site are replicated 265 times to give a total of $79 \times 265 = 20,935$ state variables. Thus even this drastic simplification defines a system that is enormously complex for analysis. We discuss below a number of ways in which this complexity can be further reduced to promote understanding of budworm/forest interactions and the management problem. These further simplifications, backed by the laboratory world of the complete simulation model, are utterly essential to successful policy design. They are predicated on the existence of a critically

TABLE 1: *Number of state and driving variables in the budworm/forest model, for each subregion.*

Density of host trees by age class	75
New foliage	1
Old foliage (retains memory of past stress)	1
Budworm	1
Weather	1
Subregion total (*other variables included implicitly*)	79
Number of subregions	265
Total number of variables in full region of 265 subregions	79 x 265 = 20,935

bounded model which explicitly leaves out everything but the essential core of the problem.

All-inclusive models cannot promote the deep understanding of inter-relationships necessary for creative policy design. Parsimony is the rule.

Causal Resolution

Myth: The goal of description is description.

If description per se were the goal of modeling, then there would be little need for a detailed understanding of causation. A multi-variate sta-tistical model would be sufficient to capture and describe historically observed patterns of behavior. In fact, that is what was done in the origi-nal analysis of the budworm problem in New Brunswick, as reported in Morris (1963). The very best of sampling procedures were applied over a 15-year period in a large number of locations, and a sophisticated statisti-cal descriptive model was developed.

However, there are two problems. The first is that ecological sys-tems often exhibit frequency behavior on the scale of decades or even centuries. As already shown in Fig. 2, the basic temporal pattern of the budworm system consists of periodicities of 30 and more years. It is hardly conceivable that there would ever be an extensive enough range of data to allow for a comprehensive description using statistical methods. At best, these can provide an effective way to mobilize whatever data are available and point to those processes or variables which contribute most to the observed variation. In addition, policies will be designed that move the system into regimes of behavior it has rarely if ever experienced during its evolutionary history. Considerable understanding of causation in terms of fundamental processes is required in order to be confident that the predicted behavior will be realistic under these novel conditions.

A certain degree of resolution in the hierarchy of causation is demanded; yet clearly one can go too far and become encumbered by details of explanation which defy comprehension. Modeling at too coarse or too fine a resolution level characteristically occurs when a system is not well understood. This can often be avoided in the modeling of ecological systems. On the basis of a rich history of experimentation, theoretical analyses and empirical field studies, the structure of key ecological processes is known not only in some detail but within a framework which has generality. This understanding can be aggregated to produce general and well-tested modules of processes such as growth, reproduction, competition and predation.

The first step in developing an explicitly causal model is to identify the component processes involved. Figure 4 shows the detailed sequence of processes (and of calculations) which occur within each 1-year period in the budworm system.

The disaggregation cannot stop here. Each of the processes shown in Fig. 4 has to be represented in the model by functions as realistic but simple as possible. Two sets of decisions must be made—first, the mathematical form of the equation describing a particular process and, second, the parameter values for that equation. If no independent information is available to identify the form of the equation, an enormous demand is placed upon the available data which must be extensive enough to define both the equation and the parameter values simultaneously. In most instances, however, there is independent knowledge of process structure that allows the equation form to be determined. With the form defined, the available data need only define the parameters and the formerly impossible demands for data become tractable. This can best be demonstrated by a specific example.

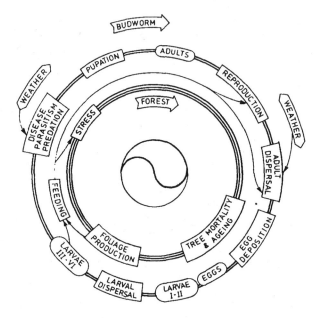

PROCESS CYCLE
OF THE BUDWORM FOREST SYSTEM

FIGURE 4: *The process cycle of the natural system. The inner ring represents the forest cycle, the outer ring the budworm cycle. Ellipses indicate the insect life stages; arrows show causal relationships among processes and budworm densities.*

One of the processes affecting survival of large larvae is predation by vertebrate predators (birds and squirrels). It has long been suspected (Morris, 1963) that vertebrate predators could play an important role in the dynamics of spruce budworm when insect populations were low. For the very reason that it is so impractical to sample low populations with any reasonable precision and accuracy, standard regression modeling approaches have been wholly inadequate to capture the effects of predation.

The predation process, however, has been analyzed sufficiently so that its variety of forms can be identified and classified (Holling, 1965). Predation is comprised of four necessary and sufficient processes—the functional response to prey density (an instantaneous rate of attack), the competition response, the development response and the numerical response. Each of these four responses has been shown to have a small number of qualitatively distinctive forms (Holling and Buckingham, 1976). The functional response to prey density, for example, can assume four and only four qualitatively different shapes. Moreover, a simple general equation has been developed whose four limiting conditions generate all these types. Equally important, the sufficient biological conditions can be precisely defined so that the most general of information is sufficient to classify any specific situation.

We have analyzed existing data on vertebrate predation of budworm in some detail. For these sorts of predators, the functional response to prey density (i.e. the instantaneous number of prey eaten per predator per unit time) is known to rise in an S-shaped manner to a plateau. The appropriate form of the general predation equation is therefore established, and it remains only to mobilize the existing data so that feasible parameter ranges can be determined. For the present purpose, only two parameters need be defined—one that determines the plateau of maximum attack rate and one that determines the rate of search.

The parameter values are influenced by the size of the predator and its searching habits. This permits the separation of budworm predator species into distinct parameter-defined groups. The birds attacking budworm are classified into three types—the small arboreal birds (e.g., warblers), the medium sized birds searching nearer the ground (e.g., white-throated sparrows) and the larger birds with a variety of searching modes (e.g., grosbeaks). Existing data on maximum consumption, sizes of predators and rates of searching establish maximum and minimum feasible values for the search and satiation parameters for each class. Together, these define a feasible range for the percentage of late instar budworm larvae that can be eaten by birds (Fig. 5). Sensitivity analysis determined how changes in parameter values within this feasible range affected the qualitative behavior of the model. It was found that, as long as the predation mortality was within the range allowed by the data in Fig. 5, the typical outbreak cycle of 35 years was generated. If less predation was introduced, the model behavior reverted to a pattern of 8–12-year cycles which is characteristic of the model with no predation. (See Holling et al. (1979) for further details.)

In summary, a good rule of thumb is to disaggregate the model first into the constituent processes that together affect growth and survival.

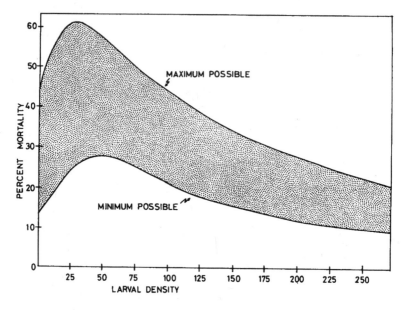

FIGURE 5: *Effect of vertebrate predators on budworm populations. Ordinate shows percent of sixth instar larval population consumed by predators as a function of larval density, given on the abscissa in units of thousands of larvae per hectare. Function shown is maximum, minimum, and range of predation rates consistent with available field data, all vertebrate predators combined (see text).*

These processes are then disaggregated one step further into their fundamental subprocesses. The principal purpose in choosing this level of causative resolution is to increase our confidence in predictions obtained under novel policies. Also, three additional benefits emerge which directly relate to our emphasis on transfer in dealing with the uncertain and unexpected. First, transfer implies that someone will be the recipient of the analysis. In many ecological problems these recipients include biologists and scientists with a sophisticated and highly detailed understanding of the mechanisms involved. Without disaggregating to the causal level suggested, the model is unlikely to be at all credible to these users. Moreover, there would be no way for the analysis to be responsive to the questions and critiques that, typically, are focused on distinct processes.

Secondly, modeling at this level of causation provides an effective way to identify factors that can be affected by policies. The qualitative behavior of the budworm/forest system is critically dependent upon vertebrate predation. Any policy which employs management acts (such as insecticides) which potentially could affect predators must therefore be evaluated in terms of possible changes in that process. It is not simply a matter of worrying about birds or other values, but rather of recognizing (in this case) the process of vertebrate predation as an integral part of potential forest management policies. A model that simply mimics past data would scarcely identify this as an issue.

Finally, some of the major advances in coping with the unexpected and unknown have applied the techniques of adaptive management (Walters and Hilborn, 1976). The key here is that, when our knowledge is tentative, management acts themselves can be designed to generate information relevant to an understanding of the underlying causal mechanisms. If our models have been conceptualized at a coarse level of resolution, the experiments of adaptive management can require extensive time or geographical areas to obtain results. This is impractical for management agencies with short time horizons and justified aversions to large scale trials. By disaggregating our models to the subprocess, or module level, "quick-and-dirty" experiments are often suggested which can yield results quickly in a localized and focused manner (Holling, 1976).

The goal of description is not description but useful causal understanding.

Invalidation
Myth: The purpose of validation is to establish whether the model is right.

There is always something about the real world which an abstract model will fail to predict, and there is always some model which can be constructed to mimic a given pattern of real world behavior. Proper validation is not a matching game, but rather an effort to explore the limits of model credibility. The establishment of these limits requires invalidation and not validation.

If our goal were to develop a micro-tactical model suitable for day-by-day predictions, then quantitative validation criteria would be in order. However, the present effort is aimed at a strategic level of regional planning, requiring projections over large spatial areas and long periods of time. Furthermore, the analysis is meant to evaluate new management situations without historical precedent. The best a quantitative validation comparison can do is assure us that we can model things that are already part of history. Confidence in the model's ability to treat new situations reliably will always be subjective to a degree. At a minimum, however, we can insist that the model generate plausible qualitative patterns of behavior in space and time under a wide range of extreme conditions. Behavior at the limits is more revealing than behavior under normal or average conditions.

Three kinds of qualitative information are available which relate to behavioral patterns of the budworm/forest system. The first concerns the "natural" or unmanaged system, and was referred to earlier when we discussed the bounding problem. The temporal pattern drawn from Blais' (1965, 1968) tree ring studies (shown in Fig. 2) suggests a natural inter-outbreak period of 30–45 years with occasional longer stretches and an outbreak duration of 3–6 years locally and 7–16 years over large geographical regions. The natural spatial pattern is reflected in extensive infestation records assembled by the Canadian Forest Insect and Disease Survey (Canada, Department of the Environment, 1938–1976), and reviewed by Greenbank (1957) and Brown (1970). In general, infestations move outwards in all directions from their areas of origin but expand most rapidly towards the east. Easterly spread rates for a new infestation may exceed 150 km per year in inland areas but are reduced to approximately 50 km per year near the Atlantic coast.

When the New Brunswick budworm forest model is run in the absence of management activities, the result is the temporal pattern of Fig. 6 and the spatial patterns of Fig. 7a and b. The temporal behavior of the model corresponds closely to the composite real-world picture of Fig. 2, generating 30–40 year inter outbreak periods and qualitatively similar levels of budworm density. The spatial pattern is also quite credible, even to the different rates of spread towards the northwest and southeast.

The second class of qualitative information concerns the behavior of the system under historical patterns of harvest and insecticide spraying. Egg density and defoliation patterns for New Brunswick between 1953 and the present are available from Webb et al. (1961) and data files of the Maritimes Forest Research Centre (C.A. Miller, personal communication, 1974). Harvest rates derived from Statistics Canada figures (averaging 4.25×10^6 m³ of wood per year) were applied to the model for this period. An average of 20% of the modeled region was "sprayed" with

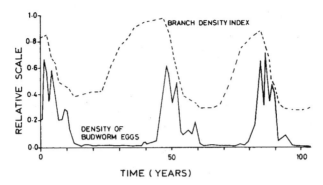

TIME (YEARS)

FIGURE 6: *Typical outbreak pattern generated by the model with no manage-ment or harvesting imposed. Ordinate is a relative scale, running from 0 to 1000 budworm eggs per 10 ft² of branch area; and from 0 to 1 for the branch density index (see text) which closely parallels the average forest age and wood volume. Initial conditions are those for New Brunswick in 1953. Data are averaged over the 265 subregions. Compare with Figures 2 and 7.*

insecticide according to the historical, state-dependent spraying rules described by Miller and Ketella (1975). In these invalidation runs, all bio-logical parameters have been determined by independent data (mostly from Morris, 1963) and remain fixed. The only "tuning" allowed is of the initial conditions (where they are ambiguous) and the management rules (harvesting trees and spraying insecticide) applied in the simula-tion model. The result is shown in Fig. 8. Initial conditions in year 0 are set as those observed in New Brunswick in 1953. Therefore, the first 23 years of this simulation run correspond to the period 1953–1975. Again, the agreement between real and simulated behavior is striking. Detailed numerical comparisons have been carried out, but in both cases the significant qualitative characteristic is a slowly eroding forest main-taining a persistent moderate level of infestation. The outbreak starts in the north, collapses there and throughout much of the province, re-emerges in the central regions and, in the early 1970's (year 21), spreads dramatically throughout the entire region (Canada, Department of the Environment, 1953–1976). The key point is that the spraying policies employed to keep the forest green, and so to preserve the forest industry, do so at the expense of maintaining semi-outbreak conditions markedly different from the "boom-and-bust" natural system (Blais, 1974).

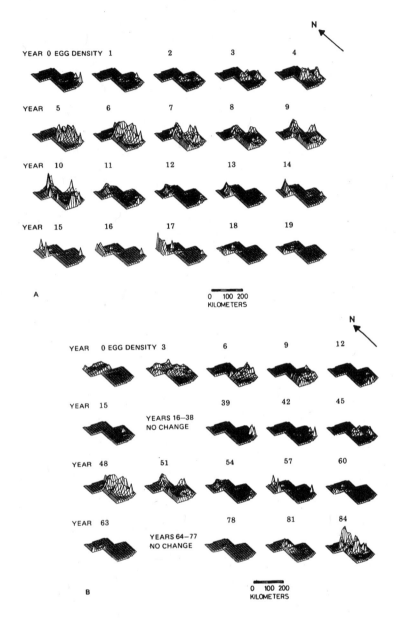

FIGURE 7: *Spatial outbreak pattern generated by the model with no management or harvesting imposed. Horizontal (x, y) coordinates define east—west and north—south spatial grid of 265 subregions within the study area (Fig. 3). The vertical (z) coordinate gives budworm egg density. (A) shows the year-by-year progression of a typical outbreak beginning in the southeast, spreading across the area, and dying out in the northwest 20 years later. (B) shows selected years from the model run for which averages were given in Figure. 6. Initial conditions are those of New Brunswick in 1953.*

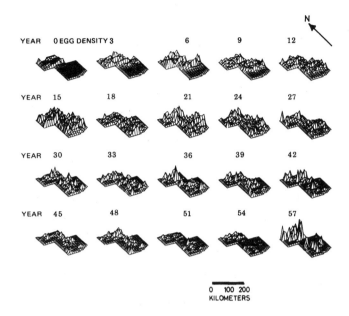

FIGURE 8: *Spatial outbreak pattern generated by the model under historical rules for spraying insecticide and harvesting trees. Initial conditions are those for New Brunswick in 1953. Compare with Fig. 7 (b), noting the loss of spatial cohesion of the outbreak due to management activities. The average time trace, analogous to Fig. 6, is given in Figure 15 (a).*

The final and, in several ways, most powerful invalidation test makes use of the fact that temporal and spatial behavior of the budworm/forest system in New Brunswick is qualitatively different from behavior in certain other regions of North America. In northwestern Ontario, for example, outbreaks are more intense (Elliot, 1960) and tend to occur at intervals of 60 or more years rather than the typical 30–45-year period observed in New Brunswick (Blais, 1968). Another pattern has been observed in Newfoundland where budworm outbreaks used to be extremely rare (Canada, Department of the Environment, 1938–1965; I. Otvos, personal communication, 1975). Now, however, moderate infestations are a more common event, perhaps because of emigration from the mainland where a persistent and spreading outbreak has been maintained by insecticide spraying (Canada, Department of the Environment, 1966–1976). The principal differences between these two regions and New Brunswick are weather conditions and forest structure. In northwestern Ontario, the proportion of susceptible host trees is less than New Brunswick while in Newfoundland it is greater. Moreover, relative to New Brunswick, the weather in northwestern Ontario is more favorable to budworm and in Newfoundland less favorable. These simple changes were introduced into the New Brunswick model. Gratifyingly, the Ontario model scenario did generate more intense outbreaks at intervals of 60 rather than 30–40 years. Similarly, the Newfoundland changes produced a world free from outbreaks except under severe invasion by external dispersers—exactly the phenomenon implicated in the recent Newfoundland infestations. These and other simple, qualitative, but highly significant tests are reported at length in Holling et al. (1975) and Clark and Holling (1979).

The three kinds of qualitative invalidation reviewed here place more rigorous demands upon the descriptive and predictive capability of the model than would any effort to fit a specific time series. By focusing on patterns in space and time, it is possible to mobilize qualitative information on a variety of behavioral modes associated with contrasting regional and historical management actions. The model's ability to reflect this broad spectrum of qualitative behavior establishes a significant degree of confidence in its utility for exploring policies which move the system far from its natural or historical conditions.

The goal for invalidation of a strategic model is to establish the limits of model credibility. A minimum requirement is agreement with a range of observed temporal and spatial patterns under a wide variety of extreme conditions. A quantitative fit to one set of data is quite insufficient.

Simplification and Compression

Myth : A complex system must be described by a complex model.

Even the most ruthlessly parsimonious but realistic ecological simulation model will be encumbered by many state variables and nonlinear functional relations. The explosive increase in the number of variables required for spatially dispersed systems presents the "curse of dimensionality" in its more intractable form. Compressions and simplifications therefore are essential, in part to encapsulate understanding and help intuition play its central role in the analysis; in part to facilitate communication in the transfer process; and in part to exploit the potential of optimization techniques that are as yet unsuitable for nonlinear stochastic systems of high dimensionality.

A powerful approach is to adopt a topological view of the system. This links the basic qualitative behavior to the number and interrelation of equilibrium states and focuses as well on our central concern for ecological resilience and policy robustness. Our first step was to use the full simulation model to generate a population growth rate or "reproduction" curve similar to that introduced by Ricker (1954) for the analysis of fish populations. This has been done in Fig. 9, where the population growth rate R (the ratio of budworm density in generation t + 1 to the density in generation t) is plotted as a function of budworm density. The growth rate curves condense all the reproduction and survival functions within the model and produce a unique curve for each state of the forest.

Interpretation of the curves is straightforward and focuses on the location and properties of the equilibrium points which occur wherever the growth rate R equals 1. These equilibria may be stable or unstable depending upon the sign of the slope of the curve as it passes through the R = 1 line. Briefly, if a slight increase in density from the equilibrium point results in further increases in the next generation (i.e. R > 1), or if a slight decrease results in further decrease (R < 1), then the equilibrium is unstable (represented as an open circle in Fig. 9). In contrast, where a slight increase in density from the equilibrium point is offset by a decrease in the next generation (R < 1), and a slight decrease is offset by a subsequent increase (R > 1), then the equilibrium is stable (shown as a solid circle in Fig. 9). The interpretation of equilibrium properties for population growth rate functions is discussed at greater length in Takahashi (1964) and Holling (1973).

Subsequent discussion will draw heavily on interpretations drawn from these growth rate curves, so it will be useful to consider their structure in some detail. The high density equilibrium points (c, d in Fig. 9) are introduced largely through intraspecific competition among budworm for foliage. Although these points are stable equilibria for budworm, they are unstable for trees. At such high budworm densities defoliation is so heavy that older trees die and are replaced by seedlings and understory growth. This shifts the system onto a lower growth rate curve (e.g. the immature forest curve in Fig. 9) where R < 1 and the insect population declines from its high level.

When the forest is immature, R < 1 for all budworm densities and no outbreak is possible. With a very mature forest, however, budworm will increase from all densities less than d. The ensuing defoliation and tree death will cycle the population back to low numbers.

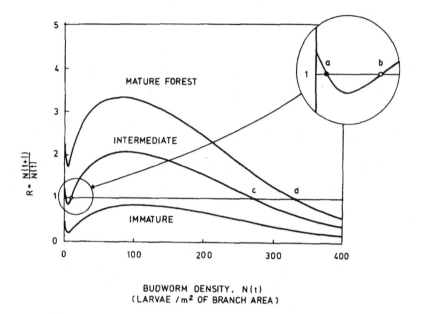

BUDWORM DENSITY, N(t)
(LARVAE / m² OF BRANCH AREA)

FIGURE 9: *Growth rate curves for budworm populations for three different forest conditions. Ordinate is population growth rate, defined as third instar budworm density in one year divided by density in the previous year. Abscissa is the previous year's third instar density. For any given forest condition, potential equilibrium budworm densities occur wherever the growth rate curve intersects the R = 1 line. Equilibria can be either stable (•) or unstable (o) (see text). Insert expands the intermediate forest curve at low budworm densities.*

When the forest is of intermediate age, two low density equilibria are introduced, one stable (point *a* in Fig. 9) and one unstable (point *b*). The additional dip in the curves at low budworm densities reflects the activity of vertebrate predators as described earlier, and is augmented to a degree by parasitism. The population may persist at density *a* until improving forest conditions move the curve's dip above the R = 1 line. An outbreak then occurs. An outbreak can occur even in an intermediate age forest if a sufficient influx of budworms disperses in from outside areas. Thus, in Fig. 9, a small influx of dispersers that increases the budworm density from equilibrium (a) to a density greater than the unstable equilibrium (b) (where R > 1) will trigger an outbreak.

The growth rate curves shown here do not include the stochastic elements of weather which affect both survival and dispersal. When these are included, there is a third outbreak trigger in the occurrence of a sequence of warm, dry summers which can raise normally low growth rates above the replacement line.

A more complete and succinct summary of these multiple equilibria can be obtained by plotting all the equilibrium points as a three-dimensional surface in a space representing condensed forms of the three key state variables—budworm, foliage condition and branch density (Fig. 10). This represents an equilibrium manifold of the kind found in topology and catastrophe theory (Thom, 1975; Jones, 1975). Such representations provide a particularly revealing way of interpreting outbreak behavior. The temporal pattern of the unmanaged system shown earlier in Figs. 2 and 6 can be understood by following the trajectory of events over this manifold as shown.

The equilibrium manifold representations also prove to be a powerful device for exploring the consequences of changes in ecological processes or management approaches. As one example, an equilibrium manifold is shown in Fig. 11 for which the foliage axis has been replaced by a predation intensity axis. When predation is at the level occurring in nature (1 on the scale), the pit responsible for the lower equilibrium is pronounced. As predation is relaxed, the pit gradually disappears along with the folded character of the manifold.

Under such conditions, the behavior of the system is radically and predictably altered, since the natural "boom-and-bust" pattern is intimately associated with the reflexive form of the manifold. Simulation runs conducted to check this topological implication of reduced predation show a world of perpetually immature forest where moderate budworm densities oscillate with foliage in a 12–16-year cycle. Since insecticides have exhibited a potential for reducing vertebrate predation directly through mortality or indirectly by affecting food availability (Pearce, 1975; Pearce et al. 1976), the significance of this result for management is obvious.

Another example is shown in Fig. 12, where the manifold is used to explore the qualitative implications of dispersal. The "immigration" axis reflects the relative intensity of immigrating budworm moths. The similarity of the dispersal manifold to that for predation is striking and significant. An increased rate of immigration clearly has qualitative properties much like those of a decreased rate of predation. This is in keeping with the earlier analysis of growth rate curves in Fig. 9, where the quantity of immigrants necessary to release a budworm population

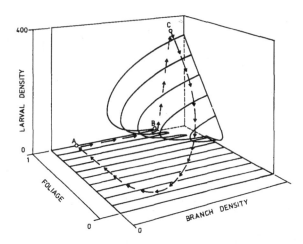

FIGURE 10: *Equilibrium manifold for budworm as a function of foliage and branch density (units are third instar larvae per 10 ft^2 of branch area, with foliage and branch density scaled 0 to 1; see text). Trajectory shows a typical unmanaged outbreak of the sort shown in Fig. 6. Segment AB is the long endemic period with budworm at low densities and the forest full of foliage and slowly growing until lower equilibrium disappears at B. Segment BC is the outbreak with budworm densities rapidly rising to upper equilibrium C. At these densities, foliage is rapidly destroyed, leading to destruction of the forest and collapse of the outbreak along CA, the final section sees the rapid recovery of foliage in a young forest, with budworm locked at their low density near A.*

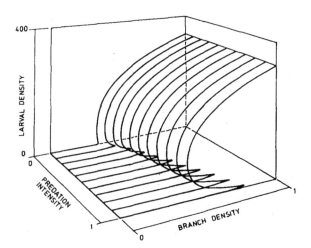

FIGURE 11: *Equilibrium manifold for budworm as a function of predation intensity and branch density. Predation scaled from 0 (no predation) to 1 (natural level of predation as included in model); other scales as in Figure 10; foliage is set at 1.*

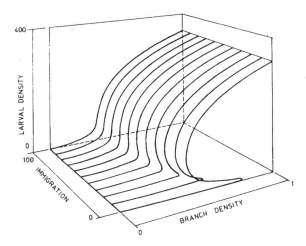

FIGURE 12: *Equilibrium manifold for budworm as a function of budworm immigration rate and branch density. Immigration scale from 0 (no immigration) to 1 (an arbitrary but feasible rate; see text). Other scales as in Fig. 10; compare with Figure. 11.*

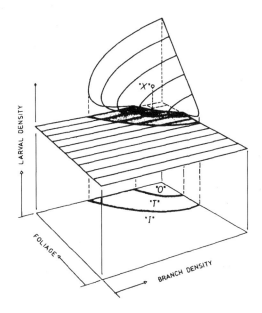

FIGURE 13: *Equilibrium manifold for budworm as used in policy analysis. See text.*

from its low density equilibrium was obviously related to the "size" of the predator-induced pit. As would be expected from the comparison of manifolds, a systematic increase in immigration affects the model very much like a systematic decrease in predation, flipping the budworm/forest system into its alternative behavioral mode of a sustained outbreak exhibiting a 12–16-year insect–foliage cycle.

The greatest payoff from the topological simplifications comes in their implications for policy. In discussing the growth rate curves of Fig. 9, we noted that a forest could be so immature that no outbreak was possible under any conditions (i.e. $R < 1$ for all budworm densities), or so mature that an outbreak would ensue if any budworm at all were present ($R > 1$ for all but extreme budworm densities). This phenomenon is reflected more clearly on the budworm–foliage–branch density manifold originally presented in Fig. 10 and redrawn in Fig. 13. The manifold has been raised above the foliage–branch density plane to illustrate critical regions of forest condition that influence budworm behavior.

Recall that an "outbreak" consists of the budworm population moving from the lower to the upper manifold surface. This is impossible if the forest is in the "inter-outbreak" region (I) of Fig. 13, since for these forest conditions no upper equilibrium surface exists. Any policy which keeps the forest in region (I) will be immune to budworm outbreak. Unfortunately, and hardly by coincidence, it will also be an unprofitably young forest, a defoliated forest, or both.

In contrast, a mature and green forest in the "outbreak" region (O) of Fig. 13 is guaranteed to produce an outbreak. Only the upper equilibrium surface exists, and any (non-zero) budworm density will increase until this surface is reached. The only management act of much significance while the forest remains in this condition is one which imposes extrinsic mortality on the budworm, keeping it from reaching its equilibrium surface where it would defoliate and kill the trees. Region (O) represents precisely the situation reached in New Brunswick in the early 1950's: the full-foliage forest matured and the inevitable outbreak began. Intensive insecticide spraying was introduced to protect the forest by killing budworm. The outbreak was indeed retarded, and the system was held near state (X) in Fig. 13—a state with moderate levels of budworm, fairly dense foliage and a mature forest.

Note, however, that point (X) does not lie on any equilibrium surface; it is manifestly an unstable and "unnatural" configuration for the system, entirely dependent on the continued external application of severe budworm mortality through insecticides. The maintenance of this state is extremely sensitive to any policy failure which decreases this mortality—be it through evolved genetic resistance, errors in spray formulation and delivery, or legal restrictions on spray dosages, targets and frequency. New Brunswick has recently discovered this in a most disagreeable way (Miller and Ketella, 1975; Baskerville, 1976).

The intermediate forest "threat" region (T) is the only one for which the outbreak issue is in doubt. Budworm densities below the reflex fold of the equilibrium surface are attracted to the lower, or endemic, surface. Densities above the fold are attracted to the upper, or outbreak surface. The availability of the lower surface immediately suggests that insecticide policies designed to "knock out" the budworm deserve consideration. On the other hand, we have already shown how dispersal and predation can affect the configuration of the fold and consequently expand or contract the range of forest conditions included in region (T). The success of any policy seeking to maintain the forest system on the lower surface of region (T) will therefore be critically sensitive to various unknowns, both those that impinge on the descriptive model which produced the manifold configuration, and those that unexpectedly change immigration or vertebrate predation intensities in the real world. These critical relationships are due particular attention in the succeeding policy design and evaluation process.

In summary, simplified and compressed versions of the simulation model—the laboratory world—can be captured in topological manifolds which focus upon the multiple equilibrium properties of that model. The manifolds can be calculated directly from the simulation model, as above, or generated from simplified, qualitative equations of the system's structure (Ludwig et al., 1978). In either case, the manifolds can then be exploited to improve understanding of system behavior and structure, and to qualitatively diagnose regions of policy sensitivity and potential. How this qualitative understanding can be put to direct use in the design and evaluation of detailed management policies will be shown in the next two sections.

Clearly, if the descriptive part of the analysis stops at the development of a complex simulation model, the clarity of understanding needed for transfer and policy design is seriously compromised. Again, creative simplification is necessary for understanding.

The Development of Policy Alternatives

Myth: The goal of policy design is the design of optimal policy.

The first responsibility of policy design is to generate and explore a strategic range of alternative approaches to the management problem. Subsequent efforts can seek to evolve a satisfactory or, more ambitiously, an optimal policy from this initial range. Unless the analysis insists from the outset on explicitly enriching the decision environment, then the policy design process will consist of little more than an unguided sequence of incremental modifications to the status quo (Braybrooke and Lindblom, 1963). Such cautious incrementalism is an appropriate procedure for a globally stable system but, as we argued earlier, it is a particularly dangerous way to manage in a world of multiple equilibria and uncertainty.

Prescription is an iterative process of specifying objectives, articulating policy and evaluating performance. We elect to enter this cycle by identifying a strategic range of alternative objectives as departure points for the analysis. The goal is not to define complex realistic objectives but to pose a strategic range of simple alternatives which encompasses the specific goals which may be sought by particular groups of individuals.

At one extreme of the range lies the classical objective of unconstrained "optimal efficiency"—for instance, maximization of present value of the forest. At the other extreme, and equally simplistic, are the objectives of the maintenance of natural dynamic variability. Table 2 lists eight of the principal alternatives used as points of departure in our budworm policy design studies.

Given a point on this strategic range of objectives, the next problem is to identify policies (i.e. sets of management rules) which will effectively promote those goals. We could, of course, seek to identify appropriate policies by simple heuristic gaming with the simulation model, but except in the most trivially simple cases, this would be a prohibitively slow, expensive, and inefficient way to develop interesting (much less optimal) policies. The number of possible policy formulations is generally so large that some formal guidance is necessary to define interesting regions in policy space. We need an organizing "brain" for the idiot simulation (Watt, 1963; Holling and Dantzig, 1977).

A variety of mathematical programming and optimization techniques have been developed to provide such guidance, but they are not yet up to

TABLE 2: *Examples of alternative objectives explored in the budworm policy analysis.*

(1)	Retain existing management approaches ("historical management")
(2)	Eliminate all human intervention, both harvest and budworm control, to recreate a natural system
(3)	Minimize budworm densities
(4)	Minimize budworm densities while eliminating insecticide applications (e.g. replacing with methods of biological control and/or forest management)
(5)	Maximize long-term profits to logging industry
(6)	Maximize long-term profits to logging industry without exceeding present industrial capacity or operational constraints, and without violating environmental standards regarding insecticide application ("constrained profit maximization")
(7)	Maximize long-term profits to logging industry subject to constraints of (5), simultaneously maximizing recreational potential of forest
(8)	Transform the system's existing temporal variability into spatial variability (i.e. develop a forest in which the budworm functions as a forest manager and the essential dynamic interplay of natural forces is retained)

the task presented by even simple ecological management problems. The high dimensionality of ecological systems cripples dynamic programming, while the essential nonlinearities and stochasticities militate against such relatively dimension-insensitive techniques as linear programming and its variants. Drastic simplification of the descriptive model is necessary to obtain any of the benefits of mathematical programming; yet with that simplification, all guarantees of real-world optimality for the resulting policies are inevitably lost.

Our response to this dilemma has been to employ first a variety of formal, and then several heuristic optimization techniques in an iterative "probing" of policy space. Extensive work has been carried out using various forms of dynamic programming (Winkler, 1975), "fixed-form" parameter search techniques (Walters and Hilborn, 1979; Stedinger, 1977), graphical nomogram approaches (Peterman, 1975, 1977b), and a combination of intuition and analysis relying heavily on the manifolds

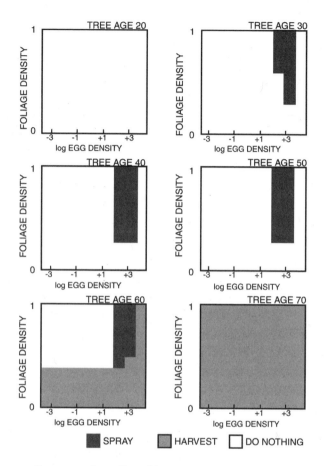

FIGURE 14: *Representative policy tables generated by the Winkler—Dantzig optimization. A separate table is provided for each age of tree (or, in practice, each mean age of stand). The table tells which management act (spray, harvest or do nothing) should "optimally" be applied as a function of field measures of trees present, foliage density (scaled as previously), and egg density (eggs per 10 ft² of branch area, here on a common logarithmic scale).*

discussed previously. Rather than try to review all of this work here (see Yorque et al., 1978), we will focus on the results obtained and lessons learned through just one sequence of optimization techniques applied to just one of the alternative objectives.

Winkler and Dantzig (Winkler, 1975; Holling and Dantzig, 1977) used dynamic programming to calculate the conditions of tree age, foliage and budworm infestation for which insecticide spraying or tree

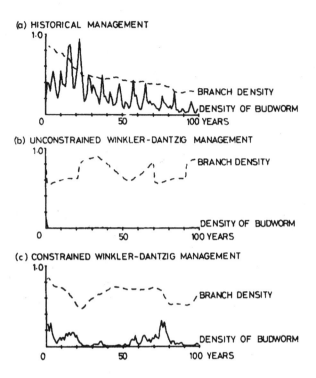

FIGURE 15: *Behavior of the budworm simulation under three management programs. Labeling conventions and initial conditions as in Fig. 6. "Historical management" rules are those used in New Brunswick in the 1960's and 1970's (spatial results for these rules are given in Fig. 8). "Unconstrained Winkler—Dantzig management" employs the rules shown in Fig. 14. "Constrained Winkler—Dantzig" rules introduce realistic constraints on total industrial harvest capacity and permissible rates of insecticide application.*

harvesting are prescribed. The dimensionality problem was resolved by viewing the forest as a collection of single trees. Movement of budworm between trees was simplified by assuming that the number leaving a tree would be balanced by the number arriving from other trees. The analysis resulted in a set of management rules ostensibly "optimal" for the (extreme) objective of maximizing long-term logging profits. These rules take the form of policy "look-up" tables telling the manager what to do each year for any possible condition of his forest (Figure 14).

It was essential to test the Winkler–Dantzig policy in the complete regional simulation model to determine whether, in spite of the simplifications necessary to perform the optimization, it still provided an interesting starting point for further investigation. The results of this test were dramatic, as can be seen in a comparison of Fig. 15 a and b. In contrast to the historical management simulation, the budworm outbreak is rapidly smothered and thereafter prohibited by the Winkler–Dantzig policy and very little budworm-induced tree mortality occurs.

These results suggest that formal optimization procedures applied to simplified models can yield starting points for policy design that are as good as or better than those developed historically. These can then provide focus for a dialogue with managers, scientists and policy people to identify modifications for subsequent heuristic exploration.

For example, there are unique management constraints in any specific region. In New Brunswick, the industrial capacity limits the annual tree harvest, insecticide spraying is only practical in large blocks, and environmental and health regulations limit insecticide dosages. Some, but not all, of these constraints could be incorporated into formal optimization but this would limit its value for other regions which have different constraints. Therefore, additional constraints are included, and the policies adapted heuristically in collaboration with the managers and policy people who have specific knowledge of the system. When the constraints mentioned above were included, the behavior shown in Fig. 15 c was generated. Clearly, the original "optimal" policy is sufficiently robust to operate reasonably well even within the more realistic operational setting of New Brunswick.

Heuristic modifications of an initial policy probe can go far beyond such tests of practicality. Totally novel rules and actions can be identified but again, dialogue is essential, and dialogue is only useful if information is generated that is directly relevant to the people affected. State variables have meaning for the analyst. But social, economic and environmental indicators have meaning for the practical policy maker. To go further in the iterative process of policy design, it is necessary to generate and evaluate indicators of policy performance. That is the focus of the next section, so that we will reserve further examples of heuristic policy development for discussion there.

We conclude that formal and simplified optimization can provide a number of useful starting points for a process of policy design and dialogue. In no sense does it guarantee an optimal or even adequate policy.

The Evaluation of Policy Performance

Myth: The goal of the evaluation process is to choose the policy best reflecting the decision makers' preferences.

Choice implies a given set of policies, one of which must be selected as "best" with respect to a given objective. The evaluation process properly includes such questions of choice but, as elsewhere, the essential goal is to promote understanding—in this case through meaningful characterization of the managed system's performance under alternative policies.

Socially relevant and responsible evaluations cannot be based upon the system's state variables we have used to this point. Rather, they require a broader set of indicators relevant to those who make, and those who endure, the ultimate policy decisions. Two classes of indicators are required, one focusing on broad questions of policy resilience and robustness, the other upon more immediate concerns of management and planning.

Resilience indicators for evaluation are a critical but as yet only partially developed aspect of the science of ecological policy design. We have explored the concept and discussed its implications and difficulties elsewhere (Holling and Clark, 1975; Holling, 1978). It would be premature to accord any of this experience the status of "lessons for policy design", and we will not dwell on it here. The important point is that resilience indicators are essentially measures of topological sensitivity. They focus upon the implications for policy performance of qualitative changes in configuration of the equilibrium manifolds governing system behavior, and not on the traditional quantitative measures of parametric and state variable sensitivity. The resulting indicator metrics are crude, but begin to provide a meaningful framework for dealing with the unexpected and unknown in policy design (Yorque et al., 1978).

The more traditional indicators of management performance are comparatively easy to generate, and can often be partitioned into categories of the sort shown in Table 3. The detailed structure of these indicators need not concern us here (see Bell, 1977b, and Yorque et al., 1978). Suffice it to note that they were defined by the New Brunswick scientists, managers and policy makers as relevant to their evaluation of alternative policies. Specific indicator values are generated by the simulation, just as it generates values for the biological variables discussed earlier.

TABLE 3: *Examples of evaluation indicators generated by the model*

Socioeconomic Indicators

(1) Total costs to the logging industry (not including insecticide)
(2) Total costs as a proportion of total sales
(3) Cost per unit volume of harvested wood
(4) Cost of insecticide spraying
(5) Employment rate reflecting proportion of mill capacity utilized

---

Resource Indicators

(1) Volume of wood in trees older than 20 years
(2) Volume of wood in trees older than 50 years
(3) Volume of wood harvested
(4) Proportion of total volume harvested
(5) Volume of wood killed by budworm
(6) Mill capacity

Environmental Indicators

(1) Damage due to visible defoliation
(2) Logging damage
(3) Age class diversity of the forest
(4) Recreational potential of affected sites
(5) Insecticide impact in terms of fraction of province sprayed

---

At an early stage in the evaluation, each decision maker can choose the particular indicators which interest him and examine the simulated time behavior of each under alternative policies. There are rigorous techniques for comparing policies through the patterns of their indicator behavior, and we will touch on these below. Sometimes visual inspection of indicator patterns is sufficient to show that one policy completely dominates another. More commonly, some of the policy trials are likely to exhibit obviously desirable behavior in others. Traditional evaluation procedures seek to provide a common denominator for ranking such complex alternatives. But we have found the opposite approach to be constructive—viewing evaluations initially as a means of highlighting differences in policy performance. These differences are then used by the policy design process to develop new management alternatives.

For example, even though indicator performance of the constrained Winkler–Dantzig policy (Fig. 17) is superior to that of the historical rules (Fig. 16), it still uses insecticides persistently and extensively. These

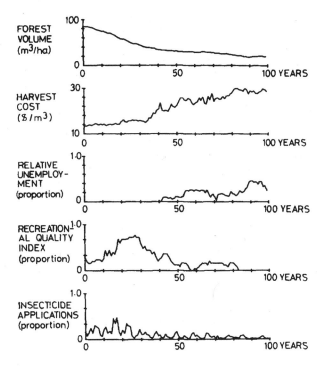

FIGURE 16: *Policy evaluation indicators under "historical management" rules (compare Fig. 15 a). Graphs show the undiscounted value of the indicators in each year. Forest volume is a density in cubic meters of merchantable timber per hectare of fully stocked forest, including only species susceptible to budworm. Harvest cost is for timber delivered to the mills, and is expressed in 1972 Canadian dollars per cubic meter of wood. It includes all stumpage, cutting, and transport costs. Unemployment is a relative rate, calculated as the proportion of mill capacity unutilized due to unavailability of suitable timber. Mill capacities are taken from historical data (1953–1975) for the first 23 years of each simulation, and fixed at 1975 values thereafter. Recreational quality reflects a complex function of logging and budworm damage, access, and forest age structure. Twenty-five subregions with high recreation potential are monitored with respect to these characteristics, and the proportion exceeding a predefined recreational quality rating in each year are recorded. Insecticide applications indicate the proportion of the study area sprayed each year.*

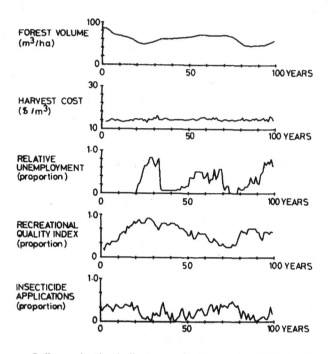

FIGURE 17: *Policy evaluation indicators under "constrained Winkler–Dantzig management" rules (compare Fig. 15 c). indicators defined as in Fig. 16.*

chemical insecticides, however useful as armaments of the forest manager, are an environmental poison that inevitably affects many species
other than the target organism. The overall consequences of their pervasive use will always be risky and uncertain, and reduction of our reliance
on them is obviously desirable. We now show one example of how the
initial Winkler–Dantzig probe can be heuristically modified to reduce
insecticide use without compromising the desirable aspects of indicator performance. Our tools are the equilibrium manifolds (Figs. 10–13)
described in the earlier section on compression and simplification.

Recall that budworm populations remain at endemic levels as long
as they are within the "pit", on the lower stable surface of the manifold.
When forest growth carries the system beyond the end of the pit, the
lower surface vanishes leaving only the upper attracting surface and
an outbreak is triggered. This manifold interpretation immediately
suggests that one way to minimize outbreaks, and hence the use of
insecticides, is to manage the forest through harvesting so that it never
achieves sufficient maturity to exit the pit (i.e. to move into region "O"

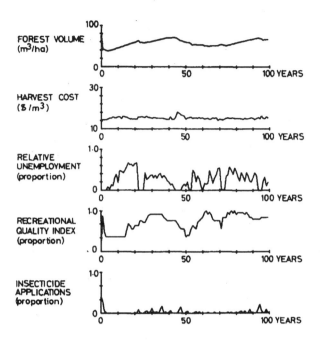

FIGURE 18: *Policy evaluation indicators under "branch density hybrid management" rules. Indicators defined as in Fig. 16.*

in Fig. 13).

Under conditions of average weather and full foliage, the end of the pit, and the loss of the lower equilibrium, is reached when the branch density index is about 0.75. Thus a new policy was designed making one addition to the constrained Winkler–Dantzig rules. If after spraying and harvesting according to Winkler–Dantzig policy, the forest's branch density remained above 0.75, a quantity of older trees were harvested so as to bring this value down to 0.75. This hybrid "branch density" policy was evaluated in the full regional model and the resulting indicator traces are shown in Fig. 18.

The objective to minimize spraying is achieved with the hybrid policy. The average area sprayed, once the initial outbreak is brought under control, drops from 25% in the original constrained Winkler–Dantzig policy (Fig. 17), to 2% when the "branch density" rule is added. Unhappily, the behavior of some other indicators has deteriorated. Total forest volume is brought to a lower value and, more important, unemployment is erratic and generally high. This is to be expected since initial overharvesting is required to bring the branch density to 0.75 which limits the supply in later periods.

The manifolds suggest several additional possibilities for improving the indicator traces, however. Rather than harvesting the forest down to the end of the pit, actions are possible which enhance the pit itself. For example, there exist a number of viruses which are highly host specific and thus potentially less damaging to the environment than insecticides. These agents have usually been treated as laboratory curiosities because they inflict only modest mortality on budworm. The manifold can be used to suggest new strategies for implementing these low mortality agents that significantly alter the dynamics of the budworm system.

The threat state (region "T" in Fig. 13) is defined as having low bud-worm densities (number of larvae less than 22 per 10 ft² of branch area) and intermediate branch densities (between 0.38 and 0.75). A forest in the threat state cannot spontaneously outbreak; it is held at the lower equilibrium until it is flooded by immigrants or until the forest matures beyond the end of the "pit". An additional mortality of only 25% applied to late instar larvae moves the end of the pit to a branch density value of 1.0. That is, the forest conditions need to be that much "better" to com-pensate for this additional mortality before a budworm outbreak begins. By applying this new control only at low densities and only in stands with branch densities nearing 0.75 we expand the threat state to a level of 1.0. The control would not be perfect as immigrants from other areas could still trigger an outbreak. Applying this mortality only at low bud-worm densities leaves the upper surface of the manifold unaltered.

This "pit enhancement" policy explicitly alters the shape of the mani-fold. However, the manifold is only a simplification of the complete model as we have not incorporated such things as stochastic weather and inter-site dispersal. Again a simulation run was generated on the complete regional model to test this new policy. The constrained Winkler–Dantzig rules were again used in conjunction with the additional provision that whenever those rules left a management area in the threat state as defined above, the pit was augmented by an additional 25% mortality, operative only under threat conditions. (This latter requirement is essential so as not to change the other equilibrium conditions which determine other kinds of qualitative behavior.) This new management act can be envi-sioned as the addition of a non-persistent and host-specific virus

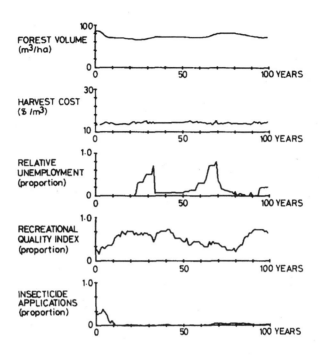

FIGURE 19: *Policy evaluation indicators under "pit enhancement hybrid management" rules. Indicators defined as in Fig. 16.*

whenever the forest reaches the state defined or, alternatively, as the enhancement of vertebrate predator populations.

The resulting indicator traces are shown in Fig. 19. Suddenly our larger objective is achieved. Once the initial outbreak is brought under control by Winkler–Dantzig insecticide spraying, the need for further spraying is almost eliminated by the threat-triggered pit enhancement. Moreover, all other indicators behave as well or better than those of the original constrained Winkler–Dantzig policy rules (compare Fig. 17).

This same procedure has been applied for a variety of objectives from the strategic range (Yorque et al., 1978). In all cases, an interesting trial policy was defined by an optimization procedure and then modified with the understanding provided by the topological analysis—each modification being evaluated using a comprehensive array of performance indicators. The procedure is a partly formal, partly heuristic process identifying new actions and rules which would otherwise elude consideration.

The evaluation of the examples discussed so far has been relatively straightforward—nearly all comparisons have offered a clear case of dominance by one indicator set over another. But the more extensive the array of indicators and the greater the number of policies to be compared, the greater is the danger of losing meaning in a wealth of numerical detail. For complex evaluation problems, some systematic approach to indicator compression is necessary. A number of concepts and techniques for multi-attribute problems are available from the field of decision analysis (e.g. Keeney and Raiffa, 1976) and several of the more practical have been brought to bear in this case study.

Preference evaluation interviews were conducted with representatives of management and research agencies, the forest industry, various provincial ministries, and the informed public. Participants identified their objectives for the forest system by specifying a set of relevant indicators and the desired levels of performance for each. In most cases, trade-off rates and dependencies among indicators were also assessed, with particular attention paid to elucidating attitudes towards risk and intertemporal trade-offs (Bell, 1977a). The result was a set of multi-attribute utility functions reflecting the preferences of the various participants (D.E. Bell, 1977b, and unpublished data).

The utility analyses have been used in the policy design studies to help articulate conflicting preferences, to provide objective functions for the optimization efforts, and to simplify comparisons of policies. These various roles are implicit in Table 4, where four of the policies developed earlier are evaluated according to the utility functions of four representative decision makers interviewed by D.E. Bell (personal communication, 1976).

The utility evaluations of Table 4 are more concise than the original uncompressed indicators. When the set of policies is very large, the compressed utilities allow rapid identification of the more promising ones. Utilities also show where decision makers really differ in the policies they prefer.

As useful as the utility simplifications are, they have severe limitations. The technical constraints are discussed by Keeney and Raiffa (1976), but there are also problems posed by the limited experience of the decision maker.

In our studies, decision makers evaluated policy alternatives consistently only when the indicator values resembled those experienced historically. The decision makers had no "feel" for hypothetical alterna-

TABLE 4: *Utility evaluations of policy alternatives. Utility functions were evaluated over the sets of indicators shown in Figs. 16–19 for four decision makers in New Brunswick. Decision maker 'A' was a senior civil servant responsible for evaluating the benefits of policies to the province as a whole; 'B' was an official in the environmental protection department; 'C' was an academic consultant; 'D' was a labor advisor concerned about the temporal pattern of employment. All utility values are for the 100-year period shown in the figures, are scaled between 0 and 100, and incorporate a 5% annual discount rate. The highest ranked policies for each decision maker are indicated by italics. Quantitative comparisons among decision makers are not meaningful, although comparisons of policy ranks can be.*

Policy alternative	Graph of original indicators	Total utility value for four decision makers			
		A	B	C	D
Historical management	Fig. 16	*97*	75	84	31
Constrained Winkler–Dantzig	Fig. 17	89	67	82	33
Branch density hybrid	Fig. 18	84	78	71	32
Pit enhancement hybrid	Fig. 19	94	*80*	*85*	*37*

tives far removed from their present experience, and attempts to compare such alternatives were therefore—literally—meaningless. Similar problems of "inconsistency" have been encountered in nearly every decision study dealing with radical alternatives (Liska, 1975).

Utility analysis presumes the existence of a comprehensive preference structure in the mind of the decision maker, and sets out to measure it. However, if meaningful preferences develop only on the basis of experience, then responses to the typical preference questions of utility analysis are more likely to reflect attributes of the question than those of the decision maker. This is precisely Lipset's (1976) finding in his retrospective analysis of public opinion surveys. Similarly, many decision studies have attributed difficulties in obtaining consistent responses to risk-taking questions to the fact that very few people have much self-conscious experience in assessing risks (e.g. Slovic and Lichtenstein, 1971).

A principal function of the policy design process is to develop meaning for a strategic range of alternative policies through the use of descriptive models and their performance indicators. This experience is a product, not a premise, of the design process. The appropriate language for most of that process will consist of carefully selected sets of indicators such as those in Figs. 16–19. Only in the latter stages—when meaning has been firmly established and simplification is desired—may it be useful to move to the highly stylized and compressed vocabulary of utility analysis.

Whether we employ utilities or simple indicators, the most profound difficulties of policy evaluation arise in attempts to deal with compression over time. The general inclination is to use various weighted averages of the indicator or utility streams (means, discounted sums, and so forth), but any time-averaging scheme necessarily implies a very particular attitude regarding the relation of present to future and the issue requires careful attention in the evaluation process.

Two time-related issues were particularly important in the budworm policy studies. The first and most novel concerned the problem of "local" time patterns. Several of the interviewed decision makers were adamant that their preference ratings for employment in one year depended upon employment levels in preceding and succeeding years. They further argued that their concern for the social dislocation induced by unemployment was such that fast rises in unemployment followed by slow drops were strongly preferred to slow rises followed by fast drops, regardless of the calendar years in which either pattern occurred. These positions made it impossible to apply any of the standard additive or multiplicative time aggregation formulae to the analysis, and also eliminated those existing approaches which permit restricted sorts of temporal interdependence (e.g. Fishburn, 1965; Meyer, 1970).

To do justice to the decision makers' stated preferences, Bell (1977a and b) found it necessary to develop a new method of deriving utility functions for time streams having inter-period dependencies. These analyses were applied where warranted in our policy evaluation work (e.g. the utility of person "D" in Table 4), and allowed subtle but significant distinctions to be made among complex indicator patterns of the sorts shown for employment in Figs. 16–19. Equally important, the unrestrictive nature of this analytical framework forced the decision makers to articulate their own attitudes towards local time patterns of indicator behavior, and to confront the implications of those attitudes. This exploratory aspect of the evaluation process emphasizes our conviction that, when used at all, utility analysis must be tailored to express the stated preferences of the decision maker rather than the convenience of the analyst. The value of the exercise lies primarily in the dialogue between analyst and decision maker, not in the ultimate utility functions.

The second important temporal issue concerned questions of "absolute" time preferences — of trade-offs between the present and the future as such. The prescriptive aspect of the absolute time question has been debated for a very long time indeed, and its resolution is well beyond our presumption or competence (but see Samuelson, 1976). Even in the descriptive treatment of intertemporal preferences, however, we were immediately confronted with apparent contradictions.

When our studies began, the New Brunswick decision makers were employing cost—benefit analyses and requesting optimization studies which utilized 5% annual discount rates for intertemporal indicator aggregation. This is not an uncommon figure in forest-related analyses (e.g. Goundry, 1960) and is apparently based upon the presumption that future returns from risky public projects should be discounted at rates equal to those at which guaranteed interest could be obtained from long-term government bonds (see Feldstein, 1964 for an exploration of the curious logic underlying this idea). But when the same decision makers were asked choice questions to reveal their implicit intertemporal trade-off rates, a much different picture emerged. None of those interviewed expressed time preference rates of less than 10% per year, and most were closer to 20% (D.E. Bell, unpublished data). Similar results were gathered independently from several different sources and via several different questioning techniques. Discount rates of the order of 15–20% are rare in the resource management literature, but are commonly used by businessmen in assessing risky enterprises. The picture was further complicated when the same decision makers discounting the future at 20% per year simultaneously insisted on "maintaining a healthy, productive forest for posterity".

The comparative evaluation of policy alternatives is highly sensitive to the discount or intertemporal preference rates, as has been demonstrated by Fox and Herfindahl (1964). Given the highly varied responses to our time preference questions, it seemed inevitable that any array of temporally aggregated utility values such as that presented in Table 4 would fail to reflect the decision makers' full range of concerns.

To ensure that meaning would not be lost in the quest for simplification, we chose to supplement the compressed utility values of Table 4 with explicit representations of the undiscounted utility time streams. Figure 20 presents an example for several of the policies and one of the decision makers.

Although the time patterns of Fig. 20 are more complex than the single values of Table 4, they also permit a direct examination of the relationships among utilities, discount rates and time horizons. The selection of a time preference rate is roughly equivalent to specifying a time horizon beyond which indicator behavior becomes largely irrelevant to the evaluation. For a plausible definition of irrelevance, this horizon might be 10 years for a 20% discount rate, about 20 years for 10%, more than 40 years for 5% and, in principle, eternity for a zero discount rate.

It is clear from Fig. 20 that the rank order of policies is related to the time horizon or discount rate selected. Simply by changing from a 10 to a 40 to a 100-year horizon, the identity of the highest ranked policy shifts from predation enhancement to historical approaches and back again. Regardless of what horizon a decision maker feels appropriate for his primary evaluation, we have found that the behavior of the utility or indicator trace in post-horizon periods is often used as an informal check to guarantee that short-term success has not been purchased at the price of "unacceptable" penalties in later years. This explicit use of utility (or individual indicator) time streams for evaluation is somewhat inelegant and in no way resolves the intertemporal aggregation issue. However, it has provided a relatively effective and unambiguous way of communicating the temporal implications of alternative policies to the decision maker.

One final caveat is necessary. We argued earlier that the descriptive model must be explicitly bounded, and that attempts to be comprehensive were counter-productive and delusory. We suspect that the same is true for prescriptive evaluation work. In any case, the evaluation results presented here are in no sense comprehensive benefit—cost or social utility calculations. Such figures would necessarily include estimates of policy implementation costs, probabilities of succes and failure, and

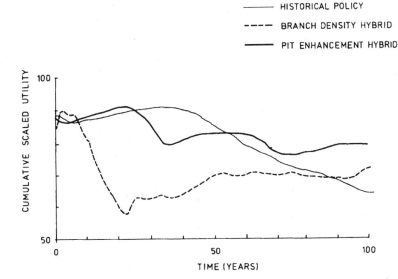

FIGURE 20: *Undiscounted utility time streams for decision maker C of Table 4. Utility value in year (t) is cumulative utility to year (t), scaled as the percent of the maximum possible utility in that year. Original evaluation indicators for the three policies shown were given in Figures 6, 18 and 19.*

so on. Whatever their possible merits for ultimate questions of choice, these "total evaluations" carry an air of finality which make them comparatively sterile and premature elements for a process of adaptive policy design.

The primary goal of the evaluation process is not to define the best policy, but rather to create a foundation of experience and critical understanding upon which informed policy designs and meaningful choices can be based.

Communication, Transfer and Implementation

Myth: A focus on generality and transferability lays sufficient groundwork for policy implementation.

We have emphasized throughout that approaches to policy design need to be transferable to a wide variety of situations. But actual implementations are made under specific circumstances, not general ones. Decisions are shaped by local constraints, by particular institutional structures, and by unique personalities. A focus on generality sets the stage, but unless it is followed by effective application to specific situations the analysis is certain to remain a naive academic exercise.

Close working ties were maintained with decision makers, managers, and their scientific advisors throughout the budworm case study. Three levels of transfer and implementation are now being explored —one involving federal and provincial agencies in New Brunswick (Baskerville, 1976), one involving key institutions from other provinces and states, and one involving Japan and European countries faced with similar problems (Norton and Holling, 1977). Our goal is not to recommend a single policy. Rather, it is to transfer concepts, modeling and evaluation techniques, and a menu of policy alternatives to the hands of those responsible for and affected by decisions. In each case, the focus has been on information packages, communication techniques and transfer workshops which can be understood, controlled, and modified by the participants.

The study reported here began in 1972 with a 5-day workshop in New Brunswick that joined scientific and managerial expertise from federal and provincial agencies, together with modeling and analysis experience from the Institute of Resource Ecology, University of British Columbia. The results were a first crude simulation model and, more importantly, the identification of critical scientific, management, and policy issues needing further consideration (Walters and Peterman, 1974). These issues were further refined in a short but intensive feasibility study that culminated in serious commitments of time, personnel and resources by concerned institutions in New Brunswick.

Cooperative linkages between institutions and individuals are necessary but are not sufficient. Truly effective transfer and implementation hinge on the identification of local leadership—of an individual with judgment, perception, and the respect of research, management, and policy groups. In New Brunswick, that key zindividual was Dr. Gordon L. Baskerville (then of the Federal Department of the Environment and subsequently of the University of New Brunswick) who became a co-equal and guiding partner in this policy design project. Our experience in this and other activities strongly suggests that without a Baskerville on the receiving end, the panoply of models and techniques will serve the management community only as glass-cased display pieces.

Our initial activities were followed by an intensive effort to expand, refine and invalidate the model, and to test the usefulness of the policy design techniques reported in this paper. This multi-disciplinary work was of necessity a cooperative effort, involving at various times 15 scientists from Canada's Department of the Environment, the University of British Columbia's Institute of Resource Ecology and Vienna's International Institute for Applied Systems Analysis (Holling et al., 1976b). A key to the venture's success has been the willingness of a variety of research and grant administrators to support the concept of an institutional network and to gamble on a long-term project that they did not control. That network spread over half the world and amongst a variety of institutions, and was inherently incapable of offering much reassurance in the way of intermediate results. Without this sort of understanding and vigorous protection, nothing the research group could have done would have kept the program from being fragmented and aborted several times over.

The real test of our approach to policy design, and its foremost opportunity for payoff, came with the initiation of a phased series of implementation workshops which constituted an essential ingredient in the transfer process (Holling and Chambers, 1973; Walters, 1974; Holling, 1978). The first were technical workshops to transfer the descriptive and prescriptive models to the computer system in New Brunswick. This phase was judged complete when the scientists and managers there had the ability to independently run, critique, and alter the models in response to local needs and priorities. It should be obvious that no transfer worth speaking of can occur until this sort of expertise and confidence in model utilization develops in the management and policy groups.

Next followed a group of workshops designed to critique the analysis and to transfer the concepts, modeling approaches and a menu of policy alternatives. The first drew together forestry scientists, budworm experts and other entomologists from throughout the United States and Canada. The descriptive models were presented and criticized component by component. Alternative hypotheses were defined, tentatively incorporated in the model, and interactively evaluated. Since the descriptive models had been defined from the start by the budworm researchers themselves, the model's residual uncertainties and approximations focused attention on critical areas for future field research. Tentative suggestions for restructuring scientific research priorities so as to be more responsive to overall system understanding were developed by the workshop participants and are summarized in Holling et al. (1975).

Subsequent workshops were held for the forest managers who will be responsible for implementing new policies, and others for the politicians, administrators and businessmen engaged in long-term policy development. Each set of workshops was followed by a period of careful review with modification to the models and design approach.

The major conclusions drawn from these transfer activities may be summarized as follows:

(1) Transfer means more than mailing the computer codes and writing a report. It also requires a program of workshops and intense "user" involvement so that the local scientists and managers end up as the real and acknowledged experts. A measure of success is the extent to which the original analysis group becomes less and less visible and the local groups more and more so as the program moves into implementation. The initiators' very strong and markedly parental inclinations to hold control too long must be resisted or transfer will fail.

(2) Vigorous institutional support and protection is necessary but not sufficient; the policy design approach can be transferred only to people, not departments. Respected local leadership of the program is necessary.

(3) The analysis must be made fully transparent and interactive. Hence the use of extensive graphical presentations (e.g. Peterman, 1975, 1977b) and an interactive computer environment which allows easy examination and modification of model assumptions. This lets cooperating scientists and managers explore their own experience and assumptions in the context of the models and is absolutely essential if a critical understanding of the strengths, weaknesses, and limitations of the analysis is to be achieved.

(4) Communication of the results must go beyond the traditional written forms. Modular slide–tape presentations describing the approach, the problem and the model can communicate the essential features vividly and rapidly without compromis-

ing content (Bunnell and Tait, 1974; Bunnell, 1976). A 4-minute motion picture of the space–time dynamics of Figs. 7 and 8 under various management regimes can better reveal that behavior than any amount of static discussion and analysis.

(5) A sequence of participatory workshops beginning with scientists, proceeding to managers, and finally involving policy makers builds a foundation of confidence and understanding on the part of all concerned. The reverse "top-down" sequence would force the technical analysis group into a premature position of prominence, alienating local experts and promoting little but suspicion.

(6) The final—and perhaps the most restrictive—requirement of effective transfer is time. The budworm policy analysis per se took less than 6 months; the full program to implementation, more than 3 years. Some of this time was spent in the workshops described above, but much was literally an incubation period. A prerequisite for effective implementation seems to be time for the analysis group to appreciate the real options and constraints, time for the local managers and scientists to become truly conversant with new concepts, and time for the policy people to credit the analysis group with relevant intent. In retrospect, we doubt that the process could be rushed without fatally prejudicing the results in one way or another. Successful implementation requires patience (Pressman and Wildavsky, 1973; Bardach, 1976).

Responsible policy choices by the decision maker are based on understanding and control of, not necessarily belief in, the technical analysis. If such understanding is not clearly communicated, if such control is not effectively transferred, then mere technique subverts political judgment as a basis for public policy decisions, with no accountability for the results. That would simply be the promulgation of another undesirable myth—the one Lewis Mumford has called the Myth of the Machine—in systems analytic disguise (Weizenbaum, 1976).

A serious commitment to effective communication, transfer and implementation is the obligation of a responsible analysis group.

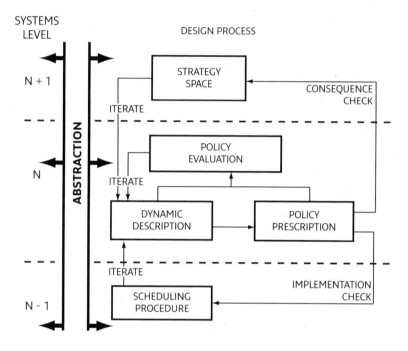

FIGURE 21: *The process of ecological policy design.*

Conclusions

Major portions of the budworm case study took place while we were at the International Institute for Applied Systems Analysis. At the same time, various groups at that institute were engaged in studies of other resource development and environmental problems. This provided the opportunity to identify elements common to a wide range of problems. The projects chosen for the comparison included the forest pest problem described here, a study of national and international aspects of Pacific salmon management (Walters, 1975b) and an analysis of global energy problems (Haefele, 1976). These covered a broad range of contemporary dichotomies of management perspective—global versus regional, ecological and social versus physical and economic, national versus international. But they proved to share a perception of the primary issue (hypotheticality and coping with the unknown) and the central organizing concept (resilience topology). Surprisingly, and significantly, they also exhibited a similar structure of analytical steps (Fig. 21).

Any particular problem, whatever its reach, whatever its extent, can be conceptualized as occurring at a systems level embedded in a hierarchy. The system level of focus for each of the case studies is arbitrarily identified as level N—the entry point for the analysis. We identify three essential steps at this level: systems description, policy prescription, and policy evaluation. The systems description provides a detailed, dynamic description of the system for which management rules or policies are to be designed. At one point it seemed that this distinction was feasible only for ecological, physical, and other "natural" problems where "Mother Nature" was a given and could be objectively modeled. However, it became clear in the energy study that there existed the same need for a detailed dynamic description of societal interactions as they related directly to energy, capital and labor. Although the mathematical languages used in the descriptive models were different (simulation in the ecological studies and differential equations in the energy models), each case study demanded a model which could act as a kind of laboratory world for the development and evaluation of policies.

Descriptive models help to answer "what if" questions concerning the likely consequences of given policies or management acts. For these models to be used effectively it is necessary to identify alternative objectives and to design policies which will achieve them. This introduces the step of policy prescription—the techniques of articulating objectives and calculating "optimal" management rules and control laws. Typically, however, the combination of high dimensionality, nonlinearity and stochastic character of the descriptive models exceeds the computational limits of existing optimization techniques. Drastically simplified variants must be developed which accommodate the limitations of mathematical programming by judiciously sacrificing realism. These simplified "optimizations" generate useful probes into policy space but carry with them no guarantee of real-world "optimality" or even sensibility.

This leads us to the third step: policy evaluation. It deals in general with the comparison of alternative policies and in particular with checks on the sensitivity of policy to surprise, error and the unknown. The simplified "optimal" policies from the prescriptive studies are now evaluated within the complete descriptive model, paying particular attention to the generation of indicators relating to issues and processes omitted in the simplifications. The sensitivity, resilience, and robustness of these policies to the simplifying assumptions, to the omitted relationships, and to the residual uncertainty can be specifically tested and quantified.

These three basic analytic steps combine to produce a set of alternative policies evaluated in terms of a range of social objectives. But no matter how effective such an analysis, and at whatever level, it must be complemented by two more significant steps, both relating to the embedding of level "N" within the overall hierarchy.

There is first the issue of execution and feasibility. Many otherwise admirable policies, when faced with specific limitations in capital and labor, and the constraints of institutional and political arrangements may be totally infeasible. A feasibility check at "$N-1$" is therefore essential, and for this a new set of techniques and approaches that identify the sequential steps of practical implementation is necessary.

Just as there is a lower systems level, so there is a higher one ($N+1$ in Fig. 21). Inevitably, the policy analysis will leave out certain social, resource, and environmental values and considerations. The best solutions at the "N" level can have disastrous consequences at "$N+1$". This higher level embedding poses difficult and critical issues which have as yet only been dealt with in a primitive but suggestive way (Haefele and Buerk, 1976). These, together with the related problems of explicit generation and measurement of systems resilience, define the present frontiers of the science of ecological policy design.

Acknowledgments

In one sense it is presumptuous to thank those directly involved in this study since they were all co-equal partners in a strange inter-institutional and interdisciplinary experiment. Nevertheless, they deserve the recognition of being as much part of this creation as the authors of this paper. K.E.F. Watt provided much of the conceptual and methodological foundation for the study, as well as anticipating many of its results. The policy people and scientists of Canada's Department of the Environment gave remarkable and consistent support throughout. In particular, Gordon Baskerville, Charles Miller, and their colleagues of the Maritimes Forest Research Centre were committed partners in the team, with their flanks admirably protected by Evan Armstrong, Dick Belyea, Murray Nielson, Dick Prentice, and John Tener.

At the International Institute for Applied Systems Analysis, an astonishing group of outstanding people gave their all to something as silly as a budworm — David Bell, George Dantzig, Myron Fiering, Carlos Winkler and Howard Raiffa.

The third institution in this effort was the Institute of Resource Ecology, University of British Columbia. Carl Walters, Ray Hilborn, Randall Peterman, Sandra Buckingham and Jeff Stander developed the initial simulation model of the budworm system, and put their results and insights at our disposal. Along with Pille Bunnell, Nick Sonntag and Zafar Rashid they carved off pieces of the subsequent analysis and resolved them at times when they saw we were faltering. The unabashed enthusiasm and competence of Ralf Yorque contributed immeasurably to this project.

Finally, thanks are due Bob Fisher, Jery Stedinger and Warren Klein for independently reviewing the computer programs; Ted Foin, Neil Gilbert, Jack McLeod, Briar Mar, John Steele and Ken Watt for comments and suggestions on an earlier draft of this report; and Ulrike Hilborn for editing, drafting, and supervising its production.

References*

Ackerman, B.A., S. Rose-Ackerman, J.W. Sawyer, Jr. and D.W. Henderson, 1974. The Uncertain Search for Environmental Quality. Free Press, New York, N.Y., 386 pp.

Bardach, E., 1976. The Implementation Game. M.I.T. Press, Cambridge, Mass., 323 pp.

Baskerville, G.L., 1975a. Spruce budworm: super silviculturist. *For. Chron.* 51:138–140.

Baskerville, G.L., 1975b. Spruce budworm: the answer is forest management — or is it?. *For. Chron.* 51:157–160.

Baskerville, G.L., Editor, 1976. *Report of the Task Force for Evaluation of Budworm Control Alternatives*, Department of Natural Resources, Fredericton, N.B., 210 pp.

Bazykin, A.D., 1974. Volterra's system and the Michaelis–Menten equation. In: V.A. Ratner, Editor, *Problems in Mathematical Genetics*, U.S.S.R. Academy of Sciences, Novosibirsk (in Russian). (Available in English as: Structural and dynamic stability of model predator–prey systems. IIASA RM-76-8.)

Bell, D.E., 1977a. A utility function for time streams having interperiod dependencies. *Oper. Res.* 25:448–458.

Bell, D.E., 1977b. A decision analysis of objectives for a forest pest problem. In: D.E. Bell, R. Keeney, and H. Raiffa, Editors, *Conflicting Objectives in Decisions*, Wiley, London, pp. 389–421.

Belyea, R., et al., 1975. The spruce budworm. *For. Chron.* 51:135–160.

Blais, J.R., 1965. Spruce budworm outbreaks in the past three centuries in the Laurentide Park. *Quebec. For. Sci.* 11:130–138.

Blais, J.R., 1968. Regional variation in susceptibility of eastern North American forests to budworm attack based on history of outbreaks. *For. Chron.* 44:17–23.

---

* In an effort to make the bibliography as current as possible, we have cited many works presently available only as publications of the International Institute for Applied Systems Analysis (IIASA) and our own Institute of Resource Ecology (IRE). These may be obtained from the following addresses:

Documents and Publications, International Institute for Applied Systems Analysis, Schloss Laxenburg, A-2361 Laxenburg, Austria.

Publications (attn. Ralf Yorque), Institute of Resource Ecology, University of British Columbia, Vancouver, B.C., V6T 1W5, Canada.

Blais, J.R., 1974. The policy of keeping trees alive via spray operations may hasten the recurrence of spruce budworm outbreaks. *For. Chron.* 50:19–21.

Branscomb, L.M., 1977. Science at the White House: a new start. *Science* 196:848–852.

Braybrooke, D., and C.E. Lindblom, 1963. A Strategy of Decision. Free Press, New York, N.Y. 268 pp.

Brewer, G., 1973. The Politician, the Bureaucrat, and the Consultant. Basic Books, New York, N.Y., 291 pp.

Brown, C.E., 1970. A Cartographic Representation of Spruce Budworm *Choristoneura fumiferana* (Clem.). *Infestation in Eastern Canada, 1909–1966.* Canadian Forestry Service Publication No. 1263.

Bunnell, P., 1976. The Spruce Budworm: an Ecosystem Problem and a Modeling Approach to its Management. An eight-part side–tape presentation (80 min), Institute of Resource Ecology, Vancouver, B.C.

Bunnell, P., and D. Tait, 1974. A Primer on Models: Why and How. A five-part slide–tape presentation (280 slides, 50 min). Institute of Resource Ecology, Vancouver, B.C.

Canada, Department of the Environment, 1938–1976. *Forest Insect and Disease Survey Annual Reports.* (Author varies.)

Clark, W.C., 1976. Mathematical Bioeconomics. Wiley, New York, N.Y., 352 pp.

Clark W.C., and C.S. Holling, 1979. Process models, equilibrium structures, and population dynamics: on the formulation and testing of realistic theory in ecology. In: U. Halbach and J. Jacobs, Editors, *Population Ecology. Fortschrifte der Zoologie* 6, Fischer Verlag, Stuttgart.

Cline, R., 1961. A Survey and Summary of Mathematical and Simulation Models as Applied to Weapon System Evaluation. *ASD Technical Report* 61–376.

Conway G.R., G.A. Norton, A.B.S. King and N.J. Small. 1975. A systems approach to the control of the sugar cane frog hopper. In: G.E. Dalton, Editor, *Study of Agricultural Systems*, Applied Science, London, pp. 193–229.

Crocker J.F.S., R.L. Ozere, S.H. Safe, S.C. Digout, K.R. Rozee and O. Hutzinger. 1976. Lethal interaction of ubiquitous insecticide carriers with virus. *Science* 192:1351–1353.

Davidson, A.G., and R.M. Prentice, Editors, 1967. *Important Forest Insects and Diseases of Mutual Concern to Canada, the United States and Mexico*, Department of Forestry and Rural Development, Canada, 248 pp.

Elliot, K.R., 1960. A history of recent infestations of the spruce budworm in northeastern Ontario, and an estimate of resultant timber losses. *For. Chron.* 36:61–82.

Feldstein, M.S., 1964. The social time preference discount rate in cost–benefit analysis. *Econ. J.* 74:360–379.

Fiering M.B., and J. Stedinger, 1976. Progress report on budworm optimization efforts. Harvard University Division of Engineering and Applied Physics, Harvard, Ill. (1976) unpublished .

Fishburn, P.C., 1965. Markovian dependence with whole product sets. *Oper. Res.* 13:238–257.

Fox, I.K., and O.C. Herfindahl. 1964. Attainment of efficiency in satisfying demands for water resources. *Am. Econ. Rev.* 1964:198–206.

Goundry, G.K., 1960. Forest management and the theory of capital. *Can. J. Econ.* 26:439–451.

Greenbank, D.O., 1957. The role of climate and dispersal in the initiation of outbreaks of the spruce budworm in New Brunswick. 2. The role of dispersal. *Can. J. Zoology* 35:385–403.

Greenbank, D.O., 1973. The Dispersal Process of Spruce Budworm Moths. Canadian Forestry Service. *Maritimes Forest Research Centre Information Report M-X-39.*

Gutierrez, A.P., D.W. Demichele and Y. Wang, 1977. New systems technology for cotton production and pest management. *Proc. 15th Int. Congr. Entomol.,* pp. 553–559.

Haefele, W., 1974. Hypotheticality and the new challenges: the pathfinder role of nuclear energy. *Minerva* 10:303–323.

Haefele, W., and R. Buerk, 1976. An Attempt of Long-range Macroeconomic Modeling in View of Structural and Technological Change. *IIASA RM-76-32.*

Haefele, W., 1976. Second Status Report of the IIASA Project on Energy Systems 1975. *IIASA RR-76-1,* 249 pp.

Hilborn, R., C.S. Holling and C.J. Walters. 1976. Managing the unknown: approaches to ecological policy design. In: J.J. Riesa, Editor, *Biological Evaluation of Environmental Impact,* President's Council on Environmental Quality.

Holcomb Research Institute, 1976. Environmental Modeling and Decision Making. Praeger, N.Y., 162 pp.

Holling, C.S., 1965. The functional response of predators to prey density and its role in mimicry and population regulation. *Mem. Entomol. Soc. Can.* 45:1–60.

Holling, C.S., 1973. Resilience and stability of ecological systems. *Ann. Rev. Ecol. Syst.* 4:1–23.

Holling, C.S., 1976. Resilience and stability of ecosystems. In: E. Jantsch and C.H. Waddington, Editors, *Evolution and Consciousness: Human Systems in Transition.* Addison-Wesley, Reading, Mass., pp. 73–92.

Holling, C.S., Editor, 1978. *Adaptive Environmental Assessment and Management,* Wiley, Chichester, 377 pp.

Holling, C.S., and S. Buckingham, 1976. A behavioral model of predator–prey functional responses. *Behav. Sci.* 21:183–195.

Holling, C.S., and A.D. Chambers, 1973. Resource science: the nurture of an infant. *BioScience* 23:13–20.

Holling, C.S., and W.C. Clark, 1975. Notes towards a science of ecological management. In: W.H. van Dobben and R.H. McConnell, Editors, *Unifying Concepts in Ecology,* W. Junk, The Hague, pp. 247–251.

Holling, C.S., and G.B. Dantzig, 1977. Determining Optimal Policies for Ecosystems. *IRE Research Report R-7-B,* 41 pp.

Holling, C.S., and M.A. Goldberg, 1971. Ecology and planning. *J. Am. Inst. Planners* 37:221–230. http://www.sciencedirect.com/science?_ob=RedirectURL&_method=outwardLink&_partnerName=3&_targetURL=http%3A%2F%2Fdx.doi.org%2F10.1080%2F01944367108977962&_acct=C000034138&_version=1&_userid=655046&md5=06f51e0670c7539296be2354914d51bb

Holling C.S., D.D. Jones and W.C. Clark, 1975. Spruce Budworm/Forest Management. In: *IRE Workshop Progress Report PR-5*, 50 pp.

Holling C.S., C.C. Huang and I. Vertinsky, 1976a. Technological change and resource flow alignments: an investigation of systems growth under alternative funding/ feedback. Int. Inst. of Management, I 76/19, Berlin, 34 pp.

Holling C.S., D.D. Jones and W.C. Clark, 1976b. Ecological policy design: a case study of forest and pest management. In: *IIASA Conference 1976* Vol. 1:139–158.

Holling C.S., D.D. Jones and W.C. Clark, 1979. Meager data and the analysis of ecological processes: predation. Unpublished.

Jones, D.D., 1975. The application of catastrophe theory to ecological systems. In: G.S. Innis, Editor, *New Directions in the Analysis of Ecological Systems*, Simulation Councils, Inc., La Jolla, Calif., pp. 133–148 (Reprinted in Simulation, 29 (1977) 1–15).

Keeney, R., and H. Raiffa, 1976. Decisions with Multiple Objectives. In: , Wiley, New York, N.Y., 569 pp.

Kiritani, K., 1977. Systems approach for management of rice pests. In: *Proc. 15th Int. Congr. Entomol.*, pp. 591–598.

Lipset, S.M., 1976. The wavering polls. *Publ. Interest* 43:70–89.

Liska, A.E., Editor, 1975. *The Consistency Controversy: Readings on the Impact of Attitude on Behavior*, Halsted Press, New York, N.Y., 324 pp.

Loucks, O.L., 1962. A forest classification for the Maritime Provinces. In: *Proc. N.S. Inst. Sci.* 25:85–167.

Ludwig, D., D.D. Jones and C.S. Holling, 1978. Qualitative analysis of insect outbreak systems: the spruce budworm and forest. *J. Anim. Ecol.* 47:315–332.

Mar, B.W., 1974. Problems encountered in multi-disciplinary resources and environmental simulation models development. *J. Environ. Manage.* 2:83–100.

Meyer, R.F., 1970. The utility of assets in an uncertain but time invariant world. In: J. Lawrence, Editor, *OR69: Proc. 5th Int. Conf. Oper. Res.* Tavistock, pp. 627–648.

Miller, C.A., and E.G. Ketella, 1975. Aerial control operations against the spruce budworm in New Brunswick, 1952–1973. In: M.L. Prebble, Editor, *Aerial Control of Forest Insects in Canada*, Department of the Environment, Ottawa, Ont.

Mitchell, R., R.A. Mayer and J. Downhower, 1976. An evaluation of three biome programs. *Science* 192:859–865.

Morris, R.F., 1963. The dynamics of epidemic spruce budworm populations. *Mem. Entomol. Soc. Can. No. 31*

Norton, G., and C.S. Holling, 1977. Editors, *Proceedings of a Conference on Pest Management* IIASA CP-77-6, 352 pp.

Noy-Meir, I., 1975. Stability of grazing systems: an application of predator-prey graphs. *J. Ecol.* 63:459–481.

Pearce, P.A., 1975. Effects on birds. In: M.L. Prebble, Editor, *Aerial Control of Forest Insects in Canada*, Department of the Environmental, Ottawa, Ont., pp. 306–313.

Pearce, P.A., D.B. Peakall and A.J. Erskine, 1976. Impact on forest birds of the 1975 spruce budworm spay operation in New Brunswick. *Canadian Wildlife Service Progress Notes No. 62:7.*

Peterman, R.M., 1975. New techniques for policy evaluation in ecological systems; methodology for a case study of Pacific salmon fisheries. *J. Fish. Res. Board Can.* 32:2179–2188.

Peterman, R.M., 1977a. A simple mechanism which causes collapsing stability regions in exploited salmonid populations. *J. Fish. Res. Board Can.* 34:1130–1142.

Peterman, R.M., 1977b. Graphical evaluation of environmental management options: examples from a forest-insect pest system. *Ecol. Modelling* 3:133–148.

Prebble, M.L., 1975. Editor, *Aerial Control of Forest Insects in Canada*, Department of the Environment, Ottawa, Ont., 330 pp.

Pressman, J.L., and A. Wildavsky, 1973. Implementation: how Great Expectations in Washington are Dashed in Oakland, University of California Press, Berkeley, Calif.

Ricker, W.E., 1954. Stock and recruitment. *J. Fish. Res. Board Can.* 11:559–623.

Ricker, W.E., 1963. Big effects from small causes: two examples from fish population dynamics. *J. Fish. Res. Board Can.* 20:257–264.

Samuelson, P.A., 1976. Economics of forestry in an evolving society. *Econ. Inquiry* 14:466–492.

Shubik M., and G. Brewer, 1972. Models, Simulation, and Games–a Survey. *Rand Report R-1060-ARPA/RC.*

Slovic P., and S. Lichtenstein, 1971. Comparison of Bayesian and regression approaches to the study of information processing in judgement. *Organizational Behav. Human Performance* 6:649–744.

Southwood, T.R.E., 1976. Bionomic strategies and population parameters. In: R.M. May, Editor, *Theoretical Ecology: Principles and Applications*, Blackwells Scientific, Oxford, pp. 26–48.

Stedinger, J., 1977. Spruce Budworm Management Models. In: *Thesis*, Harvard University, Harvard, Mass., 351 pp. mimeographed .

Takahasi, F., 1964. Reproduction curve with two equilibrium points: a consideration on the fluctuation of insect population. *Res. Popul. Ecol.* 6:26–36.

Thom, R., 1975. Structural Stability and Morphogenesis: An Outline of a General Theory of Models. Benjamin, Reading, Mass.

Walters, C.J., 1974. An interdisciplinary approach to development of watershed simulation models. *Technological Forecasting and Social Change* 6:299–323.

Walters, C.J., 1975a. Foreclosure of options in sequential resource development decisions. *IIASA RR-75-12.*

Walters, C.J., 1975b. Editor, *Proceedings of a workshop on salmon management IIASA CP-75-2.*

Walters, C.J., and R. Hilborn, 1976., Adaptive control of fishing systems. *J. Fish. Res. Bd. Can.* 33:145–159.

Walters, C.J., and R.M. Peterman, 1974. A systems approach to the dynamics of spruce budworm in New Brunswick. *Quaestions entomologicae* 10:177–186.

Watt, K.E.F., 1963. Dynamic Programming, "Look Ahead Programming", and the Strategy of Insect Pest Control. *Can. Ent.* 95:525–536.

Webb F.E., J.R. Blais and R.N. Nash, 1961. A cartographic history of spruce budworm outbreaks and aerial forest spraying in the Atlantic region of North America 1948–1959. *Can. Ent.* 93:360–379.

Weizenbaum, J., 1976. Computer power and human reason. Freeman, San Francisco, Calif.

Winkler, C., 1975. An optimization technique for the budworm forest-pest model. *IIASA RM-75-11*.

Yorque, R., Baskerville, G., Clark, W.C., Holling, C.S., Jones, D.D. and Miller, C.A. Forthcoming. Ecological policy design: a case study of forests, insects, and managers.

Qualitative Analysis of Insect Outbreak Systems

The Spruce Budworm and Forest

D. LUDWIG, D. D. JONES, AND C. S. HOLLING

D. LUDWIG, *Institute of Applied Mathematics and Statistics, University of British Columbia, Vancouver, B.C., Canada V6T 1W5.*

D. D. JONES *and* C. S. HOLLING, *Institute of Resource Ecology, University of British Columbia, Vancouver, B.C., Canada V6T 1W5.*

Reprinted with permission from *The Journal of Animal Ecology*, Vol. 47, No. 1. (1978): 315–332.

Summary

(1) A procedure has been described for the qualitative analysis of insect outbreak systems using spruce budworm and balsam fir as an example. This consists of separating the state variables into fast and slow categories.

(2) The dynamics of the fast variables are analysed first, holding the slow variables fixed. Then the dynamics of the slow variables are analysed, with the fast variables held at corresponding equilibrium values. If there are several such equilibria, there are several possibilities for the slow dynamics.

(3) In the case of the budworm, this analysis exhibits the possibility of 'relaxation oscillations' which are familiar from theory of the Van der Pol oscillator. In more modern terminology, the jumps of the system are governed by a cusp catastrophe.

(4) Such an analysis can be made on the basis of qualitative information only, but additional insight emerges when parameter ranges are defined by the kind of information typically available from an experienced biologist.

(5) At the least this can be a guide to assess subsequent priorities for both research and policy.

Introduction

As in all sciences, ecology has its theoretical and its empirical school. Perhaps because of the complexity and variety of ecological systems, however, both schools seem, at times, to have taken particularly extreme positions. And so the empiricists have viewed the theoretical school as designing misleading constructs and generalities with no relation to reality. The theoreticians, in their turn, have viewed the empirical school as generating mindless or mind-numbing analysis of specifics and minutiae.

This paper aims to apply some of the tools of the theoretician—specifically the qualitative theory of differential equations—to one of the most detailed and exhaustive empirical studies of an ecological system that has ever been attempted—the spruce budworm/forest interaction in eastern North America.

It has two purposes; the first is to demonstrate how far these mathematical tools can be pushed to give insight when information is available for a specific system at three different levels. The first level is purely qualitative and non-numerical. The second includes rough estimates of parameter values that are typically known by the informed biologist if he is asked the appropriate question. The third includes highly detailed quantitative data that, while rarely available, are provided in the extensive monograph of spruce budworm dynamics prepared by Morris (1963).

The development of the analysis described here in fact followed precisely that sequence. The first version of the equations was prepared by one of us (Ludwig) after hearing a 1-h lecture that was totally non-numerical and qualitative. Thereafter he relied on a one-half page summary description of the system (in Holling 1973, p. 14). After an afternoon discussion some modest modifications were made (particularly Step 5, in what follows) to complete the qualitative analysis as far as it could go.

We then moved to the next level of very general and easily obtained quantitative information. Our rule was to confine ourselves to guesses of parameter values that an informed entomologist or forester might reasonably be expected to have prior to the establishment of Morris' spruce budworm project.

The final step was to use the data from that detailed study to see what additional insights were added.

The second purpose emerges from that last step. Morris' detailed study has independently provided the basis for the development and rigorous testing of a simulation model (Jones 1976). Hence, the final set of differential equations, their parameter estimates, and the topological analysis could be directly compared to the functional content and behaviour of the full simulation model. The key question was to determine if there was value in compressing the detailed explanation contained in a simulation model into an analytically tractable set of three differential equations.

The paper is organized into the three levels of information. Since the approach seems to have considerable generality, we have also identified the specific steps as a kind of 'how-to-do-it' sequence.

Level I: Qualitative Information

Step 1: divide the state variables into fast and slow categories
Associated with each state variable is a characteristic time interval over which appreciable changes occur. The budworm can increase its density several hundred fold in a few years. Therefore, in a continuous representation of this process, a characteristic time interval for the budworm is of the order of months. Parasites of the budworm may be assigned a similar, or somewhat slower scale. Avian predators may alter their feeding behavior (but not their numbers) rather quickly and may be assigned a fast time scale similar to budworm. The trees cannot put forth foliage at a comparable rate, however: a characteristic time interval for trees to

completely replace their foliage is on the order of 7–10 years. Moreover, the life span of the trees themselves is between 100 and 150 years, in the absence of budworm, so that their generation time is measured in decades. We first conclude, therefore, that the minimum number of variables will include budworm as a fast variable and foliage quantity (and perhaps quality) as a slow variable.

In the case of the budworm, the main limiting features are food supply, and the effects of parasites and predators. In order to describe the former, we choose a logistic form: if B represents the budworm density, then, in the absence of predation B satisfies

$$\frac{dB}{dt} = r_B B\left(1 - \frac{B}{K_B}\right) . \tag{1}$$

The carrying capacity K_B is assumed to depend upon the amount of foliage available. The logistic equation is chosen here because it involves only two parameters. The later mathematical analysis is facilitated by this choice, but the results would be analogous if any other form of self-limited growth were assumed.

The effect of predation is included by subtracting a term $g(B)$ from the right-hand side of eqn (1). A feature of predators is that their effect saturates at high prey densities; i.e. there is an upper limit to the rate of budworm mortality due to predation. The consumption of prey by individual avian predators is limited by saturation, and the number of birds is limited by such factors as territorial behaviour. Similarly, parasites have a low searching capacity that prevents a rapid build-up of their numbers during an outbreak. Thus their impact does not appreciably increase with increasing budworm density.

We conclude that $g(B)$ should approach an upper limit β as $B \to \infty$. This limit β may depend upon the slow variables (i.e. the forest variables), but that possibility is deferred until Step 5, below. There is also a decrease in the effectiveness of predation at low budworm densities. This is a characteristic of a number of predators and arises in birds in part because of the effects of learning. Birds have a variety of alternative foods, and when one of them is scarce, that particular prey item is encountered only incidentally. As the item becomes more common, however, the birds begin to associate reward with that prey and they begin to search selectively for it. Thus we may assume that $g(B)$ vanishes quadratically as $B \to 0$. A convenient form for $g(B)$ which has the properties of saturation at a level β and which vanishes like B^2 is

$$g(B) = \beta \frac{B^2}{\alpha^2 + B^2} \cdot$$

(2)

This represents a Type-III S-shaped functional response (Holling 1959). The parameter α in eqn (2) determines the scale of budworm densities at which saturation begins to take place. The addition of vertebrate predation to eqn (1) thus produces a total equation for the rate of change of B:

$$\frac{dB}{dt} = r_B B \left(1 - \frac{B}{K_B} \right) - \beta \frac{B^2}{\alpha^2 + B^2} \cdot$$

(3)

We emphasize that this particular form was chosen to require as few parameters as possible; our final conclusions are not dependent upon the specific form of the equation, but only upon its qualitative properties.

Step 2: analyse the long-term behaviour of the fast variables when the slow variables are held fixed

In the present case, this analysis is relatively simple, since only one fast variable is considered explicitly. In more complicated situations, phase plane or other more elaborate methods might be required (Bazykin 1974). The first step in the analysis is to identify the equilibria (where $dB/dt = 0$) and determine their stability. Equilibrium values of B must satisfy

$$r_B B \left(1 - \frac{B}{K_B} \right) - \beta \frac{B^2}{\alpha^2 + B^2} = 0 \cdot$$

(4)

Clearly, $B = 0$ is one such value. If B is near zero, the first term (growth) dominates the second term (predation). The derivative dB/dt is positive for B slightly greater than zero, and therefore $B = 0$ is an unstable equilibrium. The remaining roots of eqn (4) satisfy

$$r_B \left(1 - \frac{B}{K_B} \right) - \beta \frac{B}{\alpha^2 + B^2} = 0 \cdot$$

(5)

The number of roots for eqn (5) depends upon the four parameters r_B, K_B, β, and α. The next step is to combine some of these parameters where possible by scaling.

Step 2(a): Scale the equations to reduce the number of parameters
We introduce the scaled budworm density $\mu = B/\alpha$. Equation (5) takes the form

$$r_B\left(1-\frac{\alpha\mu}{K_B}\right)-\frac{\alpha\beta\mu}{\alpha^2(1+\mu^2)}=0 .$$
(6)

We multiply through by α/B and (6) becomes

$$\frac{\alpha r_B}{\beta}\left(1-\frac{\alpha\mu}{K_B}\right)-\frac{\mu}{(1+\mu^2)}=0 .$$
(7)

Now eqn (7) involves just two combinations of the original four parameters. We set

$$R = \frac{\alpha r_B}{\beta}, Q = \frac{K_B}{\alpha}$$
(8)

and rewrite eqn (7) as

$$R\left(1-\frac{\mu}{Q}\right)=\frac{\mu}{1+\mu^2} .$$
(9)

The interpretation of eqn (9) is both simple and important. The left-hand side of eqn (9) is the *per capita* growth rate of the scaled variable μ (with respect to a scaled time $t' = \beta/\alpha t$). The right-hand side of eqn (9) is the *per capita* death rate due to predation, also in scaled variables. The points where these curves intersect are the non-zero equilibria for μ (and equivalently, B). The two sides of eqn (9) are plotted in Fig. 1. The left-hand side is a straight line, with intercepts R and Q. The right-hand side is a curve which passes through the origin and is asymptotic to the μ axis at high densities.

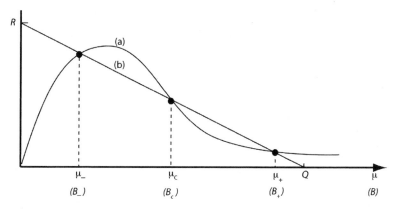

FIGURE 1: *The growth rate (a) [R(1−μ/Q)] and predation loss rate (b) [μ/(1+μ²)] of the scaled budworm density μ. Stable equilibria occur at μ₋ and μ₊; an unstable equilibrium is at μ_c.*

Step 2(b): Examine the equilibria of the fast variables as a function of the parameter values

The equilibria for the budworm variable are defined where the straight growth curve intersects the peaked predation curve (Fig. 1). The number and location of these intersections depends on the two parameters R and Q. In this section we examine the nature of this dependence.

Equation (9) provides a minimum of one and a maximum of three equilibria. A case of three is shown in Fig. 1 with the lower, middle, and upper values labelled as μ_-, μ_c, and μ_+, respectively. Although eqn (9) and Fig. 1 are, strictly speaking, in terms of the scaled variable μ we shall henceforth substitute the original variable B, keeping in mind that they differ only by a constant factor, $1/\alpha$.

It should be clear from Fig. 1 that the locations of the intersections depend upon the relative positions of the growth and predation curves. In this particular choice of scaling, both of the system parameters, R and Q, appear in the straight (growth) function. This makes it easier to visualize how changes in R and Q will change the number and location of the equilibria.

Although we call R and Q parameters, we assume that they may turn out to be functions of the slow variables of forest development. The original purpose in separating slow from fast variables was to allow us temporarily to treat the slow ones as parameters. The definition of 'fast' is synonymous with the assumption that for any (R, Q) the value of B will converge rapidly to one of the stable equilibria, either B_- or B_+.

The dynamics of the system can be visualized in Fig. 1 by imagining that initially $B = B_-$ and R is low. R is then slowly increased while keeping Q fixed. That is, the straight line is rotated clockwise about its right-hand intercept. The values of B_- and B_c will converge in an accelerating manner while B_+ increases only slightly. At the value of R where B_- and B_c coincide, the lower equilibrium is lost and the next increase in R sends the insect density to B_+. If we now reverse the path of R, the level of B_+ will decrease very slowly, even beyond the time where B_- and B_c are recreated. It is only when R assumes even lower values that B_c and B_+ coincide and the upper equilibrium is lost. Very similar geometric arguments would illuminate the effects of changing Q.

It is clear that all the action occurs when the intermediate, unstable equilibrium, B_c, coalesces with either the upper or the lower equilibrium. When this happens the density may either jump from a low value to a high one or the reverse. This behaviour is similar to the sudden outbreaks and collapse that characterize the spruce budworm populations. The equilibria B_- and B_+ correspond to budworm limitation by predators and food, respectively. As forest conditions improve, budworm growth exceeds the control by predators and an outbreak occurs. On the other hand, if the forest is destroyed far enough, the predators can again regain control. Note that the conditions under which upward and downward jumps occur may be quite different since the critical combinations of R and Q may be widely separated.

The critical values of R and Q are where two roots of eqn (9) coalesce and disappear. This translates into the two critical curves of Fig. 2. (The details of the calculation are shown in the Appendix.) The upper curve is when B_c and B_- join and the lower equilibrium is lost, and the lower curve is for $B_c = B_+$, which eliminates the upper equilibrium.

The point where both critical curves meet is that unique combination of R and Q where $B_- = B_c = B_+$.

The two critical curves define a critical region, inside of which there are three equilibria—two stable separated by one unstable. Above this region there is one (high) equilibrium and below it there is one (low). The critical region can be thought of as two overlapping surfaces of stable

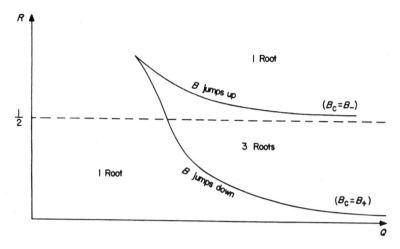

FIGURE 2: *The location of qualitative behaviour points in terms of the parameters R and Q. Regions with one root have one stable equilibrium. The region with three roots has two stable and one unstable equilibrium (as in Fig. 1). The critical curves separating these regions locate conditions where the budworm density changes radically.*

equilibria. An R-value moving upward with $B = B_-$ must pass completely through the critical region before the upward jump occurs. It must then return completely through, past the lower curve, before B collapses.

The type of phenomena we have presented readily fits into the arena of Catastrophe Theory (Thom 1975). The application of this theory to dynamical systems can be found in Zeeman (1972, 1976) and Jones (1975). The particular case of two parameters and one fast variable has been given the name of a 'cusp catastrophe'. In fact a cusp appears in Fig. 2 where the two critical curves join. The important generality provided by that body of theory is that we may use all the lessons learned from other 'cusp catastrophies' in our current case. Thom's theory says that, at the appropriate qualitative level, all such systems are the same. The equivalence, seated in deep mathematical theorems, is in harmony with our opening assertion that the exact form of our equations was not important so long as they met certain biologically necessary, qualitative criteria regarding their shape.

Step 3: Decide upon the response of the slow variables when the fast variables are held fixed

In order to characterize the state of the balsam fir forest, one ought to keep track of the age or size distribution of the trees, their foliage quantity, and their physiological condition. However, periodic budworm outbreaks synchronize the development of the trees, and the age distribution may be replaced by a single variable S, which gives the average size of the trees. S will be identified with the total surface area of the branches in a stand. Similarly, the condition of the foliage and health of the trees will be summarized in a single variable E, which may be analogously identified with an 'energy reserve'.

Since the maximum value of surface area is bounded, we shall choose a logistic form for S,

$$\frac{dS}{dt} = r_s S \left(1 - \frac{S}{K_c} \times \frac{K_E}{E} \right) \tag{10}$$

which allows S to approach its upper limit, K_S. The additional factor K_E/E is inserted into eqn (10) because S does not inevitably increase under conditions of stress; surface area may decline through the death of branches or even whole trees. However, during endemic times E will be close to its maximum value K_E and S will grow to its maximum, K_S.

We assume the energy reserve E also satisfies an equation of the logistic type:

$$\frac{dE}{dt} = r_E E \left(1 - \frac{E}{K_E} \right) - P \frac{B}{S} . \tag{11}$$

If B is small, then E will approach its maximum K_E. The second term on the right-hand side of eqn (11) describes the stress exerted on the trees by the budworm's consumption of foliage. This stress is proportional to B/S. Since B has units of number per acre and S has units of branch surface area per acre, B/S is the number of budworms per branch. This is the natural density measure for the feeding process. The factor P may be regarded as constant for our present purposes.

Step 4: Analyse the long-term behaviour of the slow variables, with the fast variables held at their corresponding equilibria

The isoclines for the systems (10), (11), are obtained by setting their left-hand sides equal to zero. Thus $dS/dt = 0$ if

$$S = \frac{K_S}{K_E} E, \quad \text{or if } S = 0$$

(12a)

and $dE/dt = 0$ if

$$S = \frac{PB}{r_E E \left(1 - \dfrac{E}{K_E}\right)} = \frac{PBK_E}{r_E} \times \frac{1}{E(K_E - E)} \, .$$

(12b)

These curves are sketched in Fig. 3, for the case when B, and therefore PBK_E/r_E, is small and there are two equilibria for S and E at C and D. The point C is a saddle point, and hence unstable. There is a single pair of trajectories which reach C and which form a separatrix (heavy arrows). If E and S start out to the right of the separatrix, then (E, S) will approach D as $t \to \infty$. If E and S start out to the left of the separatrix, they move off into the direction of $E = \infty$. While this is not realistic, it is a consequence of the form of eqns (10) and (11). In full completeness we expect E to be limited by $E = 0$ and the origin of Fig. 3 would be a stable equilibrium. In a later section we shall patch up eqns (10) and (11) for low values of E and S. For the time being it is enough to know whether the point heads towards point D or toward the left. In the latter case we argue on logical grounds that it will eventually reach the origin. Even then we could conceptualize a source term that would regenerate a new forest after the old had collapsed. This circumstance is beyond the time frame of our present model.

Now if B increases, then the U-shaped isocline will move up, and the points C and D will approach each other, and the region to the left of the separatrix will take up more and more of the (E,S) plane. Finally, the U-shaped curve can move entirely above the straight S isocline, as shown in Fig. 4.

Now every trajectory converges to the left, presumably to the origin.

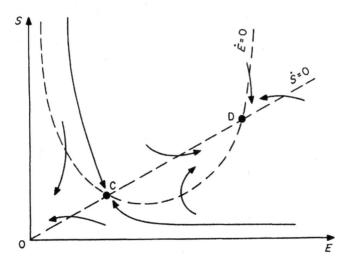

FIGURE 3: *The plane of the forest variables* S *and* E. *When there are few budworm the isoclines for no change in* S *and* E *intersect at the two points* C *and* D.

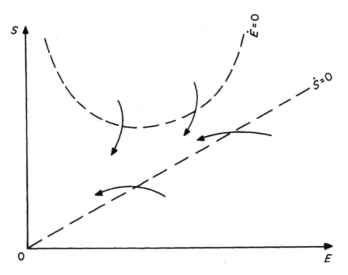

FIGURE 4: *The plane of the forest variables* S *and* E. *When there are many budworm the isoclines for no change in* S *and* E *separate and all trajectories head toward low* E *values.*

Step 5: Combine the preceding results to describe the behaviour of the complete system and identify the needs for additional coupling of the equations

Figures 2 and 3 imply the possibility of periods during which budworm populations are low and stable (at B_-). This condition holds if such populations start at low values and the system resides below the upper critical curve of Fig. 2. S and E will increase to the equilibrium condition at point D, noted in Fig. 3. The budworm will be limited primarily by predation.

However, if the upper branch of the critical curve of Fig. 2 is crossed, budworm populations will begin to increase on a fast time scale towards B_+. If this happens, Fig. 3 must give way to Fig. 4. Given a complete separation of the two isoclines as shown in Fig. 4, both E and S will decrease. If, as a consequence, the lower critical curve (Fig. 2) is crossed, there will be a rapid decline of budworm density. The particular pattern described above represents one complete outbreak/decline cycle of the budworm/balsam interaction.

A number of other possibilities also exist, however. For example, if budworm populations are partially controlled, the two isoclines of Fig. 3 might never separate to the extent shown in Fig. 4. In such a case, the complete system would reach an equilibrium, with B, S, and E all at relatively high values. This corresponds to the phenomenon of 'perpetual outbreak' that has been observed in New Brunswick as a consequence of insecticide spraying. Conditions under which this may occur will be given below.

Moreover, the above patterns are only generated if the critical curves are crossed in a particular direction. That depends on the movement across the (R, Q) plane of Fig. 2. In the present form R and Q seem to be constants. As a first departure we expect that Q would increase as the forest grew. Movement in the R direction would seem to be possible if an external driving variable changed one of its component parameters. For example, weather might increase r_B, the instantaneous rate of increase of budworm, enough to carry R across the upper curve in Fig. 2.

However, for sake of clarity, and our step-by-step format, we have left the issue of careful coupling of these equations to this point.

Let us first determine whether the budworm eqn (3) has any terms that should be expressed as functions of the slow variables. In order to do this, we refine our interpretation of B and S to represent quantities per acre of forest. Since the amount of foliage available per acre is roughly proportional to S, K_B should be proportional to S:

$$K_B = K'S. \qquad (13)$$

Thus K_B measures carrying capacity in larvae per acre, while K' measures carrying capacity in larvae per unit of branch area.

Similarly, terms in the predation rate (eqn 2) are also dependent upon the branch surface area. Predators, such as birds, search units of foliage, not acres of forest, so that the relevant density is larvae per unit of surface area. Thus the half-saturation density for B is also proportional to S:

$$\alpha = \alpha' S. \qquad (14)$$

The new parameter α' is measured in larvae per unit of branch area.

If eqns (13) and (14) are substituted into eqn (8), the result is

$$R = \frac{\alpha' r_B}{\beta} S, Q = \frac{K'}{\alpha'}. \qquad (15)$$

Note that Q is independent of S, while R is proportional to S. When the forest is young, R will be small, but Q may be quite large. Thus R and Q will be below the critical region in Fig. 2. Budworm densities will be low, not only in larvae per acre, but in larvae per branch. The densities per branch will be low because the predators will find it easy to search the small number of branches per acre. As the forest grows, control by predators becomes more uncertain, because of satiation of their appetites. Finally, the upper critical curve in Fig. 2 will be crossed and control by predators becomes ineffective as B rapidly increases to a high level (B_+) being limited now by food.

The effect of a rapid increase in B may be to change the dynamics of the slow variables from that depicted in Fig. 3 to that in Fig. 4. If this happens, the budworm outbreak will lead to a collapse of the forest. From eqns (3) and (13), we see that B/S is close to K' ($B \approx K_B$) during a budworm outbreak. Can the forest reach an equilibrium in that case? Equation (12b) may be rewritten as

$$\frac{E}{K_E}\left(1 - \frac{E}{K_E}\right) = \frac{P}{r_E K_E S}\frac{B}{S} \sim \frac{PK'}{r_E K_E}. \qquad (16)$$

Since the left-hand side of eqn (16) is dimensionless, we denote the right-hand side as a dimensionless parameter M:

$$M = \frac{PK'}{r_E K_E}.$$

(17)

M is easily interpreted as the ratio of a rate of energy consumption by budworm (by eating foliage) to the rate of energy production by the trees. The maximum possible value of the left-hand side of eqn (16) is 1/4. We conclude that no equilibrium is possible if M > 1/4. The preceding analysis indicates that an equilibrium is likely if M < 1/4.

Now, assuming that M > 1/4, a budworm outbreak must lead to a decline of the forest. Since R depends upon S, eventually the lower branch of the critical curve in Fig. 2 will be crossed, and the budworm population must collapse. However, it is not clear whether the budworm collapse will occur in time to save the forest from complete destruction. On the other hand, if the budworm density begins collapsing too soon, the effect on the forest may be reduced enough to establish a stable equilibrium for all three variables. In fact, the numerical values of the parameters in our equations will determine which of these behaviours will occur.

When realistic parameter values are substituted into our eqns (3), (10) and (11), budworm outbreaks lead to collapse of the forest. That is, E and S decline sharply during a budworm outbreak, and eventually E becomes negative. Negative values of E are unrealistic, and our model is not valid when extensive tree deaths occur. This situation can be remedied in two ways. The first (simplest) way is to recognize that the system eqns (3), (10) and (11) represents a cohort of trees and its resident budworm population. This cohort is only capable of going through one severe decline. If this were to happen, then the model must be started again with a small value of S and with E near K_E in order to generate the next outbreak. Regeneration from one cohort of trees to the next is a discontinuous process, and one might as well represent this fact explicitly in the model.

On the other hand, for mathematical convenience, it might be desirable to devise a system of differential equations which adjusts the behaviour of E and S for small values in order to simulate the growth of a new cohort. We have no need for such an adjustment in the present investigation, but we shall indicate how it might be carried out.

According to eqn (13), the carrying capacity for the budworm is independent of E, i.e. the trees put forth the same amount of foliage, regardless of their physiological condition. It would seem more reasonable that K' should depend upon E, and that K' should decline rather sharply if E falls below a certain threshold T_E. Therefore, we may replace eqn (13) by

$$K_B = K'S \frac{E^2}{T_E^2 + E^2} .$$
$$(18)$$

If T_E is small compared with K_E, then K_B will show a sharp decline near $E = T_E$. A corresponding change should also be made in eqn (11). We may set

$$K_B = K'S \frac{E^2}{T_E^2 + E^2}$$
$$(19)$$

because the stress on the trees is related to the amount of foliage consumed.

The resulting equations are as follows :

$$\frac{dB}{dt} = r_B B \left(1 - \frac{B}{K'S} \cdot \frac{T_E^2 + E^2}{E^2} \right) - \beta \frac{B^2}{(\alpha'S)^2 + B^2}$$
$$(20)$$

$$\frac{dS}{dt} = r_s S \left(1 - \frac{SK_E}{EK_S} \right)$$
$$(21)$$

$$\frac{dE}{dt} = r_E E \left(1 - \frac{E}{K_E} \right) - P' \frac{B}{S} \frac{E^2}{T_E^2 + E^2} .$$
$$(22)$$

This system appears to be much more complicated than eqns (3), (10) and (11), but its qualitative behaviour will be exactly the same, except for small values of E. If the parameter T_E is chosen properly, the system eqns (20)–(22) will exhibit a regeneration similar to that obtained by restarting eqns (3), (10) and (11). However, we shall use eqns (3), (10) and (11) in the sequel.

This is as far as we can go if we restrict ourselves exclusively to qualitative information. Some effort is now needed to quantify parameters in order to define, more precisely, the behaviour of the system. Since we are interested in determining how far one can predict with different levels of information, it is useful to identify two levels of quantitative information to add to the first qualitative level—one very general and based on estimates by experienced field naturalists and one more detailed and specific.

Level II: General Quantitative Information

In order to complete our model, we must estimate its parameters. Especially important are the combinations of parameters which form Q, R and M, eqns (15) and (17). These will determine the qualitative behaviour of the system. Most of the other parameters are rate constants which determine the speed with which certain processes occur, but do not alter the basic qualitative picture.

The parameter K' in eqn (13) measures the carrying capacity of the forest in larvae per branch. An entomologist with cursory knowledge of budworm can confidently state that from 100 to 300 larvae can be supported by an average branch of balsam foliage in good condition. The parameter α' is likely to be low. Knowing, roughly, the speed of movement and distance of perception of birds for insect prey, α' can be estimated as one to two larvae per branch. [This particular analysis has been expanded by Holling, Jones & Clark (in preparation).] These ranges then permit a calculation of the likely range for Q from eqn (15). The results range from Q = 50 to 300, and strongly suggest that the system resides in or below the critical region (Fig. 2) during the endemic phase.

But still, outbreaks will only occur if R increases above 1/2, and crosses the upper critical curve. Again, rough estimates of the elements determining the value of R (eqn [15]) can be obtained as follows:

r_B The budworm is capable of a five-fold increase in density per
    year. Since we are using a continuous time model, we set
    $e^{rB} = 5$, and conclude that $r_B = 1\cdot6$/year.

β The value of β has been estimated in the literature using the
    most general information on size of birds, their maximum
    daily consumption, the proportion of budworm in their diet,
    and their ranges of densities (Kendeigh 1947; George &
    Mitchell 1948; Mitchell 1952; Dowden, Jaynes & Carolin 1953;
    Morris et al. 1958). These estimates of the maximum consump-
    tion range from 20 000 to 36 000 larvae per acre per year.

S   A maximum value for S is K_S.

K_S A fully recovered forest contains about 24 000 average
   branches per acre.[*]

If the preceding estimates are combined (eqn [15]), we find that R, for
a mature forest, lies between 1·1 and 3·8. Because R need only be 0·5 we
have considerable leeway in the value of S below K_S that will initiate an
outbreak. A forest is fully mature at age 80, while outbreaks have a period
of about 40 years. Actually, numerical results show that an outbreak is
not immediate when the upper branch of the cusped curve is crossed. It
requires a number of years for the budworm to show an appreciable rise
in density. Although predation cannot control the budworm when
R > 0·5, the predation does appreciably slow the rate of growth. Hence, a
value of 1·1 for R at S = K_S is not unreasonable.

Now we turn to the estimation of M. As mentioned above, M is a ratio
of energy consumed by budworm to energy produced by trees. The critical value for M is 0·25 as trees collapse if M > 0·25. The time required for
such a collapse will depend upon M. A forest can withstand defoliation
for approximately 4 years, which implies that Mr_E must be approximately
0·3, since B does not in fact reach the value K'S as assumed in the derivation of M (eqn [17]).

Some rates which should be estimated are r_S and r_E. The time of regeneration of the forest after an outbreak depends on r_S. It can be adjusted to
make the period between outbreaks approximately 40 years. A value of
r_S = 0·15/year gives satisfactory results. Likewise, r_E sets the rate at which
trees recover from the stress of defoliation. Since this recovery is fairly
rapid, a value of 1/year is assumed for r_E. Since E is a synthetic variable we
can set its maximum value as K_E = 1.

The only remaining parameter is P, the rate of energy consumption by
budworm. From eqn (17)

$$P = \frac{Mr_E K_E}{K'} \approx 1·5 \times 10^{-3} \quad .$$

---

[*] The standard field measure for an 'average branch' is one that can be circumscribed by
a polygon of 10 ft² area.

TABLE 1 : *Parameter values for Level II and Level III information*

Symbol	Description	Units	Level II	Level III
r_B	intrinsic budworm growth rate	/year	1·6	1·52
K'	maximum budworm density	larvae/branch	100–300	355
β	maximum budworm predated	larvae/acre/year	20000–36000	43200
α'	½ maximum density for predation	larvae/branch	1–2	1·11
r_S	intrinsic branch growth rate	/year	0·15	0·095
K_S	maximum branch density	branches/acre	24000	25440
K_E	maximum E level	—	1·0	1·0
r_E	intrinsic E growth rate	/year	1·0	0·92
P'	consumption rate of E	/larvae	0·0015	0·00195
R	$\alpha' r_B S/\beta$	—	1·07–3·84	0·994 (S/K_S)
Q	K'/α'	—	50–300	302
M	$PK'/r_E K_E$	—	0·15–0·45	0·71

As an independent measure, it is known that 150 to 200 larvae per branch can consume approximately 25% of the foliage. Therefore

$$P = \frac{0·25}{150} = 1·7 \times 10^{-3} \ .$$

All the parameters for eqn (20) through eqn (22) are summarized in Table 1.

Level III: Empirical Quantitative Information

Level I, the development of the model, utilized only qualitative information about the system's behaviour. In Level II we made a first attempt at estimating parameter values, but restricted ourselves to general quantitative information. This is the type of information that an experienced biologist might provide without specifically examining the New Brunswick budworm.

In Level III we examine the field data that have been collected over many decades and determine the best values for the parameters as we have defined them. This step is made easier in this particular instance because much of the work has already been done in connection with the construction of a detailed simulation model of this system. That simulator is a central element in a program of ecological policy design—a program to synthesize the methodologies and concepts of systems ecology and modelling, optimization, and decision theory in a case study framework. A review of that project and some of the lessons learned can be found in Holling et al. (1976).

The primary source of data for the simulator was Morris (1963), with considerable additional expert opinion from the personnel and files of Environment Canada's Maritimes Forest Research Centre. The simulator mimics the univoltine character of the insect as a difference model with yearly time steps. As a result, its parameters are not all appropriate for a continuous model formulation without some adjustments. These adjustments could be made in the original data, but for convenience we choose to let the simulation serve as a surrogate for the real world, and we consult it for measures that are analogous to the parameters needed for our model—eqns (3), (10) and (11). The errors and discrepancies generated by going through this 'middle-man' are on the same order as those when we assume the simple continuous form that we have.

Specific details about the budworm simulation model can be found in Jones (1976). We now briefly check off the parameter values suggested by that reference and indicate the discrepancies with Level II values. First, consider the intrinsic role of growth, r_B. We find in the simulation that the maximum growth rate between generations in a mature forest with low budworm densities is 4·56. Thus, $r_B = \ln (4\cdot56) = 1\cdot52/\text{year}$, in close agreement with that found above.

The hypothetical carrying capacity per branch K' is the most difficult to interpret. The simulation has a comparable equilibrium at $K' = 335$ larvae/branch. However, numerical experience shows that there can be a transient overshoot to values of 600 or more. This wide range is a consequence of the discrete nature of the insect population. We adopt K' as 335 because of the conceptual parallel of that value, but note that the continuous model will not overshoot to the high values seen in the simulation and in nature.

The parameters of bird predation are taken from the data summarized in Holling, Jones & Clark (in preparation). That paper specifically identifies three groups of insectivorous birds that represent three distinct size classes and, to some degree, three different modes of search. This more detailed, but still qualitative, analysis of field data identifies an expected value for β of 43,200 larvae/acre/year. This maximum consumption level is significantly higher than that found in the literature and reported in Level II. That literature only considered one class of birds—the arboreal feeders (e.g. warblers). Two other classes of birds were not previously recognized as important because they are normally ground nesters (e.g. juncos and grosbeaks) and because their numbers do not increase during an outbreak. However, their large size and appetite make them at least equally as important as the smaller species and β is increased accordingly. Because the density of budworm which produces half-saturation of predation is different for each bird class we take an average for each, weighted by their contribution to the total predation. Thus, $\alpha' = 1 \cdot 11$ larvae/branch.

The parameters for tree growth are taken by fitting eqn (10) (with $E = K_E$) to a typical history of branch surface area following the collapse of an outbreak. This gives a growth rate of $r_S = 0 \cdot 095$/year and an asymptotic level of $K_S = 25 \cdot 4 \times 10^3$ branches/acre.

It was recognized early in the simulation development that something analogous to an 'energy reserve' was affecting the response dynamics of trees. However, there were insufficient data to incorporate this process adequately. The solution taken was to deputize foliage for this function, and we continue that here. As E is an intensive factor we lose no generality by defining $K_E = 1$. The value of r_E is evaluated from the rate of increase in foliage. The maximum that foliage can increase in the simulation is 1·26-fold per iteration. Thus exp $r_E = 1 \cdot 26$ gives $r_E = 0 \cdot 23$ yr^{-1}. This maximum occurs when foliage is about half its maximum; and so, as a consequence of the logistic form, $r_E = 4 \times 0 \cdot 23 = 0 \cdot 92$/year.

P is the maximum rate that an individual feeds on 'E', which is $P = 1 \cdot 95 \times 10^{-3}$/larvae.

Using the above values, the three aggregate parameters R, Q and M assume the following values. First

$$R = 1{\cdot}72(S/K_S)$$
$$Q = 302$$

(or $Q = 540$ if $K' = 600$).

Finally,

$$M = \frac{PK'}{r_E K_E} = 0{\cdot}71$$

(or $M = 1{\cdot}27$, if $K' = 600$).

The parameter values for Level III are also summarized in Table 1. There is remarkable agreement between the values found from extensive and intensive collection and analysis of field data and those estimated from a first field estimation.

One can go a lot further than traditionally assumed with informed, but qualitative, insight into ecological process. Extensive data collection efforts need not always be carried out and completed before the system is abstracted into an analysable model.

We have been emphasizing that the central and important aspect of this analysis is a process and not a product. The actual numerical integration of our model using the estimated parameters is a final, though anticlimactic, step to be performed for completeness. Because the parameter values from Level II and Level III information are so similar we adopt only Level III and use these values in our model (eqns [3], [10] and [11]). The integrated time course is shown in Fig. 5 through one outbreak cycle (ending in year 43). Fig. 6 shows a typical outbreak cycle exhibited by the simulation model.

The qualitative behaviour is similar, as this analysis predicted. The major difference in the appearance is between the graphs of surface area S. This is an expected discrepancy resulting from our attempt to mimic a 75 age class model of tree population with a single state variable. Other discrepancies between Figs 5 and 6 were also anticipated as they result from fundamental differences between discrete and continuous models. First, the maximum level of S is lower in Fig. 5 than in 6. Equivalently, the outbreak is triggered at a lower 'threshold' of S in Fig. 5. In the simulation model the process of competition between birds (not included in our differential equations) enhances their effectiveness in young forests

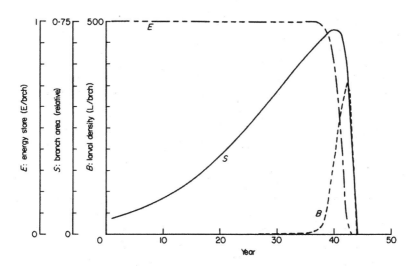

FIGURE 5: *A numerical solution of the differential equations using the parameter values of Level III information.*

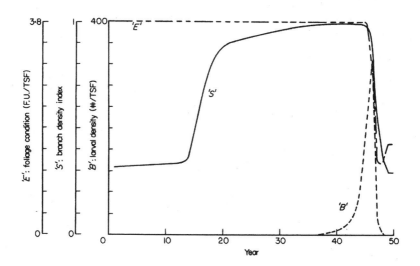

FIGURE 6: *A typical outbreak cycle from the detailed simulation model. The variables plotted are the ones most analogous to the differential equation model: 'E' is the total foliage available per unit of branch surface area in arbitrary 'foliage units'/10 ft² (F.U./TSF); 'S' is the density of surface area in TSF/acre scaled such that 'S' = 1 is equivalent to S = 0·94 in Fig. 5; 'B' is the density of third instar larvae in number 1 TSF.*

(low S). Thus S must reach a higher value in the simulation model before the control by predators is overcome (Holling, Jones & Clark, in preparation). Secondly, the differential equations as conceptualized are capable of only one outbreak while the simulation model output is periodic because the regeneration of new trees is explicitly modelled.

The differential equations could be adjusted, expanded, and otherwise 'tuned' so that their time traces more closely matched the simulation model, but we elect not to follow that procedure.

Discussion

An obvious requirement for the success of this procedure is a good understanding of the basic phenomena, to know which are the most important variables, and to know the main features of their interactions. Such knowledge is never completely available at the beginning of an investigation. In the case of the budworm we had behind us extensive field investigations and simulation experience. The present methods have been exceedingly useful when applied in conjunction with the simulation. One can narrow the reasonable parameter ranges using these procedures. Perhaps most important, the analytic model is likely to extend our understanding of the phenomena, since the full armory of mathematical techniques is available (Ludwig, Weinberger & Aronson, in preparation). This raises a final question about the level of mathematical training or ability required to carry out such a program. In principle, the methods which were applied to the budworm do not go beyond first year calculus; however, their effective use requires considerable mathematical confidence.

Acknowledgments

The authors are indebted to Dr M. Levandowsky for much advice and encouragement. Research was supported in part by NRC Grant No A9239.

References*

Bazykin, A. D. (1974). Volterra's system and the Michaelis–Menten equation (in Russian). *Problems in Mathematical Genetics* (Ed. by V. A. Ratner), pp. 103–42. U.S.S.R. Academy of Sciences, Novosibirsk. (Available in English as *Structural and Dynamic Stability of Model Predator–Prey Systems*. 1975. Institute of Resource Ecology, Report R-3-R.)

Dowden, P. B., Jaynes, H. A. & Carolin, V. M. (1953). The role of birds in a spruce budworm outbreak in Maine. *Journal of Economic Entomology*, 46, 307–12.

George, J. L. & Mitchell, R. T. (1948). Calculations on the extent of spruce budworm control by insectivorous birds. Journal of Forestry, 46, 454–5.

Holling, C. S. (1959). The components of predation as revealed by a study of small mammal predation on the European pine sawfly. *Canadian Entomologist*, 91, 293–320.

Holling, C. S. (1973). Resilience and stability of ecological systems. *Annual Review of Ecology and Systematics*, 4, 1–23.

Jones, D. S. (1975). The application of catastrophe theory to ecological systems. International Institute for Applied Systems Analysis Report RR-75-15. (To appear in 1977 in *Simulation in Systems Ecology* (Ed. by G. S. Innis), Simulation Councils Proceedings.)

Jones, D. D. (1976). *The Budworm Site Model*. Institute of Resource Ecology, Working Paper W-13.

Kendeigh, S. C. (1947). Bird population studies in the coniferous forest biome during a spruce budworm outbreak. *Biological Bulletin*, 1, Division of Research, Ontario Department of Lands and Forests.

Morris, R. F. (Ed.) (1963). *The Dynamics of Epidemic Spruce Budworm Populations*. Memoirs of the Entomological Society of Canada, No. 21. 332 pp.

Morris, R. F., Cheshire, W. F., Miller, C. A. & Mott, D. G. (1958). The numerical response of avian and mammalian predators during a gradation of the spruce budworm. *Ecology*, 34, 487–94.

Mitchell, R. T. (1952). Consumption of spruce budworm by birds in a Maine spruce-fir forest. *Journal of Forestry*, 50, 387–9.

Thom, R. (1970). Topological models in biology. *Towards a Theoretical Biology* (Ed. by C. H. Waddingtons). 3: Drafts. Edinburgh University Press.

Thom, R. (1975). *Structural Stability and Morphogenesis*. (English translation by D. H. Fowler). Benjamin, Reading, Mass.

---

*In an effort to make the bibliography as current as possible, we have cited many works available at present only as publications of the International Institute for Applied Systems Analysis (IIASA) and our own Institute of Resource Ecology (IRE). These may be obtained from the following addresses: Documents and Publications, International Institute for Applied Systems Analysis, Schloss Laxenburg A-2361, Laxenburg, Austria and Publications (att'n: Ralf Yorque), Institute of Resource Ecology, University of British Columbia, Vancouver, British Columbia, Canada V6T 1W5.

Zeeman, E. C. (1972). Differential equations for the heartbeat and nerve impulse. *Towards a Theoretical Biology* (Ed. by C. H. Waddington), 4: Essays. Edinburgh University Press.

Zeeman, E. C. (1976). Catastrophe theory. *Scientific American*, 234, 65–83.

(Received 13 June 1977)

Appendix

The double roots of equation (9)

Equation (9) has a double root if the straight line given by the left-hand side of eqn (9) is tangent to the predation term of the right side. We adopt the following expressions for the two sides of eqn (9):

$$f(\mu) = R\left(1 - \frac{\mu}{Q}\right) \tag{A1}$$

$$g(u) = \frac{\mu}{1 + \mu^2} .$$

A double root occurs if

$$f(\mu) = g(\mu)$$

and

$$\frac{df}{d\mu} = \frac{dg}{d\mu} . \tag{A2}$$

We treat eqn (A2) as a pair of simultaneous equations and solve for R and Q parametrically with respect to μ. These relationships can be written as

$$\frac{R}{Q}(Q - \mu) = \frac{\mu}{1 + \mu^2}$$

and

$$\frac{R}{Q} = \frac{\mu^2 - 1}{(1 + \mu^2)^2} .$$

It follows after some algebra that

$$R = \frac{2\mu^3}{(1 + \mu^2)^2} \tag{A3}$$

and

$$Q = \frac{2\mu^3}{(\mu^2 - 1)} \cdot \qquad \text{(A4)}$$

The cusp point of Fig. 2 where both critical curves meet (or begin) is the point where the derivatives of R and Q with respect to μ both vanish. This corresponds to the inflection point of $g(\mu)$, which occurs at $\mu = \sqrt{3} = 1 \cdot 73$. This value in eqn (A3) and (A4) gives the cusp point for (R, Q) as

$$R_c = \frac{3^{3/2}}{8} = 0.650 \qquad \text{(A5)}$$

$$Q_C = 3^{3/2} = 5.196 .$$

Further, the axis of the cusp is oriented as

$$\frac{dR}{dQ} = -1/16$$

Equations (A3) and (A4) also generate two limiting conditions. When $\mu \to \infty$, $Q \to \infty$ and $R \to 0$. As $\mu \to 1$, $Q \to \infty$, but $R \to 1/2$. Figure 2 is based upon this information.

The Evolution of an Idea—
the Past, Present, and Future
of Ecological Resilience

LANCE H. GUNDERSON, C. S. HOLLING, AND CRAIG R. ALLEN

LANCE H. GUNDERSON, Dept. of Environmental Studies, Emory University, Atlanta, GA, USA.

C. S. HOLLING, Montclair Drive, Nanaimo, BC, Canada.

CRAIG R. ALLEN, Nebraska Cooperative Fish & Wildlife Research Unit, University of Nebraska, Lincoln, NE, USA.

ECOLOGICAL RESILIENCE is a bundle of concepts or ideas developed to explain complex system dynamics (Holling 1973, Folke et al. 2002, Folke et al. 2004). Walker and Salt (2006) use the phrase "Resilience Thinking" to capture a coherent set of notions that together produce a framework for conceptualizing and explaining how systems of humans and nature behave. The resilience framework is based on observations of thresholds, abrupt or nonlinear shifts of key variables, domains or basins of attraction, and multi-stable states that characterize complex system behaviors. Resilience theory is an alternative perspective to the equilibrium-centered theories and models that guide management actions in many resource systems.

In this chapter, we attempt a synthesis of these ideas from their inception to their recent manifestations. We begin with a historical review, indicating how these ideas have evolved to date. The next section discusses the current views and applications of the resilience framework, both in theory and practice across a range of disciplines. We end with a brief set of suggestions about interesting future questions to explore. We begin with a review of how these theories have evolved, as reflected by the articles in this volume.

Theoretical Developments

Since the introduction of the idea of resilience to the ecological literature by Holling in 1973, we believe three major theoretical advances have occurred around the concepts embedded in resilience. First is the expansion and definition of the word *resilience*, which has been interpreted in many different ways. Second is the adaptive cycle, a heuristic of generic phases of ecological change, in which the property of resilience is modified through time by internal dynamics. The third development is the idea of panarchy, which involves the notion of scale and ecosystem changes.

Resilience

As noted in the introduction to this book, the word *resilience* can be traced to the Latin word *resilire*, which translates to "leap back" (Skeat 1882). Perhaps this origin has led to the interpretation of the word as the capacity to rebound or recover after a shock or event. Ecologists, including Pimm (1991) and DeAngelis (1980), among others, use the term *resilience* to describe the amount of time needed to recover following an external force or perturbation. Holling (1996) distinguished two types of resilience that have been applied by ecologists: engineering resilience and ecological resilience.

Engineering resilience is the time to recovery—how long an ecosystem takes to recover following a disturbance. In population ecology, this idea is captured by the parameter r, the intrinsic rate of population increase. Ecological resilience was first introduced by Holling (1973) to describe two different aspects of change in an ecosystem over time.

Holling's first characteristic of resilience involved the "persistence of relationships within a system" and the "ability of systems to absorb changes of state variables, driving variables and parameters, and still persist." The second defining characteristic described resilience as "the size of a stability domain or the amount of disturbance a system could take before it shifted into alternative configuration" (Holling 1973). These two views of resilience are not incompatible, with the major difference being whether or not the system of interest returns to a prior state or reconfigures into something very different. In the following subsection, one of us (Holling) recounts the origin of the idea, in reflecting on the seminal 1973 annual review paper.

Origins of Ecological Resilience (by C. S. Holling)
The idea started by blending two lines of prior research: predation and population dynamics. Nonlinear forms of the functional responses (e.g., the Type 3 S-shaped response; Holling 1959) and of reproduction responses (e.g., the Allee effect) interacted to create two stable equilibria for interacting populations, with an enclosed stability domain around one of them. It was the responses at low densities—that is, where vertebrate predators have yet to learn to locate the prey easily and where mates are too scarce to find each other easily—that were critical. Once discovered, it seemed obvious that conditions for multi-stable states were inevitable. That situation, being inevitable, suggested there were huge consequences for theory and for practice.

Up to that time, a concentration on a single equilibrium and assumptions of global stability had made ecology, as well as economics, focus on near equilibrium behavior and on fixed carrying capacity with a goal of minimizing variability. Command and control was the policy for managing fish, fowl, trees, and herds, and freedom was unlimited to provide opportunity for people.

The multi-stable state reality, in contrast, opened an entirely different direction that focused on behavior far from equilibrium and on stability boundaries. High variability, not low variability, became an attribute necessary to maintain existence and learning. Surprise and inherent unpredictability was the inevitable consequence for ecological systems. Data and understanding at low densities, rare because they are all the more difficult to obtain, were more important than those at high density. I used the word *resilience* to represent this latter kind of stability.

Hence, the useful measure of resilience was the size of stability domains or, more meaningfully, the amount of disturbance a system can take before its controls shift to another set of variables and relationships that dominate another stability region. And the relevant focus is not on constancy but on variability, not on statistically easy collection and analysis of data but on statistically difficult and unfamiliar ones. That needs a different eye to see and a different theory to perceive consequences.

About that time, I was invited to write a review article for the 1973 *Annual Review of Ecology and Systematics*. I decided to turn it into a review of the two different ways of perceiving stability and, in so doing, to highlight the significance for theory and for practice. That required finding additional, rare field data in the literature that demonstrated flips of populations from one level or state to another, as well as describing the recently discovered known nonlinearities in the processes that caused or inhibited the phenomenon. That was a big job, and I recall days when I thought it was all bunk and other days when I believed it was all real. I finished the paper on a "good" day, when all seemed pretty clear. By then, I guess I was convinced. The causal, process evidence was excellent, though the field evidence concerning population flips was only suggestive. Nevertheless, the consequences for theory and management were enormous. It implied that uncertainty was inevitable and that ecosystems, in an evolutionary time span, were momentary entities pausing in a flip to different states. As I'll describe, it took about thirty years to confirm those conclusions for others.

The 1973 paper began to influence fields outside population and community ecology: anthropology, political science, and systems science first, and then, later, ecosystem science. It became the theoretical foundation for active adaptive ecosystem management. But it was largely ignored or opposed by practitioners in the central body of ecology. What followed was the typical and necessary skepticism released by new ideas, which will be briefly described here because it is such a common foundation for developing science.

One early ecological response to the paper was by Sousa and Connell (1985). They asked the good question "Was there empirical evidence for multi-stable states?" They attempted to answer this question by analyzing published data on time series of population changes of organisms to see if the variance suggested multi-stable behavior. They found no such evidence. This analysis reinforced the dominant population ecology single-equilibrium paradigm to such an extent that the resilience concept was stopped dead in that area of science.

It seemed to be an example of evidence that refuted this new theory. But their evidence was inappropriate and the theory was not. In fact, their evidence, as is often the case, was really a model—incomplete because the collators unconsciously used an inappropriate model for choosing data that were incomplete.

There were two problems with their analysis. The first was that they did not ask any process questions (such as "Are there common nonlinear mechanisms that can produce the behavior?"). That is where the good, new hard evidence that I had discovered existed. They rightly saw the need for long time series data on populations that had high resolution. The second problem was time frame. As population and community ecologists of tradition, their view of time was a human-centered perspective in which decades were seen as being long. This view was not only small in time but small in spatial scale—referred to as a quadrat mentality. Their theory was limited to linear interactions between individuals in single species populations or between two species populations, all functioning at the same speed (e.g., predator/prey, competitors). It represented the dangers caused by inferring that "microcosm" thought and experiments have anything to contribute to the multiscale functioning of ecosystems. Steve Carpenter has a perceptive critique of that tendency (Carpenter 1996).

Multi-stable behavior can be explained only by the use of models that include at least three but not more than five variables (Holling and Gunderson 2002). Moreover, these variables need to differ qualitatively in speed from one another. The model is, therefore, inherently ecosystemic. The slowest of these variables determines how many years of data are needed for their kind of test. None of their examples had anywhere near the duration of temporal data needed. The model of the spruce budworm is a good example. The available 45 years of budworm population changes they analyzed seemed long to Sousa and Connell and to all those conditioned by single-variable behavior and the linear thinking of the times. But the relevant time scale for the multi-equilibrium behavior of budworm is set by their hosts (the trees), or the slow variable. What was needed for their tests was yearly budworm data (the fast variable) over several generations of trees (the slow variable) — that is, perhaps one and a half centuries, not 45 years. The normal boom-and-bust cycle is 40 to 60 years.

It has since taken 35 years of study of different ecosystems to develop data for appropriate tests of multistate behavior. Examples include those using paleo-ecological data covering centuries at high resolution, the deep and shallow lake studies and experiments of Carpenter (Carpenter et al. 2001) in the United States and of Marten Scheffer in Europe (Scheffer 1998), and the experimental manipulations of mammalian predator and prey systems in Australia and Africa by Tony Sinclair (Sinclair et al. 1990).

Ecologists who work in disturbance-driven ecosystems found that ecological resilience was a more applicable concept to the complex changes they were observing. These scientists observed qualitative changes in both the structure and function of ecosystems (Gunderson 2001, Scheffer et al. 2001, Folke et al. 2004) or the ecological regime or identity (Walker et al. 2004, Walker and Salt 2006).

Many examples are recorded. Walker (1981) and Dublin et al. (1990) found dramatic shifts between grass-dominated and shrub-dominated areas in semiarid rangelands that were mediated by interactions among herbivores, fires, and drought cycles. Scheffer and Carpenter (2003) described two alternative states (clear water with rooted aquatic vegetation, and turbid water with phytoplankton) in shallow lake systems. Gunderson (2001) described shifts in wetland vegetation as a result of changes in nutrient status and disturbances such as fire, drought, or frost. Coral reef system shifts between coral domination and macroalgae domination have been demonstrated (Hughes 1994, Nyström and Folke 2001, Bellwood et al. 2004). Many pathways have been documented for this phase transition, including overfishing and population decline of key grazing species, increase in nutrients, and shifts in recruitment patterns (Hughes et al. 2003). Estes and Duggin (1995) and Steneck et al. (2004) have shown how nearshore temperate marine systems shift between dominance by kelp and sea urchins as a function of the density of sea otters and other grazers. At even larger scales, the transition between the Sahara and Sahel has been described as regime shifts (Foley et al. 2003) and is driven by internal and external factors.

Carpenter's important summary paper (2001) makes the following point: multi-stable states are real and extremely important, although they are difficult to demonstrate. Surprise, uncertainty, and unpredictability are the inevitable result. Command-and-control management temporarily hides the costs, but the ultimate cost of surprises produced by managing systems that ignore multi-stable properties is too great. Active adaptive management is the only alternative management response possible. Steve Carpenter and Buz (W. A.) Brock (a great ecosystems scientist together with a wonderful "nonlinear" economist) show why in a classic paper in which a minimal model of a watershed, farming styles, and regional monitoring and regional decision making regarding phosphate control encounter the surprises created as a consequence of a multi-stable state (Carpenter et al. 1999).

Cycles of Change

Following a comparison across a number of ecosystems that had different histories of disturbance, and qualitatively different classes of dynamical change, Holling (1986) proposed an adaptive cycle (see figure 5 in Article 4 for a graphic depiction) as a metaphor to describe these observed in ecological systems. These ecological systems exhibit four distinct and usually sequential phases of change in the structures and function of a system when observed at specific scale ranges. In early stages of development (such as primary or secondary succession in ecosystems), ecosystems rapidly accumulate structure, as biomass and complexity. During this phase, competition is a scramble for resources, as winners are able to obtain and quickly convert raw materials to structure and organization. Over time, structure accumulates and the system becomes more diverse and more connections appear among the system components.

Gradually, net growth slows, as more of the acquired resources and energy are allocated to system maintenance rather than growth of new structure. Because structures and resources are accumulated and stored, this phase is the conservation phase. During the conservation phase, the system becomes increasingly connected, less flexible (more rigid), and more vulnerable to external disturbances. These phases also represent system maturation and increasing vulnerability to external variations or disturbances.

When forces external to the system stress or perturb the system, the system enters the next phase of the adaptive cycle: a period of creative destruction. This period is characterized by a release of accumulated capital or structure. This phase is also called the omega (or end) phase (Holling 1986, Holling and Gunderson 2002). Forest fires, pest out-breaks, harvesting, and hurricanes are all examples of the omega phase; they are relatively rapid periods of destruction or unraveling of previ-ously accumulated forms of capital. The destruction phase is quickly followed by a reorganization (also called the alpha or beginning) phase, where a new system emerges, leading to the growth phase of a new cycle. The new trajectory may be very similar to the previous trajectory, or it may be quite different.

The phases of creative destruction and reorganization are the phases of this cycle in which ecological resilience is expressed. The period of reorganization and renewal following a disturbance, such as fire, a drought, or even temperature variations, is when the system will flip into an alternative regime. That flip can be idiosyncratic and random, a result of what types of colonizers, invaders, entrepreneurs, or organizers establish and take hold for the next phases of growth and development. This is the period where novelty in the system is likely to emerge as new combinations of old and new elements (Allen and Holling 2008). It is also during these phases that other variables, especially slowly changing ones, can come into play. For example, in the Florida Everglades, the historic marsh vegetation has been subject to fires and droughts for thousands of years. Yet, when the soils are enriched with nutrients, a fire or drought leads to a change in vegetation, where cattails replace the native marshes (Gunderson 2001).

These first two phases correspond to system development, in which energy and resources go into building structure and connectivity, whether they are ecosystems (Holling 1986), cities (Elmqvist et al. 2004), ancient cultures (Redman and Kinzig 2003), or human organizations (Westley et al. 2002). This pattern of rapid then slowing growth, swift destruction, and reformation has been observed in many systems (Walker and Salt 2006). These include ecological examples, such as pest outbreaks and fires in temperate forests (Holling 1986), and social-ecological systems, such as water management history of the Everglades (Gunderson and Holling 2002) and aboriginal cultures (Berkes and Folke 2002, Berkes et al. 2003).

Similar state changes can be seen in organizations and management systems following perturbations, whether they are natural disasters or human-created instabilities, such as budget shortfalls, elections, or changes in personnel (Scheffer et al. 2003, Westley et al. 2002).

Panarchy

Panarchy (derived from the Greek god of nature, *Pan*, combined with *archy*, from the Greek for "rules"—hence, "rules of nature") is a term used to describe how variables at different scales interact to control the dynamics and trajectories of change in ecological and social-ecological systems (Gunderson et al., 1995, Holling 2001, Gunderson and Holling 2002). Panarchy is a theory suggesting that in ecological and other complex systems, abrupt changes occur as a result of the interaction of slow, broad variables with smaller, faster variables. Top-down control occurs when slow, broad features constrain and control the small, fast ones. For example, geology and soil types interact with climatic variables (temperature, photoperiod, rainfall) to determine the suite of plant and animal species that thrive at a given locality. In human systems, the types of climate and building materials dictate the types of structures that can be built for human dwelling. Much empirical evidence supports hierarchical or top-down controls.

Panarchy theory was proposed to suggest that both top-down and bottom-up interactions occur—that is, although top-down control does exist, many bottom-up or cascading phenomena occur. Many disturbance dynamics, such as forest fires or forest pest outbreaks, are not the result of top-down control by slower variables but are examples where faster, smaller variables appear to control the system for periods of time. A panarchy has three ingredients: (1) subsystems of adaptive cycles that represent system dynamics at a specific scale range, (2) dynamic systems that occur at different scale ranges, and (3) coupling of those systems across scales at key phases of the adaptive cycle. All of these structures are posited to change in phases described by the adaptive cycle, but each at a given scale.

Panarchy dynamics that link up scale (bottom up) have been named "revolt," suggesting that small events can cascade up to larger scales. When a level in the panarchy enters a phase of creative destruction and experiences a collapse, that collapse can cascade up to the next larger and slower level by triggering a crisis, particularly if that level is at a conservative phase where resilience is low. One example is in the dynamics of urban wildland fires, which is similar to fire in ecosystems. The lighting of a match, a strike of lightning, or a short circuit of an electrical circuit is a small, local phenomena. Under many conditions, the local fire is quickly extinguished. However, under certain conditions (such as extreme droughts or low humidity), local ignitions can create a small ground fire that spreads to the crown of a tree, then to a patch in the forest, and then to a whole stand of trees. Each step in that cascade moves the transformation to a larger and slower level. So, if not extinguished, fire can consume a house or similar structure and spread to other houses in a neighborhood. Such processes occurred in the late 1990s in central Florida and in 2003 and 2007 in southern California. Hence, part of the connotation of revolt is used to describe how fast and small events overwhelm slow and large ones, and how that effect can cascade to still higher slower levels if those levels have accumulated vulnerabilities and rigidities.

The word *remember* describes interactions from the broad to the small scale. This type of cross-scale interaction is important for recovery and renewal at a specific scale. Once a catastrophe is triggered at a level, the opportunities and constraints for the renewal of the cycle are strongly organized by capital and resources that are made available from a higher (larger) scale. After a fire in an ecosystem, for example, recovery and subsequent ecosystem development trajectory is a function of remnant resources (unburned roots and available nutrients), smaller-scale recovery processes, and the importation of resources from larger scales, such as seeds supplied from other areas. Accumulated capital, evolved structures, and other components of ecosystem memory (Berkes and Folke 2002) come into play at this stage.

The cross-scale model of panarchy can be used to understand resilience. System states (and alternative domains) exist at specific scale ranges, corresponding to levels within the panarchy. Those system states are comprised of entities (species in ecosystems, cells in organs, countries in the world) with a characteristic set of attributes or identity (Walker et al. 2004). Ecological resilience refers to the capacity of those attributes to absorb disturbances that are a result of cross-scale interactions, as described in the previous paragraphs. Other work (Holling 1992, Allen and Holling 2008) indicates that the cross-scale identities or attributes of complex systems are discontinuously distributed in space and time—that is, the cross-scale attributes are patterned in clumps (or lumps) that are separated by gaps. Two of the articles in this volume (Peterson et al. 1998, Allen et al. 1999) indicate how this lumpy structure and cross-scale interactions confer resilience.

Peterson et al. (1998) and Allen et al. (1999) contend that species interact with scale-dependent sets of ecological structures and processes that determine functional opportunities. At a particular ecological scale, the function of species may overlap but they will tend to differ as species evolve to avoid interspecific competition, and this will increase diversity at a particular ecological scale. Across ecological scales, there is more overlap in ecological function because species are less likely to face competition from species that interact with the environment (e.g., forage, compete, disperse, or defend territories) at different spatial and temporal scales. The combination of within-scale diversity of ecological function and cross-scale redundancy (i.e., reinforcement) adds to ecological resilience. The ecological function of a species loss at a particular scale can be offset by similar species that interact with the environment at a different scale (Walker et al. 1999). High within-scale diversity and cross-scale redundancy are predicted to produce ecosystems that are capable of resisting minor ecological disruptions (e.g., invasion by nonnatives) and regenerating after major disturbances (e.g., hurricanes, fire). Low within-scale diversity leads to low resilience. Allen et al. (1999) found that shifts in state—such as extinction, endangerment, or invasion of species—occur at the edge of lumps, near gaps.

Applying Resilience Theory

The application of resilience theory to the management of natural resources has created fundamentally different approaches (Holling 1973, 1986). Indeed, the theory proposed by Holling (1973) resulted in the development of adaptive resource management as described by Holling (1978), Walters (1986), Lee (1993), and Gunderson et al. (1995). The theory has shifted focus away from managing for particular equilibria to the management of regimes, as described below.

Managing Regimes

Adaptive capacity has been defined in the ecological literature as the ability to manage resilience (Gunderson 2000, Walker et al. 2004). Humans manipulate ecological systems to secure goods and services and in doing so leave the system more vulnerable to change, by eroding ecological resilience (Holling and Meffe 1996). Ecological resilience is difficult to assess and measure *a priori* and is often known only after the fact—that is, the complexities, nonlinearities, and self-organized processes that generate regime shifts or ecological phase transitions are generally understood only after a shift has occurred, and then only partly. Even so, humans do manage for adaptive capacity. Those management actions can be categorized as those that are aimed at buffering the impact of disturbances (Berkes and Folke 1998, 2002), those that accelerate recovery and renewal, and those that attempt to choose and manage transitions among alternative regimes.

Regime management has two key components that must be actively managed. Quite simply, they revolve around two basic questions: (1) "What kind of system do we want?" and (2) "What kind of system can we get?" (Clark and Munn 1986). The first question is beguilingly simple but actually very complex. The issue of desirability is at the core of many social-ecological problems. For example, endangered species legislation states that the population size and trend of a species that is in danger of extinction is not a desired state. Similarly, the frameworks of multiple objective or ecosystem management were developed in acknowledgment that different people have different expectations or desires of system state. Take, for example, the history of water management in the Everglades.

During the twentieth century, the Everglades region went through four distinct management eras, each reflecting different expectations or desires around the water resources (Light et al. 1995). At the beginning of the century, flooding and a new administration at the state level led to the construction of canals to drain the wetland (Light et al. 1995). When flooding in 1947 overwhelmed the capacity of the early drainage system, more control of the water resources was sought by a massive federal state partnership. This led to the current geographical array of land uses (agricultural, conservation, and urban development) that persists to date. For the first half of the century, flood control was the paradigm, with the desired state one in which unwanted, excess water would be removed in agriculture and urban areas. In 1971, a drought and an increase in human populations led to another regime. This time, the desired system was one in which water could be both removed during times of flood and retained during times of drought. In the mid-1980s, nutrients associated with agricultural runoff were leading to unwanted regime changes in the remnant wetlands of the Everglades. The ecosystem regime shift from sawgrass to cattails signaled a loss of ecological resilience and invoked a new management regime to stop the spread of nutrients (Gunderson 2001).

The point is that we develop many different institutions to work through these issues about desirable and undesirable regimes. The question of what is desirable can change over time, as indicated by the history of water management in the Everglades (Light et al. 1995). Yet the design and implementation of those institutions remains an open and ongoing activity, as social values change as well as the structure and function of management institutions (Olsson et al. 2006, Ostrom 2005).

When Resilience May Not Be a Good Thing

While many social-ecological systems exhibit rhythms of change (Walters 1986, Gunderson et al. 1995, Berkes and Folke 1998, 2002), many others do not. Systems can become trapped when they cannot or do not change or adapt to new conditions nor escape from a trajectory

TABLE 1: *Level of Each of the Three Variables That Characterize Four System Traps*

	Capital/Potential	Connectivity	Resilience
Rigidity Trap	High	High	High
Poverty Trap	Low	Low	Low
Lock-in Trap	Low	High	High
Isolation Trap	High	Low	Low

Source: Modified from Holling and Gunderson 2002, Allison and Hobbs 2004.

toward an undesired regime. Trapped systems exist in narrow management regimes, with few or no options for the future. The following paragraphs describe some examples of trapped systems; some are perversely resilient, whereas others are pathologic or maladaptive (Holling and Gunderson 2002).

At least four different types of resource management traps can be identified (Table 1). These different types are each defined by a combination of three properties: (1) capital or potential, (2) degree of connectivity, and (3) level of resilience (Holling and Gunderson 2002). Although it is possible to have eight combinations of these three properties, four combinations have been identified and are listed in Table 1. A system in a rigidity trap has high capital, connectivity, and resilience (Holling and Gunderson 2002). A system in a poverty trap has low amounts of these three properties. A system caught in a lock-trap has low capital but high levels of connectivity and resilience. The fourth trap—an isolation trap—is the least well understood of the four. It has high capital or potential but is not tightly coupled nor is it resilient.

Rigidity Trap

The Everglades management system has been and continues to be in a management trap (Gunderson and Holling 2002, Gunderson and Light 2006). This is a type of social trap (Rothstein 2005), defined as a system configuration or regime that persists over time despite being subjected to a wide range of shocks or perturbations (Allison and Hobbs 2004). It is a very resilient system (sensu Holling 1973) that is maintained by considerable infusions of money, which are tied to the conventional bureaucratic system. This system is governed by rules and procedures that are no longer appropriate to accomplish a highly complex and multiobjective mission. The result is that, for the sake of consistency, Everglades restoration remains in a policy straitjacket.

In addition to the command-and-control culture (Holling and Meffe 1996) mentioned above, rigidity traps have other characteristics, including (1) avoidance of learning (from past mistakes), (2) lack of trust among management institutions and stakeholders, and (3) strong feedbacks that maintain core elements of the status quo. Certainly, hundreds of millions of dollars have been put into Everglades research over the past three decades. Yet, understanding is and always will be incomplete and partial. While adaptive experiments were designed in workshops held in the early 1990s and suggested in scientific articles (Walters et al. 1992, Walters and Gunderson 1994), discussions still persist as to what should be done. There are fiscal and political costs to experimentation. Moreover, the reasons that more experiments have not been conducted are related to the fear of risking conflict and the fear of failure to produce desired or even meaningful results. This is compounded by the inability of current bureaucracies to comprehend the value received from learning now when compared to the costs of inaction.

Unfortunately, current practices have government agencies supporting large scientific endeavors that focus on modeling and data collection rather than on using experiments to reduce uncertainties and explore new options. A recent National Academy of Sciences panel (2003) indicated that ongoing and future research should move away from self-serving, piecemeal studies to ones that are more synthetic and integrative. To do so will require scientists to become motivated to pursue collective learning. Perhaps the main reason for the rigidity trap is a lack of social capital and trust fostered by institutional power imbalances in the region (Rothstein 2005). Special interests and resource managers who feel that experimentation would supplant an opportunity to secure water options for the park and conservation interests have stymied attempts at adaptive experiments. Rather than acknowledge that it is currently unknown what it would take to restore the lost environmental values, some chose to replace scientific uncertainty with political certitude, as false as it may be.

Poverty Trap

Systems in a poverty trap are not very resilient and are characterized by little capital or connectivity. Open-water pelagic systems are an ecological example. In these systems, productivity (ability to convert sunlight to plant material) is very low. These blue-water systems are limited by nutrients and hence build little, if any, structure. Because of a fluid and changing environment, few connections develop among parts of the system. Other systems, where capital (in any form) has been mined or used can also fall into a poverty trap. Areas where soils are degraded through poor management (organic soil oxidation in the Everglades, soil salinization in many arid regions, or loss of topsoil in tropical regions) are in poverty traps, as they lack the resources for renewal and connections and are vulnerable to change into many different states.

Lock-in Trap

The agricultural region of Western Australia (wheat belt) has gone through boom-and-bust cycles over multiple decades (Allison and Hobbs 2004). Currently, it is in a type of trap where capital is low but connectivity and resilience remain relatively high. The system has lost many natural resources (such as native biodiversity), primarily through land conversion. Over the past few decades, pollution has increased and social structures (such as towns) have declined. Yet, wheat production has been maintained by improved agricultural practices and crop varieties in spite of increases in soil salinity. The system of farming and agro-business maintains a tight connection. This has led to characterizing the system as being in a lock-in trap. It is locked in, because the supports (external and policy) and connections maintain the current agricultural system. It is trapped, as social capital (in the form of communities, churches, schools, and so forth) and natural capital have degraded.

Isolation Trap

Remnant plant or animal populations (including human cultures) can be in an isolation trap. This trap is characterized by relatively high capital or potential capital, low connectivity, and low resilience. Many isolated populations are classified (and receive special management consideration) as threatened or endangered because they are very vulnerable to perturbations (i.e., they are not very resilient). One such example is found in the Colorado River in the United States, where the endangered humpback chub (*Gila cypha*) persists in areas that are remnants of former range. Because of dams that prohibit movement, former meta-populations are now isolated populations. The isolated populations have the potential to increase in number, and the numbers are low enough to warrant special designation and management. But with increased preda-tion and temperature changes, among other factors, the populations are not very resilient and the species is in danger of extinction (an alternate and undesirable state).

Escaping traps poses some of the most difficult and frustrating issues facing resource managers. These problems are best confronted by active adaptive management approaches that attempt to unravel the complex-ity of these traps by attempting management actions designed to help understand as much as meet other social objectives.

Ecological Resilience and Sustainable Futures

Ecological resilience is an idea that has come of age. It has taken more than three decades to mature, during which time it has withstood attempts of refutation and narrow interpretations of the concepts. The literature in this volume and other outlets documents how the concept has been used in hundreds of cases to explain complex, unpredictable changes in ecosystems. A growing number of examples worldwide are beginning to use the idea of ecological resilience as a way of confronting the complexities of seeking sustainability.

Two key examples of resilience and sustainable futures involve the man-agement of alternative regimes in lake systems (Scheffer and Carpenter 2003, Carpenter et al. 1999) and in coral reef systems (Hughes 1994, Hughes et al. 2003, Olsson et al. 2006). In both of these cases, it has taken decades to develop an understanding of system indicators that can be used to signal changes in system resilience and system state and to iden-tify different forms of capital that contribute to resilience. Both cases identify a variety of forward-looking mechanisms (such as markets or policies) that manage various forms of capital (especially natural capital) needed to maintain ecological resilience. In both cases, management focuses on keeping the system within the bounds or thresholds of a sin-gle beneficial and desired regime—that is, society has made a choice, through a variety of mechanisms, about which regime among many pos-sible alternative ecological configurations, is valued and preferred.

Multiple or alternative ecological domains can be viewed and valued in a range of ways by society. Simply, these domains can be viewed in a normative sense as positive (desired state), negative (undesired state), or neutral. The two cases of lakes and reefs in the previous paragraph are ones in which the desire is to maintain a particular ecological state — that is, the scientists and practitioners are attempting to maintain ecological resilience in order to preserve the desired current state. They are attempting to avoid a change of state and the corollary flow of ecosystem goods and services and economic activities associated with an undesired, alternative state (e.g., a turbid lake or algae-covered reefs).

Other examples exist in resource systems where (1) it may not matter to the social or economic systems as to the ecological state, or (2) resilience may be a negative attribute (e.g., the system is in a social trap), or (3) the system has already flipped to an alternative state and people have adapted to the new regime. For example, wool production in New South Wales, Australia, has been stable for over a century, in spite of a change of ecological state in the rangeland (Abel et al. 2006). In the wheat belt of western Australia, the social-ecological system is very resilient to change, exhibiting a lock-in trap (Allison and Hobbes 2004), from which indicators or forward-looking market mechanisms have not helped lead to escape. Finally, many of the largest conservation and environmental issues of our time occur in places where the system is already in an undesired domain. In the Everglades of the United States, for example, endangered species populations and vegetation conversions are both examples where the system has already crossed a threshold into an undesired state (Gunderson 2001). In these cases, infusions of capital (human, fiscal, social, and political) are needed to help restore or recover from a degraded state (Olsson et al. 2006).

The issue of regime management is central to sustainability. The management and maintenance of key forms of capital are crucial to ecological resilience through the application of forward-looking economic and institutional structures. Integrating ecologic, economic, and other social dimensions is needed to build adaptive capacity to deal with known and unknown futures. It is our ability to learn and confront the uncertainties of resilience of coupled systems of humans and nature that will help as we navigate a path to sustainability.

Literature Cited

Abel, N., D. H. M. Cumming, and J. M. Anderies. 2006. Collapse and reorganization in social-ecological systems: Questions, some ideas, and policy implications. *Ecology and Society* 11 (1):17. http://www.ecologyandsociety.org/vol11/iss1/art17/.

Allen, C. R., and C. S. Holling. 2008. *Discontinuities in ecosystems and other complex systems.* New York: Columbia University Press.

Allen, C. R., E. A. Forys, and C. S. Holling. 1999. Body mass patterns predict invasions and extinctions in transforming landscapes. *Ecosystems* 2:114–21.

Allison, H. E., and R. J. Hobbs. 2004. Resilience, adaptive capacity, and the "lock-in trap" of the Western Australian agricultural region. *Ecology and Society* 9 (1):3. http://www.ecologyandsociety.org/vol9/iss1/art3/.

Bellwood, D. R., T. P. Hughes, C. Folke, and M. Nyström. 2004. Confronting the coral reef crisis. *Nature* 429:827–33.

Berkes, F., and C. Folke, eds. 1998. *Linking social and ecological systems: Management practices and social mechanisms for building resilience.* Cambridge: Cambridge University Press.

Berkes, F., and C. Folke. 2002. Back to the future: Ecosystem change, institutions and local knowledge. In *Panarchy: Understanding transformations in human and natural systems,* ed. L. H. Gunderson and C. S. Holling, 103–19. Washington, DC: Island Press.

Berkes, F., J. Colding, and C. Folke, eds. 2003. *Navigating social-ecological systems: Building resilience for complexity and change.* Cambridge: Cambridge University Press.

Carpenter, S. R. 1996. Microcosm experiments have limited relevance for community and ecosystem ecology. *Ecology* 77 (3):677–80.

———. 2001. Alternate states of ecosystems: Evidence and its implications. In *Ecology: Achievement and challenge,* ed. M. C. Press, N. Huntly, and S. Levin, 357–83. London: Blackwell.

Carpenter, S., W. Brock, and P. Hanson. 1999. Ecological and social dynamics in simple models of ecosystem management. *Conservation Ecology* 3 (4). http://www.consecol.org/vol3/iss2/art4.

Carpenter, S. R., J. J. Cole, J. R. Hodgson, J. F. Kitchell, M. L. Pace, D. Bade, K. L. Cottingham, T. E. Essington, J. N. Houser, and D. E. Schindler. 2001. Trophic cascades, nutrients and lake productivity: Whole-lake experiments. *Ecological Monographs* 71:163–86.

Clark, W. C., and R. E. Munn. 1986. *Sustainable development of the biosphere.* Cambridge: Cambridge University Press.

DeAngelis, D. L. 1980. Energy flow, nutrient cycling, and ecosystem resilience. *Ecology* 61:764–71.

Dublin, H. T., A. R. E. Sinclair, and J. McGlade. 1990. Elephants and fire as causes of multiple stable states in the Serengeti-Mara woodlands. *Journal of Animal Ecology* 59:1147–64.

Elmqvist, T., J. Colding, S. Barthel, A. Duit, S. Borgström, J. Lundberg, E. Andersson, K. Ahrné, H. Ernstson, J. Bengtsson, and C. Folke. 2004. The dynamics of social-ecological systems in urban landscapes: Stockholm and the National Urban Park, Sweden. *Annals of New York Academy of Sciences* 1023:308–22.

Estes, J. A., and D. O. Duggins. 1995. Sea otters and kelp forests in Alaska: Generality and variation in a community ecology paradigm. *Ecological Monographs* 65:75–100.

Foley, J. A., M. T. Coe, M. Scheffer, and G. Wang. 2003. Regime shifts in the Sahara and Sahel: Interactions between ecological and climatic systems in Northern Africa. *Ecosystems* 6:524–32.

Folke, C., S. Carpenter, T. Elmqvist, L. Gunderson, C. S. Holling, and B. Walker. 2002. Resilience and sustainable development: Building adaptive capacity in a world of transformations. *Ambio* 31:437–40.

Folke, C., S. Carpenter, B. Walker, M. Scheffer, T. Elmqvist, L. Gunderson, and C. S. Holling. 2004. Regime shifts, resilience and biodiversity in ecosystem management. *Annual Review of Ecology and Systematics* 35:557–81.

Gunderson, L. H. 2000. Resilience in theory and practice. *Annual Review of Ecology and Systematics* 31:425–39.

———. 2001. Managing surprising ecosystems in southern Florida. *Ecological Economics* 37:371–78.

Gunderson, L. H., and C. S. Holling. 2002. *Panarchy: Understanding transformations in human and natural systems.* Washington, DC: Island Press.

Gunderson, L. H., and S. S. Light. 2006. Adaptive management and adaptive governance in the Everglades ecosystem. *Policy Sciences* 9 (4):323–34.

Gunderson, L. H., C. S. Holling, and S. S. Light. 1995. *Barriers and bridges to the renewal of ecosystems and institutions.* New York: Columbia University Press.

Holling, C. S. 1959. Some characteristics of simple types of predation and parasitism. *Canadian Entomologist* 91:385–98.

———. 1973. Resilience and stability of ecological systems. *Annual Review of Ecology and Systematics* 4:1–23.

———, ed. 1978. *Adaptive environmental assessment and management.* Chichester: Wiley.

———. 1986. The resilience of terrestrial ecosystems: Local surprise and global change. In *Sustainable development of the biosphere*, ed. W. C. Clark and R. E. Munn, 292–317. Cambridge: Cambridge University Press.

———. 1992. Cross-scale morphology, geometry and dynamics of ecosystems. *Ecological Monographs* 62:447–502.

———. 1996. Engineering resilience vs. ecological resilience. In *Engineering within ecological constraints*, ed. P. C. Schulze, 31–43. Washington, DC: National Academy Press.

———. 2001. Understanding the complexity of economic, ecological, and social systems. *Ecosystems* 4:390–405.

Holling, C. S., and L. H. Gunderson. 2002. Resilience and adaptive cycles. In *Panarchy: Understanding transformations in human and ecological systems*, ed. L. H. Gunderson and C. S. Holling, 25–62. Washington, DC: Island Press.

Holling, C. S., and G. K. Meffe. 1996. Command and control and the pathology of natural resource management. *Conservation Biology* 10:328–37.

Hughes, T. P. 1994. Catastrophes, phase shifts, and large-scale degradation of a Caribbean coral reef. *Science* 265:1547–51.

Hughes, T. P., A. H. Baird, D. R. Bellwood, M. Card, S. R. Connolly, C. Folke, R. Grosberg, et al. 2003. Climate change, human impacts, and the resilience of coral reefs. *Science* 301:929–33.

Lee, K. 1993. *Compass and gyroscope: Integrating science and politics for the environment.* Washington, DC: Island Press.

Light, S. S., L. H. Gunderson, and C. S. Holling. 1995. The Everglades: Evolution of management in a turbulent environment. In *Barriers and bridges to the renewal of ecosystems and institutions,* ed. L. H. Gunderson, C. S. Holling, and S. S. Light, 94–132. New York: Columbia University Press.

National Academy of Sciences. 2003. *Science and the greater Everglades ecosystem restoration: An assessment of the critical ecosystem studies initiative panel to review the critical ecosystem studies initiative.* Washington, DC: National Research Council, National Academy of Sciences.

Nyström, M., and C. Folke. 2001. Spatial resilience of coral reefs. *Ecosystems* 4:406–17.

Olsson, P., C. Folke, and T. Hahn. 2004. Social-ecological transformation for ecosystem management: The development of adaptive co-management of a wetland landscape in southern Sweden. *Ecology and Society* 9 (4):2. http://www.ecologyandsociety.org/vol9/iss4/art2/.

Olsson, P., L. H. Gunderson, S. R. Carpenter, P. Ryan, L. Lebel, C. Folke, and C. S. Holling. 2006. Shooting the rapids: Navigating transitions to adaptive governance of social-ecological systems. *Ecology and Society* 11 (1):18. http://www.ecologyandsociety.org/vol11/iss1/art18/.

Ostrom, E. 2005. *Understanding institutional diversity.* Princeton, NJ: Princeton University Press.

Peterson, G. C. Allen, and C. S. Holling. 1998. Ecological resilience, biodiversity, and scale. *Ecosystems* 1:6–18.

Pimm, S. L. 1991. *The balance of nature?* Chicago: University of Chicago Press.

Redman, C. L., and A. P. Kinzig. 2003. Resilience of past landscapes: Resilience theory, society, and the longue durée. *Conservation Ecology* 7 (1):14. http://www.consecol.org/vol7/iss1/art14/.

Rothstein, B. 2005. *Social traps and the problem of trust.* Cambridge: Cambridge University Press.

Scheffer, M. 1998. *The ecology of shallow lakes.* London: Chapman and Hall.

Scheffer, M., and S. R. Carpenter. 2003. Catastrophic regime shifts in ecosystems: Linking theory to observation. *Trends in Ecology and Evolution* 18:648–56.

Scheffer, M., S. R. Carpenter, J. Foley, C. Folke, and B. Walker. 2001. Catastrophic shifts in ecosystems. *Nature* 413:591–696.

Scheffer, M., F. Westley, and W. Brock. 2003. Slow response of societies to new problems: Causes and costs. *Ecosystems* 6:493–502 .

Sinclair, A. R. E., P. D. Olsen, and T. D. Redhead. 1990. Can predators regulate small mammal populations? Evidence from house mouse outbreaks in Australia. *Oikos* 59:382–92.

Skeat, W. W. 1882. *A concise etymological dictionary of the English language.* Oxford: Clarendon.

Sousa, W. P., and J. H. Connell. 1985. Further comments on the evidence for multiple stable points in natural communities. *American Naturalist* 125:612–15.

Steneck, R. S., J. Vavrinec, and A. Leland. 2004. Accelerating trophic level dysfunction in kelp forest ecosystems of the western North Atlantic. *Ecosystems* 7:323–31.

Walker, B. H. 1981. Is succession a viable concept in African savanna ecosystems? In *Forest succession: Concepts and application*, ed. D. C. West, H. H. Shugart, and D. B. Botkin, 431–47. New York: Springer-Verlag.

Walker, B., and D. Salt. 2006. *Resilience thinking*. Washington, DC: Island Press.

Walker, B., A. Kinzig, and J. Langridge. 1999. Plant attribute diversity, resilience and ecosystem function: The nature and significance of dominant and minor species. *Ecosystems* 2:95–113.

Walker, B., C. S. Holling, S. R. Carpenter, and A. Kinzig. 2004. Resilience, adaptability and transformability in social-ecological systems. *Ecology and Society* 9 (2):5. http://www.ecologyandsociety.org/vol9/iss2/art5/.

Walters, C. J. 1986. *Adaptive management of renewable resources*. New York: McGraw-Hill.

Walters, C. J., and L. H. Gunderson. 1994. Screening water policies for ecosystem restoration. In *The Everglades: The ecosystem and its restoration*, ed. S. Davis and J. Ogden, 489–519. Delray Beach, FL: St. Lucie.

Walters, C. J., L. H. Gunderson, and C. S. Holling. 1992. Experimental policies for water management in the Everglades. *Ecological Applications* 2:189–202.

Westley, F., S. R. Carpenter, W. A. Brock, C. S. Holling, and L. H. Gunderson. 2002. Why systems of people and nature are not just social and ecological systems. In *Panarchy: Understanding transformations in human and natural systems*, ed. L. H. Gunderson and C. S. Holling, 103–19. Washington, DC: Island Press.

Selected Bibliography

HERE WE PROVIDE A SHORT BIBLIOGRAPHY of important papers on resilience that are not referenced in the commentaries for this volume. These works have been influential in defining, refining, and expanding our understanding of resilience. Although the concept of ecological resilience took decades to be integrated into both theory and practice, within the past decade the number of published articles and books with the word *resilience* in the title has increased exponentially. The Resilience Alliance maintains a bibliography that, though not exhaustive, is extensive. The Alliance Web site can be accessed at http://www.resalliance.org/, and we invite readers to visit the site for updates.

Adaptive Capacity

Adger, W. N., and K. Vincent. 2005. Uncertainty in adaptive capacity. (IPCC Special Issue on "Describing Uncertainties in Climate Change to Support Analysis of Risk and Options.") *Comptes Rendus Geoscience* 337:399–410.

Allison, H. E., and R. J. Hobbs. 2004. Resilience, adaptive capacity, and the "lock-in trap" of the Western Australian agricultural region. *Ecology and Society* 9 (1):3. http://www.ecologyandsociety.org/vol9/iss1/art3.

Brooks, N., W. N. Adger, and P. M. Kelly. 2005. The determinants of vulnerability and adaptive capacity at the national level and the implications for adaptation. *Global Environmental Change* 15:151–63.

Folke, C., S. Carpenter, T. Elmqvist, L. Gunderson, C. S. Holling, and B. Walker. 2002. Resilience and sustainable development: Building adaptive capacity in a world of transformations. *Ambio* 31:437–40.

Smith, J. B., R. J. T. Klein, and S. Huq. 2003. *Climate change, adaptive capacity and development.* London: Imperial College Press.

Adaptive Management

Armitage, D., F. Berkes, and N. Doubleday, eds. 2007. *Adaptive co-management: Collaboration, learning, and multi-level governance.* Vancouver, BC: University of British Columbia Press.

Berkes, F., J. Colding, and C. Folke. 2000. Rediscovery of traditional ecological knowledge as adaptive management. *Ecological Applications* 10:1251–62.

Folke, C., T. Hahn, P. Olsson, and J. Norberg. 2005. Adaptive governance of social-ecological systems. *Annual Review of Environment and Resources* 30:441–73.

Gunderson, L. H. 1999. Resilience, flexibility and adaptive management—antidotes for spurious certitude? *Conservation Ecology* 3 (1):7. http://www.consecol.org/vol3/iss1/art7.

Hahn, T., P. Olsson, C. Folke, and K. Johansson. 2006. Trust-building, knowledge generation and organizational innovations: The role of a bridging organization for adaptive co-management of a wetland landscape around Kristianstad, Sweden. *Human Ecology* 34:573–92.

Olsson, P., C. Folke, and F. Berkes. 2004. Adaptive co-management for building resilience in social-ecological systems. *Environmental Management* 34:75–90.

Tompkins, E. L., and W. N. Adger. 2004. Does adaptive management of natural resources enhance resilience to climate change? *Ecology and Society* 9 (2):10. http://www.ecologyandsociety.org/vol9/iss2/art10.

Walters, C., L. Gunderson, and C. S. Holling. 1992. Experimental policies for water management in the Everglades. *Ecological Applications* 2:189–202.

Biodiversity

Chapin, F. S., E. S. Zavaleta, V. T. Eviner, R. L. Naylor, P. M. Vitousek, H. L. Reynolds, D. U. Hooper, S. Lavorel, O. E. Sala, S. E. Hobbie, M. C. Mack, and S. Diaz. 2000. Consequences of changing biodiversity. *Nature* 405:234–42.

Colding, J., C. Folke, and T. Elmqvist. 2003. Social institutions in ecosystem management and biodiversity conservation. *Tropical Ecology* 44:25–41.

Dornelas, M., S. R. Connolly, and T. P. Hughes. 2006. Coral reef diversity refutes the neutral theory of biodiversity. *Nature* 440:80–82.

Lundberg, J., and F. Moberg. 2003. Mobile link organism and ecosystem functioning—implications for ecosystem resilience and management. *Ecosystems* 6:87–98.

Norberg, J. 2004. Biodiversity and ecosystem functioning: A complex adaptive systems approach. *Limnology and Oceanography* 49:1269–77.

Walker, B., A. Kinzig, and J. Langridge. 1999. Plant attribute diversity, resilience and ecosystem function. *Ecosystems* 2:95–113.

Discontinuities and Resilience

Allen, C. R. 2006. Predictors of introduction success in the South Florida avifauna. *Biological Invasions* 8:491–500.

Ernest, S. K. M. 2005. Body size, energy use, and community structure of small mammals. *Ecology* 86:1407–13.

Garmestani, A., C. R. Allen, and K. M. Bessey. 2005. Time series analysis of clusters in city size distributions. *Urban Studies* 42:1507–15.

Holling, C. S. 1988. Temperate forest insect outbreaks, tropical deforestation and migratory birds. *Memoirs Entomological Society of Canada* 146:22–32.

Kerkhoff, A. J., and B. J. Enquist. 2007. The implications of scaling approaches for understanding resilience and reorganization in ecosystems. *BioScience* 57:489–99.

Lambert, W. 2006. Functional convergence of ecosystems: Evidence from body mass distributions of North American late Miocene mammal faunas. *Ecosystems* 9:97–118.

Governance

Adger, W. N., K. Brown, J. Fairbrass, A. Jordan, J. Paavola, S. Rosendo, and G. Seyfang. 2003. Governance for sustainability: Towards a "thick" analysis of environmental decision-making. *Environment and Planning A* 35:1095–110.

Adger, W. N., K. Brown, and E. L. Tompkins. 2005. The political economy of cross-scale networks in resource co-management. *Ecology and Society* 10 (2):9. http://www.ecologyandsociety.org/vol10/iss2/art9/.

Carpenter, S. R., and W. A. Brock. 2004. Spatial complexity, resilience and policy diversity: Fishing on lake-rich landscapes. *Ecology and Society* 9 (1):8. http://www.ecologyandsociety.org/vol9/iss1/art8.

Lebel, L., J. M. Anderies, B. Campbell, C. Folke, S. Hatfield-Dodds, T. P. Hughes, and J. Wilson. 2006. Governance and the capacity to manage resilience in regional social-ecological systems. *Ecology and Society* 11 (1):19. http://www.ecologyandsociety.org/vol11/iss1/art19/.

Ostrom, E. 2005. *Understanding institutional diversity.* Princeton, NJ: Princeton University Press.

Ostrom, E., and M. Janssen. 2004. Multi-level governance and resilience of social-ecological systems. In *Globalisation, poverty and conflict: A critical "development" reader,* ed. M. Spoor, 239–59. Dordrecht: Kluwer Academic Publishers.

Westley, F., B. Zimmerman, and M. Q. Patton. 2006. *Getting to maybe: How the world has changed.* Toronto: Random House Canada.

Multiple Stable States, Regime Shifts

Done, T. J. 1992. Phase shifts in coral reef communities and their ecological significance. *Hydrobiologia* 247:121–32.

Hanski, I. 1995. Multiple equilibria in metapopulation dynamics. *Nature* 377:618–21.

Hare, S. R., and N. J. Mantua. 2000. Empirical evidence for North Pacific regime shifts in 1977 and 1989. *Progress in Oceanography* 47:103–45.

Knowlton, N. 1992. Thresholds and multiple stable states in coral-reef community dynamics. *American Zoologist* 32:674–82.

Scheffer, M., and S. R. Carpenter. 2003. Catastrophic regime shifts in ecosystems: Linking theory to observation. *Trends Ecology and Evolution* 18:648–56.

Scheffer, M., S. H. Hosper, M.-L. Meijer, B. Moss, and E. Jeppesen. 1993. Alternative equilibria in shallow lakes. *Trends in Evolutionary Ecology* 8:275–79.

Resilience (Synthetic)

Carpenter, S. R., B. H. Walker, J. M. Anderies, and N. Abel. 2001. From metaphor to measurement: Resilience of what to what? *Ecosystems* 4:765–81.

Chapin, F. S., G. Peterson, F. Berkes, T. V. Callaghan, P. Anglestam, M. Apps, et al. 2004. Resilience and vulnerability of northern regions to social and environmental change. *Ambio* 33:344–49.

Folke, C. 2003. Reserves and resilience—from single equilibrium to complex systems. *Ambio* 32:379.

Gunderson, L. H., and L. Pritchard. 2002. *Resilience and the behavior of large scale ecosystems*. Washington, DC: Island Press.

Holling, C. S. 1978. Myths of ecological stability: Resilience and the problem of failure. In *Studies in crisis management*, ed. C. F. Smart and W. T. Stanbury, 97–109. Montreal: Butterworth.

Nyström, M., and C. Folke. 2001. Spatial resilience of coral reefs. *Ecosystems* 4:406–17.

Socioecological Systems

Abel, N., D. H. M. Cumming, and J. M. Anderies. 2006. Collapse and reorganization in social-ecological systems: Questions, some ideas, and policy implications. *Ecology and Society* 11 (1):17. http://www.ecologyandsociety.org/vol11/iss1/art17/.

Adger, W. N. 2000. Social and ecological resilience: Are they related? *Progress in Human Geography* 24:347–64.

Adger, W. N., T. P. Hughes, C. Folke, S. R. Carpenter, and J. Rockström. 2005. Social-ecological resilience to coastal disasters. *Science* 309:1036–39.

Berkes, F., and C. Folke. 1998. *Linking social and ecological systems: Management practices and social mechanisms for building resilience*. Cambridge: Cambridge University Press.

Berkes, F., and C. S. Seixas. 2005. Building resilience in lagoon social-ecological systems: A local-level perspective. *Ecosystems* 8:967–74.

Matutinovic, I. 2001. The aspects and the role of diversity in socioeconomic systems: An evolutionary perspective. *Ecological Economics* 39:239–56.

Peterson, G. D. 2000. Political ecology and ecological resilience: An integration of human and ecological dynamics. *Ecological Economics* 35:323–36.

Social Systems

Adger, W. N. 2003. Social capital, collective action and adaptation to climate change. *Economic Geography* 79:387–404.

Berkes, F., J. Colding, and C. Folke. 2003. *Navigating social-ecological systems: Building resilience for complexity and change*. Cambridge: Cambridge University Press.

Burton, I, R. Kates, and G. White. 1993. Natural extremes and social resilience. In *The environment as hazard*, ed. I. Burton, R. Kates, and G. White, 219–40. New York: Guilford.

Levin, S. A., S. Barrett, S. Aniyar, W. Baumol, C. Bliss, B. Bolin, P. Dasgupta, et al. 1998. Resilience in natural and socioeconomic systems. *Environment and Development Economics* 3:222–34.

Perrings, C. 1998. Resilience in the dynamics of economy-environment systems. *Environmental and Resource Economics* 11:503–20.

Sustainability

Arrow, K. J., P. Dasgupta, and K.-G. Mäler. 2003. Evaluating projects and assessing sustainable development in imperfect economies. *Environmental and Resource Economics* 26:647–85.

Brown, K., W. N. Adger, E. Tompkins, P. Bacon, D. Shim, and K. Young. 2001. Trade-off analysis for marine protected area management. *Ecological Economics* 37:417–34.

Clark, W. C., and R. E. Munn, eds. 1986. *Sustainable development of the biosphere.* Cambridge: Cambridge University Press.

Janssen, M. A., T. A. Kohler, and M. Scheffer. 2003. Sunk-cost effects and vulnerability to collapse in ancient societies. *Current Anthropology* 44:722–28.

Ludwig, D., B. Walker, and C. S. Holling. 1997. Sustainability, stability, and resilience. *Conservation Ecology* 1 (1):7. http://www.consecol.org/vol1/iss1/art7.

McMichael, A. J., C. D. Butler, and C. Folke. 2003. New visions for addressing sustainability. *Science* 302:1919–20.

Permissions and Original Sources

The editors gratefully acknowledge permission to reprint articles from the following sources.

Article 1, "Resilience and Stability of Ecological Systems," was originally published by *Annual Reviews*. C. S. Holling, "Resilience and Stability of Ecological Systems," *Annual Review of Ecological Systems* 4 (1973): 1–23.

Article 2, "Engineering Resilience versus Ecological Resilience," is reprinted with permission from the National Academies Press, Copyright 1996, National Academy of Sciences. C. S. Holling, "Engineering Resilience versus Ecological Resilience," in *Engineering within Ecological Constraints*, ed. Peter C. Schulze, 31–44 (Washington, DC: National Academies Press, 1996).

Article 3, "The Resilience of Terrestrial Ecosystems: Local Surprise and Global Change." Source: W. C. Clark and R. E. Munn, eds., *Sustainable Development of the Biosphere* (Cambridge: Cambridge University Press, 1986).

Article 4, "Regime Shifts, Resilience, and Biodiversity in Ecosystem Management," was originally published by *Annual Reviews*. Carl Folke, Steve Carpenter, Brian Walker, Marten Scheffer, Thomas Elmqvist, Lance Gunderson, and C. S. Holling, "Regime Shifts, Resilience, and Biodiversity in Ecosystem Management," *Annual Review of Ecology, Evolution, and Systematics* 35 (2004): 557–81.

Article 5, "Biological Diversity, Ecosystems, and the Human Scale," is reproduced by permission of the Ecological Society of America. Carl Folke, C. S. Holling, and Charles Perrings, "Biological Diversity, Ecosystems, and the Human Scale," *Ecological Applications* 6, no. 4 (1996): 1018–24.

Article 6, "Ecological Resilience, Biodiversity, and Scale," is reprinted with kind permission from Springer Science+Business Media: Garry Peterson, Craig R. Allen, and C. S. Holling, "Ecological Resilience, Biodiversity, and Scale," *Ecosystems* 1 (1997): 6–18.

Article 7, "Catastrophes, Phase Shifts, and Large-scale Degradation of a Caribbean Coral Reef." Source: T. P. Hughes, "Catastrophes, Phase Shifts, and Large-scale

451

Degradation of a Caribbean Coral Reef," *Science* 265 (1994):1547–51. Reprinted with permission from AAAS.

Article 8, "Sea Otters and Kelp Forests in Alaska: Generality and Variation in a Community Ecological Paradigm," reproduced by permission of the Ecological Society of America. James A. Estes and David O. Duggins, "Sea Otters and Kelp Forests in Alaska: Generality and Variation in a Community Ecological Paradigm," *Ecological Monographs* 65, no. 1 (1995):75–100.

Article 9, "Body Mass Patterns Predict Invasions and Extinctions in Transforming Landscapes," is reprinted with kind permission from Springer Science+Business Media: Craig R. Allen, Elizabeth A. Forys, and C. S. Holling, "Body Mass Patterns Predict Invasions and Extinctions in Transforming Landscapes," *Ecosystems* 2 (1999):114–21.

Article 10, "Resource Science: The Nurture of an Infant," originally published in *BioScience*, copyright American Institute Biological Sciences. C. S. Holling and A. D. Chambers, "Resource Science: The Nurture of an Infant," *BioScience* 23, no. 1 (1973): 13–20.

Article 11, "Lessons for Ecological Party Design: A Case Study of Ecosystem Management." Reprinted from *Ecological Modeling*, 7, William C. Clark, Dixon D. Jones, and C. S. Holling, "Lessons for Ecological Policy Design: A Case Study of Ecosystem Management," 1–52, Copyright (1979) with permission from Elsevier.

Article 12, "Qualitative Analysis of Insect Outbreak Systems: The Spruce Budworm and Forest," by D. Ludwig, D. D. Jones, and C. S. Holling, *Journal of Animal Ecology* 47, no. 1. (1978): 315–32. Reprinted with kind permission of Wiley-Blackwell.

Index

Page numbers in italics indicate figures and tables.

Island Press | Board of Directors

ALEXIS G. SANT *(Chair)*
Summit Foundation

DANE NICHOLS *(Vice-Chair)*

HENRY REATH *(Treasurer)*
Nesbit-Reath Consulting

CAROLYN PEACHEY *(Secretary)*
President
Campbell, Peachey & Associates

S. DECKER ANSTROM
Board of Directors
Comcast Corporation

STEPHEN BADGER
Board Member
Mars, Inc.

KATIE DOLAN
Eastern New York
 Chapter Director
The Nature Conservancy

MERLOYD LUDINGTON LAWRENCE
Merloyd Lawrence, Inc.
 and Perseus Books

WILLIAM H. MEADOWS
President
The Wilderness Society

DRUMMOND PIKE
President
The Tides Foundation

CHARLES C. SAVITT
President
Island Press

SUSAN E. SECHLER

VICTOR M. SHER, ESQ.
Principal
Sher Leff LLP

PETER R. STEIN
General Partner
LTC Conservation Advisory
 Services
The Lyme Timber Company

DIANA WALL, PH.D.
Professor of Biology
 and Senior Research Scientist
Natural Resource Ecology
 Laboratory
Colorado State University

WREN WIRTH
President
Winslow Foundation